P9-ELO-639

NOBEL LECTURES CHEMISTRY

NOBEL LECTURES

INCLUDING PRESENTATION SPEECHES
AND LAUREATES' BIOGRAPHIES

PHYSICS

CHEMISTRY

PHYSIOLOGY OR MEDICINE

LITERATURE

PEACE

PUBLISHED FOR THE NOBEL FOUNDATION
BY
ELSEVIER PUBLISHING COMPANY
AMSTERDAM - LONDON - NEW YORK

NOBEL LECTURES

INCLUDING PRESENTATION SPEECHES
AND LAUREATES' BIOGRAPHIES

CHEMISTRY

1963-1970

PUBLISHED FOR THE NOBEL FOUNDATION
IN 1972 BY
ELSEVIER PUBLISHING COMPANY
AMSTERDAM - LONDON - NEW YORK

ELSEVIER PUBLISHING COMPANY
335 JAN VAN GALENSTRAAT
P.O. BOX 211, AMSTERDAM, THE NETHERLANDS

AMERICAN ELSEVIER PUBLISHING COMPANY, INC.
52 VANDERBILT AVENUE
NEW YORK, N.Y. 10017

LIBRARY OF CONGRESS CARD NUMBER 63-22071
ISBN 0-444-40987-4

WITH 131 ILLUSTRATIONS
AND 16 TABLES

PRINTED IN THE NETHERLANDS BY
KONINKLIJKE DRUKKERIJ G.J. THIEME N.V., NIJMEGEN
BOOK DESIGN: HELMUT SALDEN

Foreword

The Nobel Lectures, which according to the statutes of the Nobel Foundation are to be delivered by each Laureate within six months of receiving the Prize, subsequently appear in the annual publication of the Foundation, *Les Prix Nobel.*

It was felt, however, that a collection of lectures within each Prize group would be of interest for many readers not only as a source of highly qualified information for specialists and research workers in the pertinent field, but also because it offers a unique historical perspective of the scientific and cultural development in the areas in question. This idea materialized some ten years ago in the form of the Nobel Lecture Series published by the Elsevier Publishing Company, Amsterdam, each volume containing the lectures, presentation speeches and updated biographies. The first volumes cover physics, chemistry and physiology or medicine in the years 1901–1962, literature 1901–1967 and peace 1901–1970.

The public interest in these carefully presented volumes has been encouragingly high. In order to make subsequent lectures available, it has been agreed with Elsevier to continue the series by new volumes at suitable intervals. It is hoped that the continued series will meet with an equally favourable reception and serve the same purpose of focusing interest on recent and past major achievements in those fields which were selected by Alfred Nobel.

Ulf von Euler – President, Nobel Foundation

Publisher's Note

The translations of the lectures which were delivered in a language other than English have been taken from *Scientific American*, which we gratefully acknowledge here.

The lectures that did not appear in *Scientific American* in an English version, as well as the presentation speeches that were only available in French or German, were translated by the Express Translation Service, London.

The biographies were completed with the cooperation of many of the laureates themselves.

To all these persons we would like to offer our sincerest thanks in appreciation of their efforts.

Elsevier Publishing Company

Contents

Chemistry 1963

KARL ZIEGLER

GIULIO NATTA

«for their discoveries in the field of the chemistry and technology of high polymers»

Chemistry 1963

Presentation Speech by Professor A. Fredga, member of the Nobel Committee for Chemistry of the Royal Swedish Academy of Sciences

Your Majesties, Royal Highnesses, Ladies and Gentlemen.

Our epoch has witnessed the gradual replacement of traditional materials by synthetic ones. We have all seen that plastics can often substitute glass, porcelain, wood, metals, bones, and horn, the substitutes being frequently lighter, less fragile, and easier to shape and work. It has in fact been said that we live in the Age of Plastics.

Plastics consist of very large molecules or macromolecules often forming long chains of thousands of atoms. They are made by joining together normal-size molecules constituting the basic units. These molecules must be reactive, but some outside help is also necessary to make them combine. This outside assistance often used to be supplied by free radicals, added to trigger off the reaction of polymerization. The term «free radical» may conjure up political connotations, and indeed free radicals have much in common with revolutionaries: they are full of energy, difficult to control, and have an unpredictable outcome. Thus, free-radical reactions give polymer chains with branches and other anomalies.

However, Professor Ziegler has found entirely new methods of polymerization. Studying organometallic compounds, he discovered that organo-aluminium compounds, which are easy to prepare, are particularly suitable for work on the industrial scale. Peculiar electrical forces operate around an aluminium-carbon bond in a hydrocarbon chain: reactive molecules are drawn in and sandwiched between the carbon atom and the aluminium atom, thus increasing the length of the chain. All this happens much more quietly than in free-radical reactions. When the chain is long enough, we detach the aluminium and thus stop the further growth of the molecule. The combination of aluminium compounds with other metallic compounds gives Ziegler catalysts. These can be used to control polymerizations and to obtain molecular chains of the required length. However, many systematic experiments – and indeed some accidental findings – were necessary to reach this stage. Ziegler catalysts, now widely used, have simplified and rationalized polymerization processes, and have given us new and better synthetic materials.

The individual molecules strung together to form polymers are often so built that the resulting chain exhibits small side groups or side-chains at certain points, generally one at every other carbon atom. But the picture is more complicated, since these side groups can be oriented either to the left or to the right. When their orientations are randomly distributed, the chain has a spatially irregular configuration. However, Professor Natta has found that certain types of Ziegler catalysts lead to stereoregular macromolecules, *i.e.* macromolecules with a spatially uniform structure. In such chains, all the side groups point to the right or to the left, these chains being called isotactic. How is this achieved when the microstructure of the catalyst is probably highly irregular? The secret is that the molecular environment of the metal atom, at which new units are stuck on to the chain as mentioned before, is so shaped that it permits only a definite orientation of the side groups.

Isotactic polymers show very interesting characteristics. Thus, while ordinary hydrocarbon chains are zigzag-shaped, isotactic chains form helices with the side groups pointing outwards. Such polymers give rise to novel synthetic products such as fabrics which are light and strong at the same time, and ropes which float on the water, to mention only two examples.

Nature synthesizes many stereoregular polymers, for example cellulose and rubber. This ability has so far been thought to be a monopoly of Nature operating with biocatalysts known as enzymes. But now Professor Natta has broken this monopoly.

Towards the end of his life, Alfred Nobel was thinking of the manufacture of artificial rubber. Since then, many rubber-like materials have been produced, but only the use of Ziegler catalysts enables us to synthesize a substance that is identical with natural rubber.

Professor Ziegler. Your excellent work on organometallic compounds has unexpectedly led to new polymerization reactions and thus paved the way for new and highly useful industrial processes. In recognition of your services to Science and Technology, the Royal Academy of Sciences has decided to award you the Nobel Prize. It is my pleasant duty to convey to you the best wishes of the Academy.

Professor Natta. You have succeeded in preparing by a new method macromolecules having a spatially regular structure. The scientific and technical consequences of your discovery are immense and cannot even now be fully estimated. The Swedish Royal Academy of Sciences wishes to express its appreciation by awarding you the Nobel Prize. Please accept the best wishes of the Academy. I would also like to express the admiration of the Academy for the intensity with which you are continuing your work in the face of difficulties.

Professor Ziegler. In the name of the Academy, I now ask you to accept the Nobel Prize from His Majesty the King.

Professor Natta. In the name of the Academy, I now ask you to accept the Nobel Prize from His Majesty the King.

KARL ZIEGLER

Consequences and development of an invention*

Nobel Lecture, December 12, 1963

The awarding of the Nobel Prize for Chemistry for the year 1963 is related to the precipitous expansion of macromolecular chemistry and its industrial applications, which began precisely ten years ago at my Max-Planck-Institute for Coal Research, in Mülheim/Ruhr. The suddenness with which this began, and the rapidity with which it was propagated are comparable to an explosion. The energy carriers in this case were the ingenuity, activity, creative imagination and bold concepts of the many unnamed chemists, designers and entrepreneurs in the world who have fashioned great industries from our humble beginnings.

If today I stand with my colleague Natta, who has been particularly effective in promoting this explosive wave, in the limelight of distinction, and do wish to manifest, with this address, my appreciation for the honor bestowed upon me, I must begin by thanking these many anonymous persons. They, too, deserve this distinction.

The extent of this «explosion» may be illustrated by two charts[1], in which the location of newly-established plants is indicated. The places marked by black circles refer to the production of high molecular weight materials, the crosses to new production facilities which, though concerned with low molecular weight materials, nevertheless also have some connection with the address I am delivering today (Figs. 1 and 2).

The new development had its inception near the end of 1953, when I, together with Holzkamp, Breil and Martin[2], observed - during only a few days of an almost dramatic course of events - that ethylene gas will polymerize very rapidly with certain catalysts that are extremely easy to prepare, at 100, 20 and 5 atmospheres and, finally, even at normal pressure, to a high molecular weight plastic.

I would like to first describe our normal-pressure polymerization experiment, which actually takes about an hour but which has been condensed in the film[3] to a few minutes (not shown here).

* This translation of Prof. Ziegler's Nobel Lecture is reproduced with some modifications, by permission of the publishers, from *Rubber Chem. Technol.*, 38 (1965) xxiii.

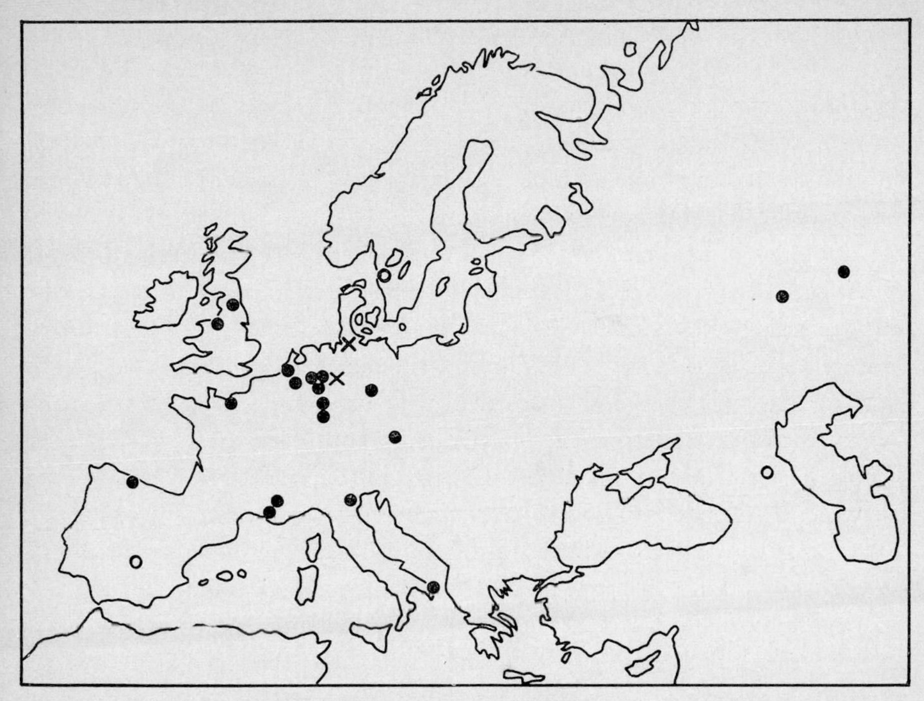

Fig. 1. Location of industrial applications of the Mülheim processes in Europe (as of 1963). On the figure: ●, High molecular weight materials; ×, Aluminium alkyls and low molecular weight materials; ○, Under construction or planned.

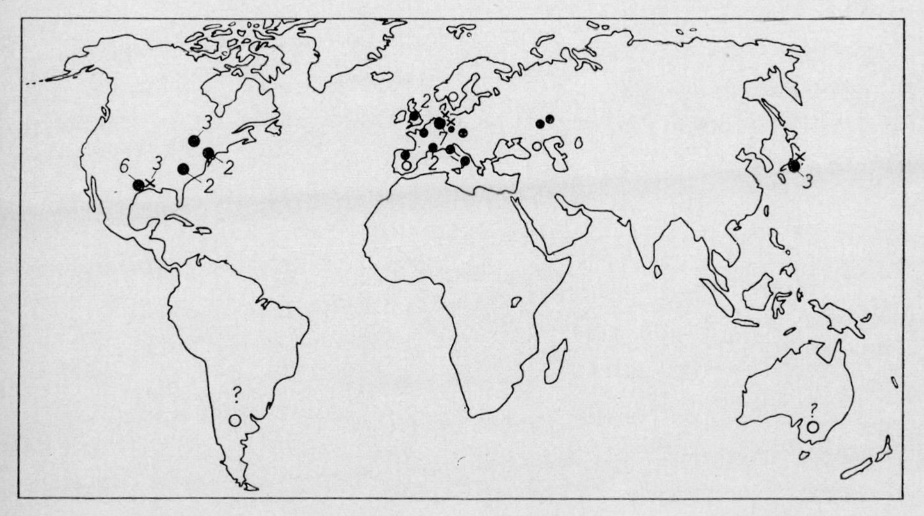

Fig. 2. Location of industrial applications of the Mülheim processes in the world (as of 1963). Symbols as in Fig. 1. Numbers indicate the number of factories.

The catalyst is prepared simply by simultaneously pouring, with exclusion of air, two liquid materials into about two liters of a gasoline-like hydrocarbon, after which ethylene is introduced, while stirring. The gas is absorbed quickly; within an hour one can easily introduce 300–400 liters of ethylene into the two liters of liquid. At the same time, a solid substance precipitates, in such a way that after approximately one hour the material becomes doughy and can scarcely be stirred any more. If the brown catalyst is then destroyed, by the addition of some alcohol and by the introduction of air, the precipitate becomes snow-white and can be filtered off. In its final state it will accumulate, in amounts of 300–500 g, as a dry, white powder.

The results of this experiment greatly surprised us, and, later on, many others, since up to that time ethylene had been considered extremely difficult to polymerize. The «polythene» of the Imperial Chemical Industries, a product which had been known for some seventeen years, was being prepared under pressures of 1000–2000 atmospheres, and at a temperature of 200°C. Our experiment thus destroyed a dogma. It led, in addition, to a polyethylene which differed quite markedly from the high-pressure product. Low-pressure polyethylene not only has a better resistance to elevated temperatures and a higher density, but is also more rigid. This is easily demonstrated by holding in one hand two similar objects made of the two materials, and press-

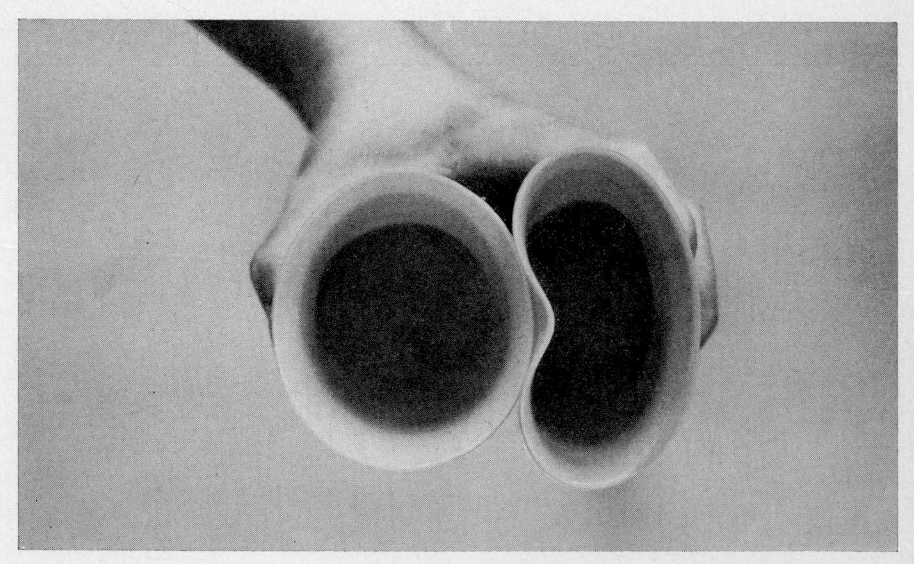

Fig. 3. Comparison between the rigidity of two beakers, one of low-pressure, one of high-pressure polyethylene.

ing them together (Fig. 3). Low-pressure polyethylene can be drawn without difficulty to form fibers or ribbons of high tensile strength. This cannot be done at all with high-pressure polyethylene, or at best only an indication of drawing is obtained. We established these facts immediately after our discovery, with test specimens which were still quite primitive[1].

The differences can be attributed to the fact that in our process molecules of ethylene are joined together linearly, without interruption, whereas in the high-pressure process chain growth is disturbed, so that a strongly branched molecule results (Fig. 4).

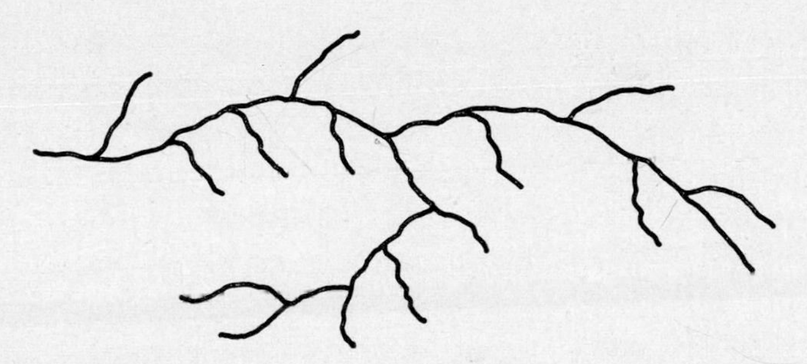

Fig. 4. High-pressure polyethylene, structural principle.

The low-pressure process found immediate acceptance in industry. By 1955, 200 metric tons of this new type of plastic had been produced; in 1958 it was 17000 tons, and in 1962 some 120000 tons. The much higher figures occasionally cited for this and other plastics have resulted from the confusion of available, but unused, capacities with actual production. The increase in dimensions can be indicated by comparison of our first test specimens, prepared ten years ago with rather primitive means, with containers that are twenty cubic meters in capacity, the largest now being made from polyethylene. A subsequent figure shows the lightness of the material, since a very large container can easily be carried by only a few men.

The catalyst employed in the experiment described was prepared by mixing aluminium triethyl, or diethyl aluminium chloride, with titanium tetrachloride. However, this is only one example, taken from the countless series of «organometallic mixed catalysts». Most generally they will form, as we found, whenever standard organometallic compounds, preferably those of aluminium, but also many of other metals, are brought into contact with compounds of certain heavy metals. Those of titanium, zirconium, vanadium, chromium, molybdenum, cobalt and the like are especially effective. Since there are many different metal alkyls and many different heavy metal compounds, and since, furthermore, components can be mixed together in varying proportions, and by different methods, and because all this can have an effect, often a truly decisive effect, on the nature of the catalytic activity, it is easy to understand why this field has grown to practically limitless proportions.

In place of the metal alkyls, one can also use metal hydrides, or the metals themselves, whereas metal alkyls probably will still form during the catalyzed processes.

Our catalysts then became known, at the turn of the year 1953/4, to our friends in industry and to their foreign colleagues, in Frankfurt, the Ruhr, Manchester, and – last but not least – Milan. Shortly thereafter this knowledge jumped over to the U. S. A. as well, and ultimately our findings became available to all. The consequences have been characterized, elsewhere, by the statement that revelation of the Mülheim catalysts had the same effect as the starting gun of a race in which the laboratories of the interested industries had been entered[4]. However, representatives of purely scientific chemistry also participated.

Because of the magnitude of the new field, arrival at further stages, or the order of such arrivals, was necessarily dependent upon contingencies. Indeed, many important observations were made within short spaces of time, independently of one another, and at different places. Let me illustrate this with two examples: It was pure chance that in November of 1953 the first of the catalysts in which our invention was clearly recognizable happened to be a relatively weak-acting combination of an aluminium alkyl with a zirconium compound, by which ethylene could be polymerized only under a few atmospheres pressure, and with which propylene, already tested the day after our critical experiment with ethylene, would not polymerize at all. Then, for a number of weeks, we were absorbed in experimenting with normal-pressure polymerizations of ethylene by means of titanium-containing catalysts. Early in 1954 we recognized the possibility of copolymerizing ethylene and propylene, after which we succeeded, at Mülheim, in polymerizing propylene with more effective catalysts, but – and this we did not know at the time – a short while after my colleague Natta of Milan had already observed this. In a first substantiation of his observation, and in an act of fairness, Natta had referred to the catalyst used as a «Ziegler catalyst», and that is how this expression found its way into the literature[5]. It is surely understandable that I myself prefer to speak of them as «Mülheim catalysts».

The second example: Near the close of 1955 work was being done in many places on the polymerization of butadiene with our catalysts. But no one had observed that in addition to the desired high polymers, a very interesting trimer of butadiene, namely 1,5,9-cyclododecatriene, was being produced. Günther Wilke, of my institute, became aware of this, and showed how one can guide the reaction entirely in this new direction. While endeavouring to explain the formation mechanism of cyclododecatriene, Wilke discovered a way to redirect this reaction at will, either toward a dimerization to an eight-membered ring, or – by a co-reaction with ethylene – toward a co-oligomerization to a ten-membered ring[6]. The result was that the Mülheim catalysts also achieved importance for polycondensation plastics such as Nylons 8, 10 and 12, into which the ring compounds can be transformed.

These cyclizations constituted the third surprising development afforded the scientific community by the organometallic mixed catalysts, if I assign number one to the new polyethylene process. I saved the second surprise for

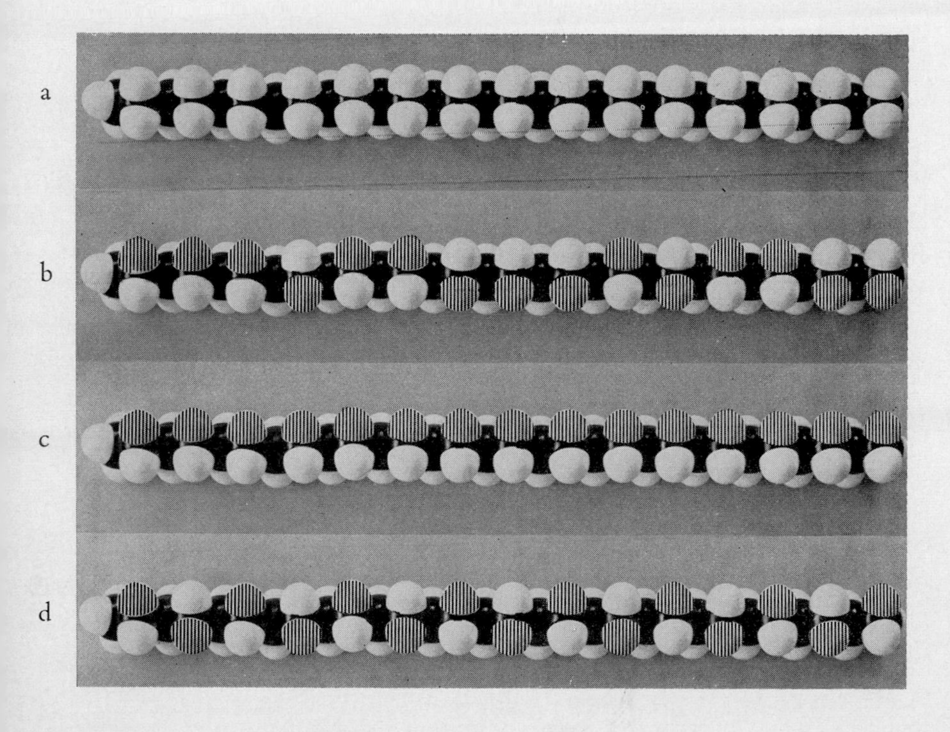

Fig. 5. Portions of the chains of (a) polyethylene, (b) atactic, (c) isotactic and (d) syndiotactic polypropylene. The methyls in the polypropylene are striped, and are actually much larger than shown.

later, and I must go into that now. From the middle of 1954 on, it began to be obvious that the Mülheim catalysts were capable of polymerizing in a structurally specific, as well as a stereo-specific manner. This realization is an essential contribution of my colleague Natta. He had often pondered over the mechanism of the polymerization, and very successfully strove to «train» the catalysts in such a way that they would possess extremely high specificity. Without wishing to anticipate Natta[7] in any way, I nevertheless feel obliged, for the sake of completeness, to explain briefly what this is all about.

The chain of linear polyethylene in the model, at an enlargement of fifty million, has approximately the following shape: (Fig. 5a). If a substituted ethylene, for example propylene, is polymerized, only the two doubly-bound carbon atoms of the olefin molecule will participate in the chain formation. The substituents, as side chains, will remain on the outside. If they combine in a purely random fashion the resultant product will show an entirely arbitrary distribution of the substituents along the two sides. Previously it had been believed that only those polymers could be formed which Natta–so far as I know at the suggestion of his wife–later called «atactic» (Fig. 5b). In stereo-specific polymerizations, polymers with highly regular structure are produced, with all the substituents on one side – isotactic (Fig. 5c) according to Natta–or with the substituents in a regular right-left sequence–syndiotactic (Fig. 5d) according to Natta. Both these terms were again inspired by Mrs. Natta. The particularly favorable properties of the products correspond to the regularity of the structure.

Analogous phenomena were encountered when our catalysts were used for polymerization of butadiene. In this instance, either only one of the two double bonds present can take part in the polymerization process. The result is a configuration comparable to that of polypropylene and containing, instead of methyl groups, only the unsaturated residues of ethylene, the so-called vinyl groups C_2H_3, in which case it can still be isotactic, atactic, or syndiotactic. This is a so-called 1,2-polymer of butadiene (Fig. 6, upper). Or, all four of the C-atoms can enter into the long chain of the polymer, in 1,4-polymerizations, so that in the middle of each individual C_4 structure unit a new double bond is formed, which was not present previously at that particular site (Fig. 6, lower).

In addition, because of the double bonds, and from their aspect, the valences of the two adjacent carbon atoms point either both toward one side, or to opposite sides. The first is the *cis* configuration, and the other is the *trans* (Fig. 7).

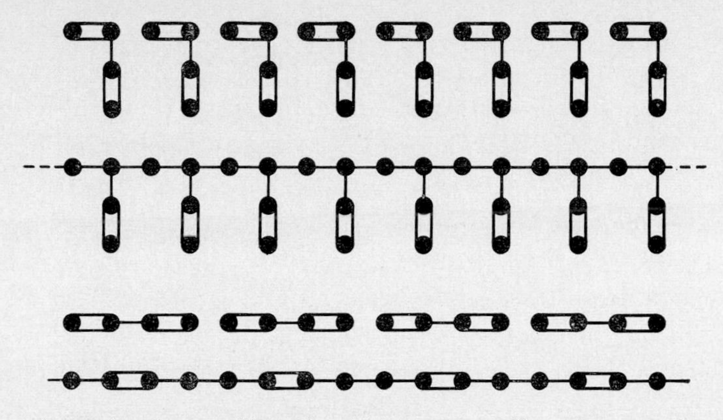

Fig. 6. «1,2-» (upper) and «1,4-» (lower) polymerizations of butadiene. Hydrogen atoms are not shown.

Natural rubber is a *cis*-1,4-polybutadiene, in which, at every double bond, the hydrogen atom has been replaced by a methyl group. Another important natural substance, guttapercha, corresponds to the *trans*-1,4-polymer (Fig. 7). The difficulty with all earlier attempts to synthesize rubber or rubberlike materials was that it was not possible to steer the polymerization of the basic materials – butadiene, isoprene – uniformly into the one or the other configuration. For this reason synthetic products contained a chaotic array of 1,4-*cis*, 1,4-*trans* and 1,2 structural units, even in the individual molecules. Although they resembled the natural product to some extent, none of them ever corresponded to it completely.

With the aid of the easily prepared Mülheim catalysts it is now possible to synthesize all these types uniformly, as desired, in a structure-specific or stereospecific manner. For example, 1,2-polybutadiene is formed by using a

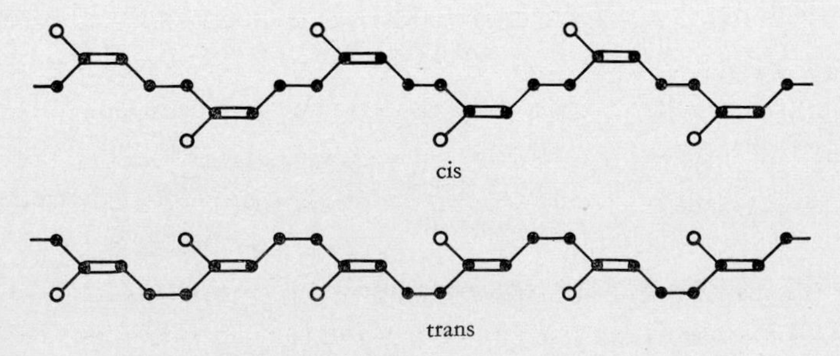

Fig. 7. Structural principle of natural rubber (*cis*) and guttapercha (*trans*). White circles: methyl groups. Hydrogen atoms are not shown.

catalyst made from titanium acid ester and 3 aluminium triethyl. With the catalysts obtained from $TiCl_4 + 0.5\ Al(C_2H_5)_2Cl$, *trans*-1,4-polybutadiene can be produced, and with those derived from $1\ TiI_4 + 1\ Al(C_2H_5)_3$ or $1\ CoCl_2 + 1\ Al(C_2H_5)_2Cl$, *cis*-1,4-polybutadiene will be formed, Finally, I would like to add that an increase of the Al:Ti ratio in the catalyst, to 5:1, will lead to cyclododecatriene.

A group from the B. F. Goodrich Research Center in the U. S. A. first made these observations with *cis*-1,4-polyisoprene, the synthetic «natural rubber», a few weeks after their company had learned about the essential features of our catalysts[8]. Actually, this represented only the final, closing stages of a 50-year effort to synthesize «genuine» rubber. Corresponding polymers of butadiene itself were then intensely studied, in a number of places, and *cis*-1,4-polybutadiene is today considered to be of great technological importance.

I will close this short survey with a discussion of recent developments pertaining to the rubber-like copolymers of ethylene and propylene, particularly those obtained with vanadium-containing organometallic mixed catalysts, and to the so-called terpolymers, into whose molecules certain diolefins – (dicyclopentadiene, or, again as discovered by Natta and coworkers[9], our cyclooctadiene-1,5) – have been incorporated.

Large quantities of all these new synthetic materials, discovered in connection with low-pressure polyethylene, are already being produced throughout the world, and production is sure to continue rising at a substantial rate.

With this I have shown, in broad outline, what has resulted in the course of ten years from our early experiments with organometallic mixed catalysts. In order to make the sequence of events which led to such a fruitful invention more understandable, I shall have to go back exactly forty years. Shortly after my graduation, having been a student of Karl von Auwers at the University of Marburg/Lahn, I began my independent scientific work with experiments for testing the theory of so-called free radicals. I incidentally found, in 1923, a new method for the formation of organic compounds of the alkali metals potassium and sodium[10], which brought my attention to the metal alkyls as an interesting, highly diversified field, that has continued to fascinate me, over and over again, up to the present. The new catalysts grew out of this, as a side sprout, in 1953. Permit me now to pursue the unbroken chain of causal relationships that links Then and Now by using special block schemes (Fig. 8, *0* → *1*, Figs. 9–12).

A few years later, in 1927, Bähr and I[11-13] made the discovery – important for the further development – that alkali alkyls can be added with ease to butadiene or styrene, at room temperature (Fig. 8,*2*). Repetition of the process leads first to oligomers, in a «stepwise organometallic synthesis», and finally to polymers and high polymer reaction products (Fig. 8,*4*).

$$R-A + CH_2=CH-CH=CH_2 \rightarrow R-CH_2-CH=CH-CH_2-A$$

$$\downarrow$$

$$R-(C_4H_6)_n-A \leftarrow R-CH_2-CH=CH-CH_2-CH_2-CH=CH-CH_2A$$

(arbitrarily designated as pure 1,4 structures)
Oligomers when $n = 2,3,4$, etc.
Polymers when n is considerably higher.

This first contact of mine with «macromolecular chemistry» later gave impetus to many investigations by third parties, and recently butyl lithium has also been suggested for industrial polymerizations of isoprene. At first, however, another, indirect result of our work was of more importance. Secondary observations suggested the conclusion that metallic lithium should be amenable to a reaction analogous to the one by which Grignard compounds are formed from magnesium:

$$RCl + Mg = RMgCl$$

$$RCl + 2\,Li = RLi + LiCl$$

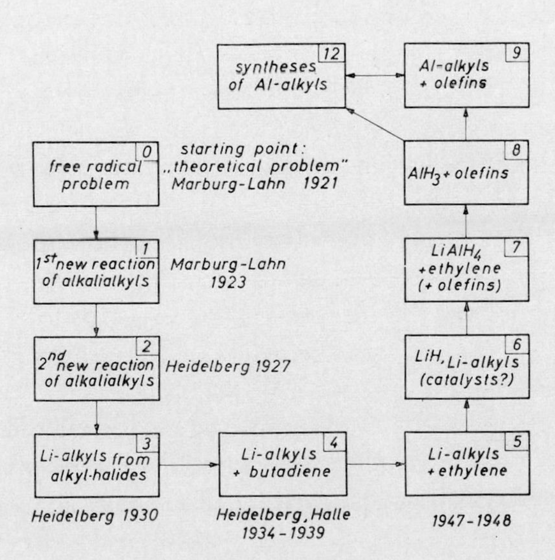

Fig. 8. Preliminary work (Marburg/Lahn, Heidelberg, Halle/Saale) 1921–1939. First results in Mülheim/Ruhr.

With Colonius[14], I was able to confirm this in 1930, and that is how the organolithium compounds became easily accessible (Fig. 8,*3*).

In Mülheim/Ruhr, where I have been working since 1943, Gellert and I succeeded in transferring the technique of a «stepwise organometallic synthesis» from butadiene to ethylene[15]. In this instance the reaction leads from lithium alkyl directly to the higher straight-chain lithium alkyls, and hence also to alcohols, carboxylic acids, and the like (Fig. 8, *5*).

$$C_2H_5-Li \xrightarrow{C_2H_4} C_2H_5-CH_2-CH_2-Li \xrightarrow{C_2H_4} C_2H_5-(CH_2-CH_2)_n-Li$$

$$\xrightarrow{O_2} C_2H_5-(CH_2-CH_2-)_nOH \xrightarrow{CO_2} C_2H_5-(CH_2-CH_2)_nCO_2H$$

Contrary to what holds true for butadiene, however, there is a limit here to the growth which a chain can undergo, since, for ethylene addition, the temperatures required are such that the lithium alkyls will readily decompose to lithium hydride + olefin. This certainly seemed to justify the following conclusion: If, in such decompositions, it is a question of a reversible reaction, as we had reason to believe, then lithium hydride and lithium alkyls should, under proper conditions, function as catalysts for the polymerization, or rather, oligomerization, of ethylene to the higher α-olefins (Fig. 8, *6*).

We did find such a reaction in principle, but it was so complicated by secondary and subsequent reactions that we could do nothing with it. Then when I had already decided to give up these efforts, my coworker, H. G. Gellert, conducted one more experiment–and the last, he was convinced–with the just recently discovered lithium aluminium hydride. This led immediately to the desired higher α-olefins (Fig. 8, *7*). As the decisive turning point, this resulted in the realization that the alkali metal was not the crucial issue at all, and that everything we already knew about the lithium alkyls, and all that we had anticipated besides, with respect to the chemistry of the olefinic hydrocarbons, could be achieved with a great deal more ease through use of organoaluminium compounds[16]. That is:

(*1*) There are genuine equilibria aluminium alkyl $\rightleftharpoons$ aluminium hydride + olefin lying, as a rule, entirely to the left, so that, in reverse of the situation with lithium, it is possible to synthesize the aluminium alkyls from hydride + olefin (Fig. 8, *8,9*).

(*2*) In the case of aluminium, too–and this came as a real surprise–at moderately high temperatures a stepwise organometallic synthesis, or as we now call it, a «growth» or propagation reaction takes place, leading to the higher aluminium alkyls; thus, a synthesis of the higher straight-chain primary monofunctional aliphatic compounds, particularly the fatty alcohols (Fig. 9, *10,11*), became possible.

(*3*) Furthermore, we have, from about 150° on, a catalytic oligomerization of the ethylene to higher α-olefins (Fig. 9, *21*).

Here the organometallic synthesis appears as the partial reaction of a completely understood, homogeneous intermediate reaction catalysis: After a certain number of addition steps the intermediate product decomposes to a hydride and an olefin whereupon, after the addition of ethylene to the hydride, the cycle is repeated.

$$C_2H_5al + (n-1)C_2H_4 = C_2H_5(C_2H_4)_{n-1}\text{-}al$$

$$C_2H_5(C_2H_4)_{n-1}\text{-}al = alH + C_2H_5(C_2H_4)_{n-2}\text{-}CH{=}CH_2$$

$$alH + C_2H_4 = alC_2H_5$$

$$\overline{\qquad nC_2H_4 = (C_2H_4)_n \qquad} \qquad al = 1/3\,Al$$

Such a reaction was encountered in its most primitive form with propylene, for which the homogeneous catalysis leads, without supplementary chain growth, almost exclusively to a well-defined dimer[17] (Fig. 9, *19*):

$$CH_3{-}CH_2{-}CH_2{-}al \xrightarrow{C_3H_6} \begin{array}{c} CH_3{-}CH_2{-}CH_2 \\ | \\ CH_3{-}CH{-}CH_2{-}al \end{array}$$

$$\downarrow$$

$$\uparrow\; C_3H_6 \text{——} alH + \begin{array}{c} CH_3{-}CH_2{-}CH_2 \\ | \\ CH_3{-}C{=}CH_2 \end{array}$$

$$\begin{array}{c} CH_3{-}CH_2{-}CH \\ \| \\ CH_3{-}C{-}CH_3 \end{array} \longleftarrow$$

$$\downarrow$$

$$CH_4 + \begin{array}{c} CH_2{=}CH \\ | \\ CH_2{=}C{-}CH_3 \end{array}$$

Recently this reaction has achieved significance for high molecular weight chemistry as well, since the cracking of isohexene, following the shifting of the double bond, produces isoprene, in addition to methane[18] (Fig. 9, *20*).

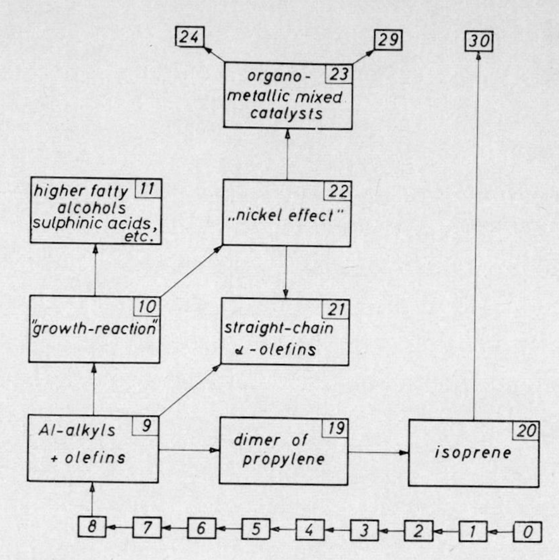

Fig. 9. Course of the Mülheim Experiments, Part I.

The transition of all these reactions into industrial applications was finally accomplished by the so-called «direct synthesis» of aluminium alkyls from aluminium, hydrogen and olefins, discovered by us at approximately the same time as the new polyethylene process. Aluminium hydride, from which the aluminium trialkyls are quite easy to obtain through the addition of olefins, cannot be prepared directly from the metal and hydrogen. However, in already-prepared aluminium trialkyls, aluminium will dissociate with hydrogen to dialkyl aluminium hydrides which, with ethylene, will give 1.5 times the original amount of aluminium triethyl,

$$\mathrm{Al} + 2\,\mathrm{Al(C_2H_5)_3} + 1\tfrac{1}{2}\,\mathrm{H_2} = 3\,\mathrm{HAl(C_2H_5)_2}$$

$$3\,\mathrm{HAl(C_2H_5)_2} + 3\,\mathrm{C_2H_4} = 3\,\mathrm{Al(C_2H_5)_3}$$

$$\mathrm{Al} + 1\tfrac{1}{2}\,\mathrm{H_2} + 3\,\mathrm{C_2H_4} = \mathrm{Al(C_2H_5)_3}$$

so that any amount of aluminium trialkyl can be prepared without difficulty[19] (Fig. 10, *12,14*).

In the charts shown at the beginning of this address, the location of the industries engaged in the production of aluminium alkyls and their low molecular weight applications were included. This area of the industrial development initiated by Mülheim is likewise in a state of continuous evolution, though it has been less rapid than that of the high molecular weight phase.

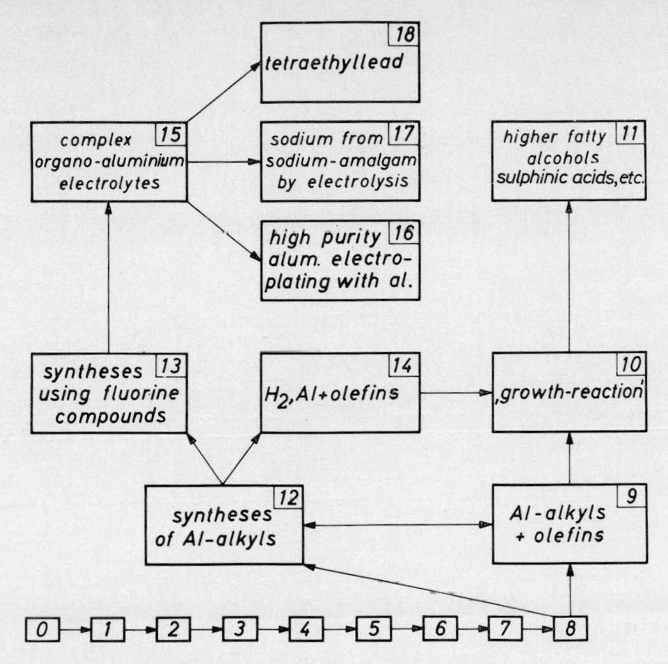

Fig. 10. Course of the Mülheim Experiments, Part II.

Up to the year 1952 we had frequently conducted «growth» reactions based on aluminium triethyl, and thought we were thoroughly acquainted with such reactions. But when, together with E. Holzkamp[20], I attempted to apply this type of reaction to aluminium tripropyl the formation of chains, to our great surprise, did not materialize at all. On the contrary, we obtained propylene–from the propyl aluminium–in addition to aluminium triethyl and α-butylene. Even starting from aluminium triethyl our reaction now yielded nothing but butylene and the unchanged aluminium compound.

The explanation was obvious: A catalyst in trace amounts must have gotten into this series of experiments, leading to an uncommonly rapid acceleration of the displacement reactions:

$$C_3H_7al + C_2H_4 = C_2H_5al + C_3H_6$$

and

$$C_4H_9al + C_2H_4 = C_2H_5al + C_4H_8$$

Under such conditions the alkyl had to be forced out at the aluminium, immediately after the first propagation step, as butylene. It is now generally known that we detected a tiny trace of metallic nickel as the disturbing element[2] (Fig. 9, *22*). Thus our attention was again directed to the problem to polymerize also ethylene just as, years ago, we had been able to do with butadiene and styrene, to produce a genuine macromolecule with the aid of metal alkyls, in this case aluminium alkyls in particular.

Our growth reaction must lead to a genuine polyethylene, if we succeeded in adding about 1 000 ethylene units to the aluminium triethyl. For this, with our reaction, only about 100–200 atmospheres pressure, instead of the 1 000–2 000 atmospheres used heretofore, would be required. Nevertheless, in properly performed experiments we had obtained only waxy products, because the chain at the aluminium was prematurely split off–apparently by a displacement reaction–as an olefin, with the re-formation of ethyl at the aluminium, an occurrence known to chemists working in the high molecular weight field as a «chain transfer reaction»:

$$C_2H_5\text{-}(CH_2\text{-}CH_2)_n\text{al} + C_2H_4 = C_2H_5\text{-}(CH_2\text{-}CH_2)_{n-1}\text{-}CH{=}CH_2 + C_2H_5\text{al}$$

To the extent that catalyst traces, as we now might well suspect, had been involved here also in effecting the displacement, there existed the prospect that completely «aseptic» procedures could eventually lead to a true polyethylene. In order to provide the essentials for the «asepsis», we began, in the middle of 1953, to systematically investigate substances which have effects somewhat similar to those of nickel. We found, instead, the polymerization-promoting organometallic mixed catalysts, and, in particular, we achieved a low-pressure polymerization of ethylene, and with this I have again arrived at my starting point (Fig. 9, *23*).

Fig. 11, which follows, once again shows, in schematic form, all that has resulted from the discovery of organometallic mixed catalysts. In this connection, isoprene is doubly concerned in our work: because of the aforementioned synthesis, and for polymerization purposes.

Finally, I would like to present the following scheme (Fig. 12), in order to show the entire development. The areas in two types of hatching indicate the important transitions (from Li to Al, from the Al-alkyls to the mixed catalysts), and also a third transition to an electrochemical side branch, which I cannot go into at this time (*cf.* Fig. 10, *15*→ *18*). The important consequences of the discovery of organometallic mixed catalysts, evident even to the layman, have led many to regard me, nowadays, as a «macromolecular chemist», and, in fact, even as a plastics expert. I have intentionally set my work in this field within a much broader framework, to show you that I am a «macro-

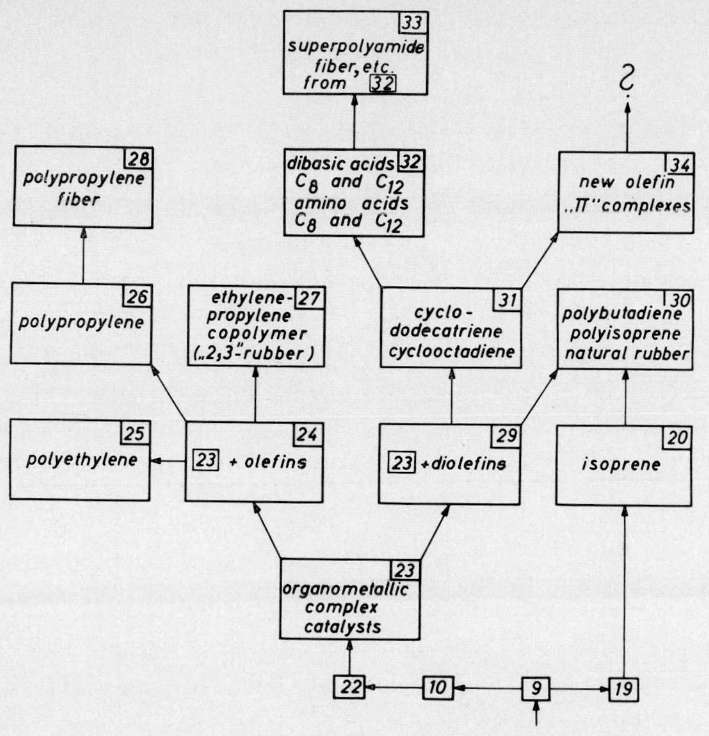

Fig. 11. Course of the Mülheim Experiments, Part III.

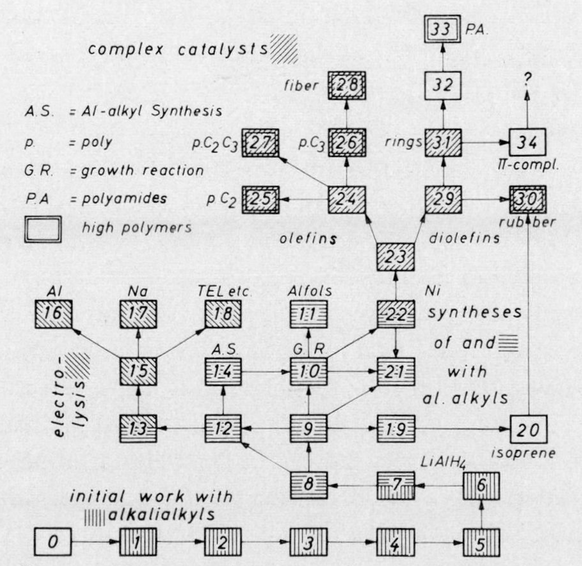

Fig. 12. The Mülheim experiments (1948–1963), overall aspect. Numbers are as in Figs. 8–11.

molecular chemist» only peripherally, and that I am not at all a plastics expert. Rather, I have always looked upon myself as a pure chemist. Perhaps that is also why the impact of the invention has been so enduring. The new knowledge has, after all, not come from macromolecular chemistry. It is the metal alkyls that have insinuated themselves into the chemistry of macromolecules to effectively fertilize this field. Typical of the course I have followed from those early beginnings of forty years ago until today, is the fact that I have never started with anything like a formally presented problem. The whole effort developed quite spontaneously, from a beginning which was actually irrational in nature, through an unbroken causal series of observations, interpretations of findings, rechecking of the interpretations by new experiments, new observations, etc. My method resembled a meandering through a new land, during which interesting prospects kept opening up, during which one could frequently view part of the road to be traveled, but such that one never quite knew where this trip was actually leading. For decades I never had the slightest notion that successful technological and industrial applications were also to be encountered during the journey.

Twice this path seemed seriously blocked. The first time was before the transition from lithium hydride to lithium aluminium hydride and from the lithium to the aluminium compounds had been accomplished. The second was when our growth reaction suddenly, in a truly mysterious way, refused to go any more. In both cases, a capitulation in the face of these difficulties would undoubtedly have broken the red thread of continuity, which can now be followed clearly.

But a much more formidable impediment might have presented itself. In order to illustrate this, I must elaborate on the paradox that the critical concluding stages of the investigations I have reported took place in an institute for «coal research».

When I was called to the Institut für Kohlenforschung in 1943, I was disturbed by the objectives implied in its name. I was afraid I would have to switch over to the consideration of assigned problems in applied chemistry. Since ethylene was available in the Ruhr from coke manufacture, the search for a new polyethylene process, for example, could certainly have represented such a problem. Today I know for certain, however, and I suspected at the time, that any attempt to strive for a set goal at the very beginning, in Mülheim, would have completely dried up the springs of my creative activity. As a matter of fact: Giving up my preoccupation with organometallic compounds in favor of the other, «bread-and-butter» problems of coal chemistry–many of my colleagues were of the opinion, at the time, that this would be the natural consequence of my removal to Mülheim–would have cut the leads which I already held invisibly in my hand, and which were to lead me safely to the results that proved of such importance also for the Ruhr industry.

As a condition of my transfer to Mülheim I stipulated that I was to have complete freedom of action in the entire field of the chemistry of carbon compounds, without regard to whether any direct relation to coal research was or was not recognizable. The acquiescence of my stipulation was in accord with the principles of the then Kaiser-Wilhelm, and now Max-Planck Society, of which my institute is a part. As far as the German coal mining industry which supported my institute is concerned, this was an act of great foresight on their part which did in fact provide the conditions for everything that occurred, particularly the present circumstance that my institute, and I with it, have now received this very great distinction.

The institute, however, –what is it, in its distinctive spiritual and intellectual substance, other than the totality of its active people. I began this address with an expression of gratitude to the many people in the world whom I know only slightly, or not at all, and who have developed great industries from our beginnings. I will end the address by expressing my heartfelt thanks to the many, very well known members of my institute who have stood by me faithfully throughout all these years, and who share with me the prize for which I have been singled out.

1. Figs. 1–11 are taken from K. Ziegler, *Arbeitsgemeinsch. Forsch. des Landes Nordrhein-Westfalen*, 128 (1964) 33.
2. K. Ziegler, E. Holzkamp, H. Breil and H. Martin, *Angew. Chem.*, 67 (1955) 541.
3. The experiment corresponding to ref. 1, Fig. 1 was shown as a movie in Stockholm.
4. *Hercules Chemist*, 46 (1963) 7.
5. K. Ziegler, E. Holzkamp, H. Breil and H. Martin, *Angew. Chem.*, 67 (1955) 426.
6. G. Wilke, *Angew. Chem.*, 75 (1963) 10; *Angew. Chem., Intern. Ed.*, 2 (1963) 105.
7. G. Natta, following lecture; also *Angew. Chem.*, 76 (1964) 553.
8. S. E. Honer *et al.*, *Ind. Eng. Chem.*, 48 (1956) 784.
9. G. Natta *et al.*, *Chim. e Ind.* (*Milan*), 45 (1963) 651.
10. K. Ziegler and F. Thielmann, *Ber.*, 56 (1923) 1740.
11. K. Ziegler and K. Bähr, *Ber.*, 61 (1928) 253.
12. K. Ziegler and H. Kleiner, *Ann. Chem.*, 473 (1929) 57.
13. K. Ziegler, F. Dersch and H. Wollthan, *Ann. Chem.*, 511 (1934) 13.
14. K. Ziegler and H. Colonius, *Ann. Chem.*, 479 (1930) 135.
15. K. Ziegler and H.-G. Gellert, *Ann. Chem.*, 567 (1950) 195.

16. K. Ziegler, *Angew. Chem.*, 68 (1956) 721, 724.
17. K. Ziegler, *Angew. Chem.*, 64 (1952) 323, 326.
18. V. F. Anhorn *et al.*, *Chem. Eng. Prog.*, 57 (1961) 43.
19. K. Ziegler *et al.*, *Ann. Chem.*, 629 (1960) 1.
20. K. Ziegler *et al.*, *Ann. Chem.*, 629 (1960) 121, 135.

Biography

Karl Ziegler was born in Helsa near Kassel in Germany, on November 26, 1898. He graduated in 1920 under Prof. von Auwers at the University of Marburg/Lahn, and qualified as a lecturer in 1923. After working for a short period at the University of Frankfurt/Main, he spent 10 years as a lecturer at Heidelberg.

His research work in the field of radicals with trivalent carbon and his syntheses of multi-membered ring systems earned him the Liebig Medal in 1935. In 1936 he became Professor and Director of the Chemisches Institut at the University of Halle/Saale. In the same year he lectured as a visiting professor at Chicago University.

From 1943 until 1969 he was Director of the Max-Planck-Institut für Kohlenforschung (formerly known as the Kaiser-Wilhelm-Institut für Kohlenforschung) in Mülheim/Ruhr. He continues his active association with the Institute in his capacity of Scientific Member. After the war he was instrumental in the foundation of the Gesellschaft Deutscher Chemiker, whose president he was for five years. From 1954 until 1957 he was president of the Deutsche Gesellschaft für Mineralölwissenschaft und Kohlechemie.

His research work at the above-mentioned institute over the past 20 years on syntheses and reactions in the chemistry of organoaluminium compounds, his discovery of organometallic mixed catalysts for the polymerization of olefins (*e.g.* the synthesis of high-density polyethylene) – all these are widely known.

Many honours have been bestowed upon him. He holds honorary doctorates at the Technische Hochschulen of Hannover and Darmstadt, and of the Universities of Heidelberg and Giessen. He has received the Liebig Medal from the Verein Deutscher Chemiker, the Carl Duisberg Plakette from the Gesellschaft Deutscher Chemiker, the Carl Engler Medal from the Deutsche Gesellschaft für Mineralölwissenschaft und Kohlechemie, and the Lavoisier Medal from the Société Chimique de France. The Werner von Siemens Foundation awarded him the Siemens Ring. His Nobel prize was followed by the award of a distinguished order by the German Federal Government, the Swinburne

Medal by the Plastics Institute, London, the International Synthetic Rubber Medal by *Rubber and Plastics Age*, London, and, in 1971, the Carl Dietrich Harries Plakette by the Deutsche Kautschuk Gesellschaft, as well as the Wilhelm Exner Medal by the Österreichischer Gewerbeverein, Vienna. He is an honorary senator of the Max-Planck Gesellschaft, founder president (1970–1972) of the Rheinisch-Westfälische Akademie der Wissenschaften, as well as member or honorary member of various German and foreign scientific societies and academies. In 1971, The Royal Society, London, elected him as a Foreign Member.

On the death of Otto Hahn, Karl Ziegler was appointed the latter's successor to the Order «Pour le mérite für Wissenschaften und Künste».

Karl Ziegler has been married to Maria Kurtz, since 1922. His daughter, Marianne Witte, is a doctor of medicine and is married to the chief physician of a children's hospital in the Ruhr. His son, Dr. Erhart Ziegler, is a physicist and patent attorney. Karl Ziegler has five grandchildren by his daughter and five by his son.

GIULIO NATTA

From the stereospecific polymerization to the asymmetric autocatalytic synthesis of macromolecules

Nobel Lecture, December 12, 1963

Introduction

Macromolecular chemistry is a relatively young science. Though natural and synthetic macromolecular substances had long been known, it was only between 1920 and 1930 that Hermann Staudinger placed our knowledge of the chemical structure of several macromolecular substances on a scientific basis[1]. In the wake of Staudinger's discoveries and hypotheses, macromolecular chemistry has made considerable progress.

Very many synthetic macromolecular substances were prepared both by polymerization and by polycondensation; methods were found for the regulation of the value and distribution of molecular weights; attempts were made to clarify the relationships existing among structure, chemical regularity, molecular weight, and physical and technological properties of the macromolecular substances. It was far more difficult to obtain synthetic macromolecules having a regular structure from both the chemical and steric points of view.

An early result in this field, which aroused a certain interest in relation to elastomers, was the preparation of a polybutadiene having a very high content of *trans*-1,4 monomeric units, in the presence of heterogeneous catalysts[2].

A wider development of this field was made possible by the recent discovery of stereospecific polymerization. This led to the synthesis of sterically regular polymers as well as to that of new classes of crystalline polymers.

Before referring to the stereospecific polymerizations and to their subsequent developments, I wish to make a short report on the particular conditions that enabled my School to rapidly achieve conclusive results on the genesis and structure of new classes of macromolecules. I also wish to describe the main stages of the synthesis and characterization of the first stereoregular polymers of α-olefins.

The achievement of these results has also been helped by the research I did in 1924 when I was a trainee student under the guidance of Professor Bruni. At that time I began to apply X-ray study of the structures of crystals to the resolution of chemical and structural problems[3].

At first, investigations were mainly directed to the study of low-molecular-weight inorganic substances and of isomorphism phenomena; but, after I had the luck to meet Professor Staudinger in Freiburg in 1932, I was attracted by the study of linear high polymers and tried to determine their lattice structures.

To this end I also employed the electron-diffraction methods which I had learned from Dr. Seemann in Freiburg and which appeared particularly suitable for the examination of thin-oriented films[4]. I applied both X-ray and electron-diffraction methods also to the study of the structure of the heterogeneous catalysts used for certain important organic industrial syntheses, and thus had the possibility of studying in the laboratory the processes for the synthesis of methanol[5] and the higher alcohols[6], and also of following their industrial development in Italy and abroad.

In view of the experience I had acquired in the field of chemical industry, certain Italian Government and industrial bodies entrusted me in 1938 with the task of instituting research and development studies on the production of synthetic rubber in Italy.

Thus the first industrial production of butadiene–styrene copolymers was realized in Italy at the Ferrara plants, where a purely physical process of fractionated absorption was applied for the first time to the separation of butadiene from 1-butene[7].

At that time I also began to be interested in the possible chemical applications of petroleum derivatives, and particularly in the use of olefins and diolefins as raw materials for chemical syntheses such as oxosynthesis[8] and polymerization[9].

The knowledge acquired in the field of the polymerizations of olefins enabled me to appreciate the singularity of the methods for the dimerization of α-olefins that Karl Ziegler described[10] in a lecture delivered in Frankfurt in 1952; I was struck by the fact that in the presence of organometallic catalysts it was possible to obtain only one dimer from each α-olefin, while I knew that the ordinary, cationic catalysts previously used yielded complex mixtures of isomers with different structures.

At this time I also became acquainted with Ziegler's results on the production of strictly linear ethylene oligomers, obtained in the presence of homogeneous catalysts. My interest was aroused, and in order to understand better the reaction mechanism[11], concerning which very little was known, I started the kinetic study of such polymerizations. In the meantime Ziegler discovered the process for the low-pressure polymerization of ethylene[12]. I then decided to focus attention on the polymerization of monomers other than ethylene; in particular I studied the α-olefins, which were readily available at low cost in the petroleum industry.

At the beginning of 1954 we succeded in polymerizing propylene, other α-olefins, and styrene; thus we obtained polymers having very different properties from those shown by the previously known polymers obtained from these monomers[13]. I soon observed that the first crude polymers of α-olefins and of styrene, initially obtained in the presence of certain Ziegler catalysts ($TiCl_4$ + aluminium alkyls), were not homogeneous, but consisted of a mixture of different products, some amorphous and non-crystallizable, others more or less crystalline or crystallizable. Accordingly, I studied the separation of the different types of polymer by solvent extraction and the structures of the single separated products. Even if the more soluble polymers were amorphous and had a molecular weight lower than that of the crystalline, but far less soluble, polymers deriving from the same crude product, I observed that some little-soluble crystalline fractions had a molecular weight only a little higher than that of other amorphous fractions. Therefore, convinced of the well-known saying *natura non facit saltus*, I did not attribute crystallinity to a higher molecular weight, but to a different steric structure of the macromolecules present in the different fractions[14].

In fact all vinyl polymers may be regarded as built from monomeric units containing a tertiary carbon atom. Thus in a polymer of finite length, such a carbon atom can be considered asymmetric, and hence two types of monomeric units may exist, which are enantiomorphous[13,15].

Since all the polymers of vinyl hydrocarbons previously known, even those recognized as having a head-to-tail enchainment like polystyrene, were amorphous, we examined the possibility that the crystallinity we observed was due to a chemically regular (head-to-tail) structure, accompanied by regular succession of steric configurations of the single monomeric units. Indeed, X-ray analysis permitted us to determine the lattice constants of crystalline polypropylene[16] and polystyrene[17]. The identity period along the chain axis in the fiber spectra was of about 6.5 Å and might be attributed to a chain segment containing three monomeric units[18]. This led us to exclude the idea that the crystallinity was due to a regular alternation of monomeric units having opposite steric configuration. Thus it could be foreseen, as was in fact later proved by more accurate calculations of the structure factors, that the polymeric chains consisted of regular successions of monomeric units, all having the same steric configuration[14].

In the subsequent study of the butadiene polymers, prepared by us in the presence of organometallic catalysts (for example, catalysts containing chromium[19]) that have 1,2-enchainment, two different types of crystalline polymers were isolated and purified.

The X-ray and electron-diffraction analyses of these products enabled us to establish that the structure of one of them is analogous to the structures of poly-α-olefins[20] – that is, characterized by the repetition of monomeric units having the same configuration. We also established that the other crystalline product is characterized by a succession of monomeric units, which are chemically equivalent but have alternately opposite steric configuration[21], as confirmed by a thorough X-ray analysis of the structure. In order to distinguish these different structures I proposed the adoption of terms coined from the

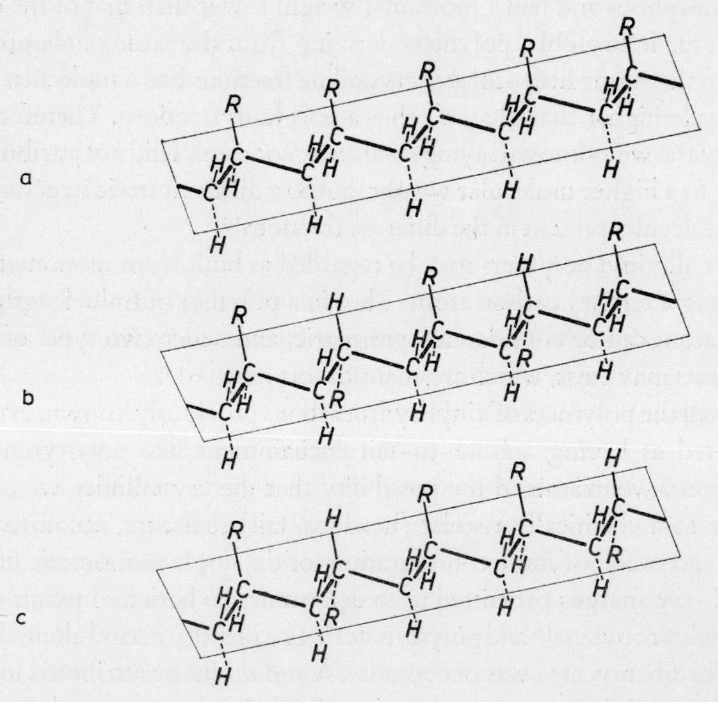

Fig. 1. Models of chains of head-to-tail vinyl polymers supposed arbitrarily stretched on a plane, having, respectively, isotactic (a), syndiotactic (b), and atactic (c) successions of the monomeric units.

ancient Greek, and these are now generally used[22]; that is, *isotactic*[14] and *syndiotactic*[21].

Fig. 1 shows the first device we used for an easy distinction of the different types of stereoisomerism of vinyl polymers; the main chains have been supposed arbitrarily stretched on a plane.

By accurate examination of the structure of isotactic polymers on fiber spectra, we could establish that all crystalline isotactic polymers have a helical structure, analogous to that found by Pauling and Corey[23] for α-keratin (Fig. 2); in fact only the helix allows a regular repetition of the monomeric units containing asymmetric carbon atoms, as was foreseen by Bunn[24].

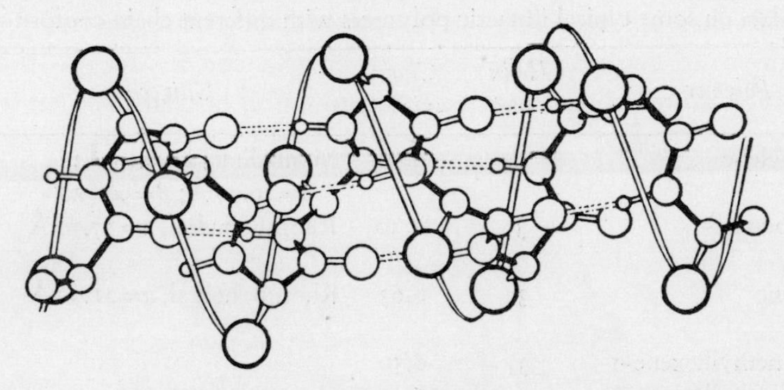

Fig. 2. Model of chain of α-keratin, according to Pauling and Corey.

Soon after the first polymerizations of α-olefins we realized the importance and vastness of the fields that were opened to research, from both the theoretical and the practical points of view.

Our efforts were then directed to three main fields of research: (*1*) To investigate the structures of the new polymers in order to establish the relationships existing between chemical structure, configuration, and conformation of the macromolecules in the crystalline state. (*2*) To find the conditions that allowed the synthesis of olefinic polyhydrocarbons having a determined type of steric structure, with high yields and high degree of steric regularity[25], as well as to study the reaction mechanism, and regulation of the molecular weight. (*3*) To attempt the synthesis, possibly in the presence of nonorganometallic catalysts, of stereoregular polymers corresponding to other classes of monomers having a chemical nature different from that of α-olefins.

I. *Crystalline Structure of High Stereoregular Polymers*

1. Homopolymers

The synthesis of new classes of crystalline macromolecules and the X-ray analysis of their structures led to the formulation of some general rules which determine the structure of linear macromolecules[26]. Table 1 summarizes some data concerning the structure of isotactic polymers; the data indicate that four-fold or higher order helices exist besides the three-fold ones already mentioned.

Table 1

X-Ray data on some typical isotactic polymers with different chain conformations.

Polymer	*Helix type*[a]	*Chain axis (Å)*	*Unit cell*	*Space group*
Polypropylene	3_1	6.50	Monoclinic, a=6.65 Å; b=20.96 Å; β=99° 20′	C2/c
Poly-α-butene[b]	3_1	6.50	Rhombohedral, a=17.70 Å	R3c or R$\bar{3}c$
Polystyrene	3_1	6.63	Rhombohedral, a=21.90 Å	R3c or R$\bar{3}c$
Poly-5-methylhexene-1	3_1	6.50		
Poly-5-methylheptene-1	3_1	6.40		
Poly-3-phenylpropene-1	3_1	~6.40		
Poly-4-phenylbutene	3_1	6.55		
Poly-o-methylstyrene	4_1	8.10	Tetragonal, a=19.01 Å	$I4_1cd$
Poly-α-vinylnaphthalene	4_1	8.10	Tetragonal, a=21.20 Å	$I4_1cd$
Polyvinylcyclohexane	4_1	6.50	Tetragonal, a=21.76 Å	$I4_1a$
Poly-3-methylbutene-1	4_1	6.84		
Poly-4-methylpentene-1	7_2	13.85	Tetragonal, a=18.60 Å	P$\bar{4}$
Poly-4-methylhexene-1	7_2	14.00	Tetragonal, a=19.64 Å	
Poly-m-methylstyrene	11_3	21.74	Tetragonal, a=19.81 Å	

[a] It is to be understood that, besides the right-handed X_n helix, the left-handed X_{X-n} helix also exists.

[b] Modification 1.

The conformation assumed by the single macromolecules in the lattice always corresponds to the conformation, or to one of the conformations, of the isolated molecule that shows the lowest internal energy content, the intramolecular Van der Waals forces being taken into account. The mode of packing of the polymer chains in a crystalline lattice takes place, as in the case of molecular crystals of low-molecular-weight substances, so as to fill the space in the best possible way.

If the polymer chain assumes a helicoidal conformation in the crystalline state, and if it does not contain asymmetric carbon atoms, it can be expected that either helices of the same sense, or, in equal ratio, helices of opposite sense are represented in the lattice.

Analogous to the case of nonenantiomorphous low-molecular-weight crystalline substances, so also in polymers that do not contain asymmetric carbon atoms, right- and left-handed helices are usually represented in the lattice in equal amount.

On the other hand, in the case of isotactic polymers containing asymmetric carbon atoms, the space group will not contain symmetry elements involving inversion, as, for instance, centers of symmetry or mirror or glide planes.

A racemic mixture of antipode macromolecules can be an exception. Furthermore, it is interesting to note that the chain symmetry is often maintained in the space group to which the unit cell of the polymer belongs.

With regard to the occurrence of enantiomorphous space groups, typical examples are represented by some isotactic poly-1-alkylbutadienes, in the

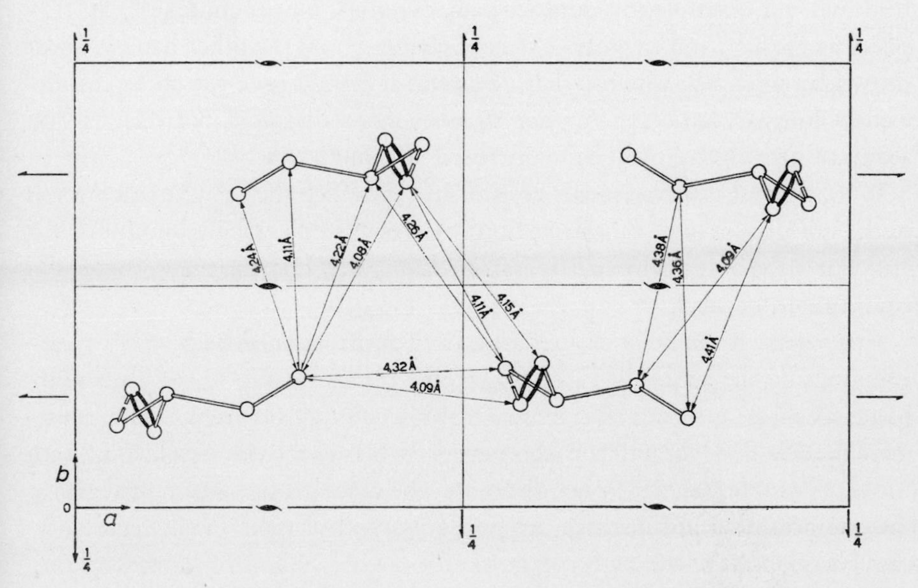

Fig. 3. Model of packing of isotactic *trans*-1,4-poly-1-ethylbutadiene in the crystalline state, projected on the (001) plane (space group $P2_12_12_1$).

crystalline lattice of which macromolecules with helices of exclusively one sense, right or left, exist for each crystal[27] (Fig. 3). Also in the case of isotactic poly-*tert.*-butylacrylate, the helices in the lattice seem to be all of the same sense[28].

If the chain symmetry is maintained in the crystal lattice, the possible occurrence of different space groups is considerably restricted. Where equal amounts of enantiomorphous macromolecules are contained in the lattice, we must distinguish two cases concerning the relative orientation of side groups of enantiomorphous macromolecules facing one another, which can be either isoclined or anticlined.

In the first case, possible symmetry operators for the covering of near macromolecules are either a mirror plane or a glide plane, parallel to the chain axis.

It is, however, known that good packing is generally obtained more easily with a glide plane than with a mirror plane, especially in the case of bodies having periodical recesses and prominences, as in the case of spiralized polymer chains.

In the case of a three-fold helix, each right-handed helix will be surrounded, because of the existence of the glide plane, by three isoclined left-handed helices, and *vice versa*; the space group will be $R3c$ (Fig. 4). This lattice is shown, for example, by isotactic polystyrene[29], by polybutene[30], by 1,2-polybutadiene[31], and by poly-*o*-fluorostyrene[32]; on the other hand it is not shown by isotactic polypropylene, because it would give rise to an insufficiently compact lattice, if Van der Waals contact distances, between carbon atoms of near chains, must be maintained[3] around 4.2 Å.

In the second case previously considered, in which the relative orientation of the side groups of enantiomorphous macromolecules facing one another is anticlined, the only symmetry operator relating neighboring macromolecules is a symmetry center.

And again, if the helix is threefold, each right-handed helix will be surrounded, by the action of three symmetry centers at 120°C, by three left-handed helices, and *vice versa*; the macromolecules are oriented so as to minimize the length of the unit cell axes perpendicular to the three-fold axis, with the best possible Van der Waals distances: the space group, which probably is the one presented, for instance, by polyvinylmethyl ether[33] and by poly-*n*-butylvinyl ether[34], will be $R\bar{3}$ (Fig. 5).

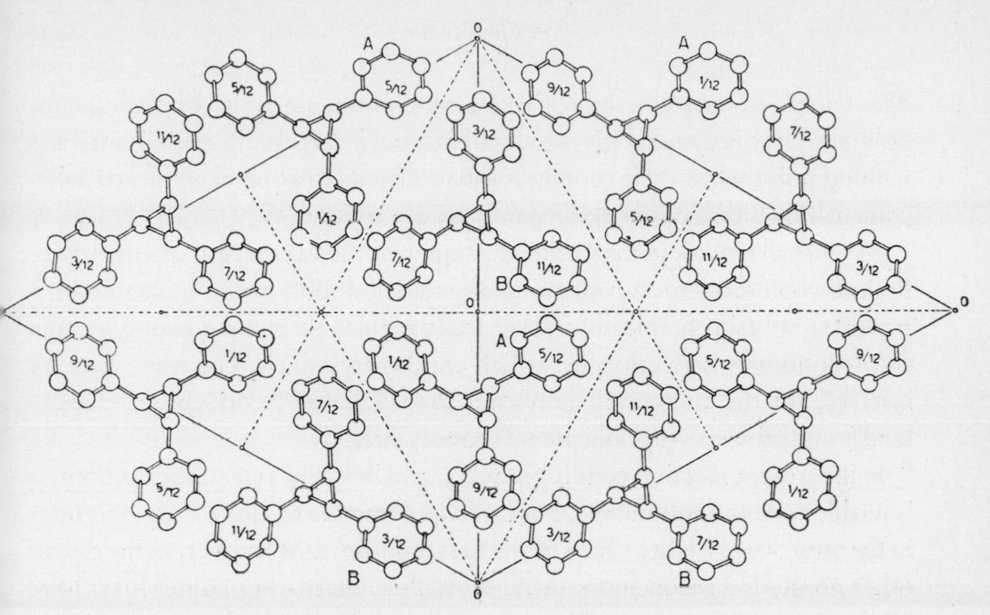

Fig. 4. Model of packing of isotactic polystyrene in the crystalline state, projected on the (001) plane (space group R3*c*).

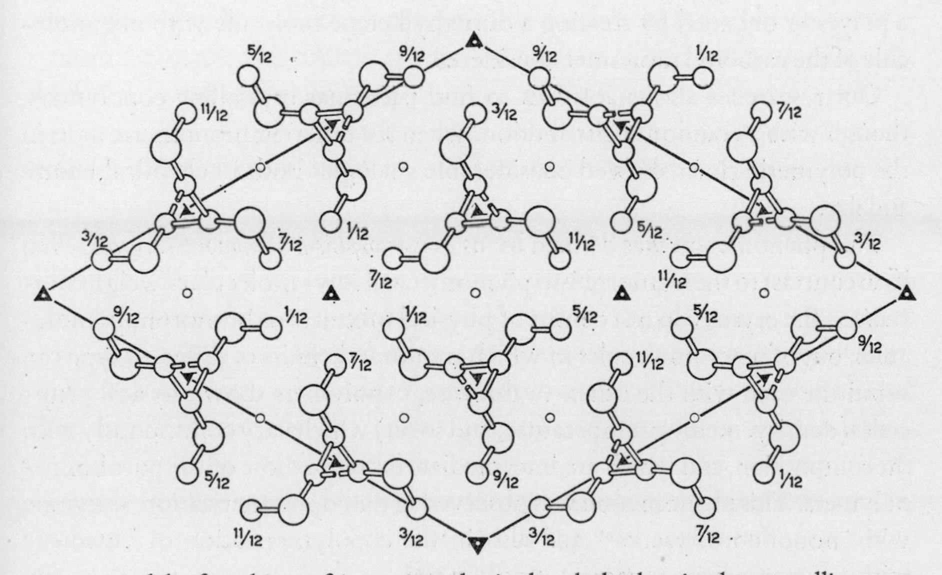

Fig. 5. Model of packing of isotactic polyvinylmethyl ether in the crystalline state, projected on the (001) plane (space group R3).

2. *Copolymers*

The «random» introduction of different monomeric units in a crystalline polymer by copolymerization generally causes a decrease in crystallinity and melting point when their content is lower than 20 to 25 percent, but at higher content values the copolymer is generally amorphous.

As we shall remark in the section dealing with the stereoregular polymers of hydrocarbon monomers containing an internal double bond, it is sometimes possible to obtain chemically and sterically regular alternating copolymers of these monomers with ethylene, which are also crystalline. This is the case, for instance, for the alternating ethylene–*cis*-2-butene[35], ethylene–cyclopentene[36], and ethylene–cycloheptene[37] copolymers.

In these cases, reaction conditions were used in which one of the monomers is unable to homopolymerize, but can copolymerize to alternating polymers in the presence of a large excess of the first monomer. Moreover, in the case of other nonhydrocarbon monomers, crystalline alternating copolymers have been obtained[38] from two different monomers that are both very reactive in the presence of stereospecific catalysts (for example, in the copolymerization of dimethylketene with higher aldehydes[39]), when the values of the relative copolymerization rates are much higher than those of homopolymerization. In the cases mentioned above, the repeating structural unit has the structure of a polyester obtained by treating a dimethylketene molecule with one molecule of the carbonyl monomer considered.

Our researches also enabled us to find particular crystalline copolymers, though with a «random» distribution, when the different monomeric units in the polymeric chain showed considerable analogies both in chemical nature and size.

This phenomenon was defined by us as *isomorphism of monomeric units*, even if, in contrast to the isomorphism phenomena of low-molecular-weight substances, the crystals do not consist of physical mixtures of isomorphous molecules, but of macromolecules in which monomeric units of different type can substitute each with the other. In this case, copolymers show physical properties (density, melting temperature, and so on) which vary continuously with the composition, and which are intermediate between those of the pure homopolymers. This phenomenon was observed in the copolymerization of styrene with monofluorostyrenes[40] and also in the copolymerization of butadiene with 1,3-pentadiene to *trans*-1,4 polymer[41].

Crystalline copolymers of a completely different type are obtained by successive polymerization of different monomers in the presence of catalysts able to homopolymerize both of them. These are linear copolymers constituted by successive blocks, each consisting of a chemically and sterically regular succession of units of the same type.

In some of these cases X-ray analysis reveals both the crystallinities corresponding to the single homopolymers[42].

II. *Stereospecificities in Polymerization Processes of Hydrocarbon Monomers*

The importance of the stereospecific polymerization–from the standpoint of both theory and practical applications–is due to the fact that in most cases (even if not always) the stereoregularity of linear polymers determines crystallinity. When the glass transition temperature and the melting temperature are very different, the physical and especially the mechanical properties are very different from those of the corresponding stereoirregular polymers. Due to such properties, these materials have very interesting practical applications, either as plastics and textiles when the melting point is high or as elastomers when the melting point does not considerably exceed the temperature of use.

The knowledge acquired in these last 10 years in the field of the stereospecificity of the polymerization processes shows that stereoregular and, in particular, isotactic polymers can be obtained in the presence of suitable catalysts acting through an ionic (both anionic and cationic) coordinated mechanism; however, they cannot generally be obtained by processes characterized by radical mechanism.

The catalysts having a higher degree of stereospecificity are characterized by the presence of metal atoms able to coordinate the monomer molecules in a stage immediately preceding that of insertion of the monomeric unit between the end of the growing chain and the catalyst[43–45].

In fact, a stereospecific action is shown either by the catalysts containing metal atoms, the coordinating properties of which are due to their charge and to their small ionic radius (aluminium, beryllium, lithium)[44], or by compounds of the transition metals[46,47].

Some authors[48] were led to believe that the steric structure of the last monomeric unit, or units, of the growing chain played an important role in the steric regulation of the polymerization processes. However, the low degree of stereospecificity observed in the radical processes shows that this factor alone cannot exert a determining action. In any case stereoregularity in these last processes is of the syndiotactic type and may be attributed also to thermodynamic factors, according to the strong increase in stereospecificity with decrease in temperature.

The first highly stereoregular isotactic polymers were obtained in the presence of heterogeneous catalysts; however, it soon became clear that the heterogeneity of the catalytic system is an essential factor for the polymerization of aliphatic olefins to isotactic polymers, but not for the polymerization of other types of monomers. In fact it was found that aliphatic aldehydes and certain monomers containing two electron-donor functional groups able to be coordinated (for example, conjugated diolefins, vinyl ethers, alkenyl ethers, acrylic monomers, styrenes that are substituted differently in the benzene ring, vinyl pyridine, and so on) can be polymerized in the stereospecific way also in the presence of soluble catalysts.

It must be borne in mind that, even if the most typical highly stereospecific catalysts for the polymerization of α-olefins contain organometallic compounds, some classes of monomers (for example, vinyl ethers) can be polymerized to isotactic polymers in the presence of cationic catalysts without the presence of organometallic compounds[49].

The stereospecificity of the polymerization processes not only depends on the catalytic system but is a property of each monomer-catalyst system. This is particularly evident in the case of the polymerization of some conjugated homologs of diolefins, in which the variation of the monomer changes both the degree of stereospecificity of the process and, in some cases, the type of stereoregularity of the polymer obtained[50].

Therefore, in order to attain a general view of the present state of the stereospecific polymerization, it is helpful to examine the most important results obtained in each class of monomers.

1. α-Olefins

This is the most studied branch of stereospecific polymerization. As already mentioned, isotactic polymers of α-olefins have been obtained so far only with the use of heterogeneous catalysts.

High stereospecificity is observed only when one employs organometallic catalysts containing a particular crystalline substrate, such as that deriving from the violet α, γ (ref. 51), and δ (ref. 52) modifications of $TiCl_3$, having a layer lattice[42,53,54]. The use of the β modification of $TiCl_3$ (ref. 55), which does not correspond to layer lattices, or of other heterogeneous catalysts (for example, catalysts containing a substrate formed by metal oxides) which also yield linear polymers of ethylene, leads to the formation of catalysts having little stereospecificity in the polymerization of α-olefins[53,56].

The study of the catalysts prepared from organometallic compounds containing aromatic groups[56] or labeled carbon enabled us to determine the ionic coordinated mechanism of such polymerization and the number of active centers on the surface of the heterogeneous catalysts[57].

Chemical and kinetic studies led to the conclusion that the stereospecific polymerization of propylene is a polyaddition reaction (stepwise addition), in which each monomeric unit, on its addition, is inserted on the bond between an electropositive metal and the electronegative terminal carbon atom of the growing polymeric chain. This study revealed also that some organometallic catalysts, which contain only titanium as metal atoms, could be stereospecific (ref. 58). The first reaction step corresponds to a coordination of the monomer molecule to the transition metal belonging to the active center[43,45].

The reaction chain generally does not show a kinetic termination[59], the length of the single macromolecules being determined by the rate of the processes of chain transfer either with the monomer[60] or with the alkyls of the organometallic compounds present[61]; these transfer processes allow, after the formation of a macromolecule, the start of another macromolecule on the same active center[56,62].

The single-rate constants of the different concurrent processes of chain growth and termination have been determined for some typical catalysts[63]. Later on, the study of homogeneous catalysts based on vanadium compounds and on alkyl aluminium monochloride permitted us to synthesize crystalline polypropylenes with a nonisotactic structure. The detailed development of this study led to the preparation of catalysts, obtained by treating hydrocarbon-soluble vanadium compounds (acetylacetonates or vanadium tetrachloride) with dialkyl aluminium monochloride. These catalysts yield, at low temperature, more or less crystalline polymers, free, however, from isotactic crystallizable macromolecules[64].

X-Ray analysis, applied to the fiber spectra, permitted us to establish that this is a syndiotactic polymer; its lattice structure has an identity period of 7.4 Å, corresponding to four monomeric units[65]. The comparison between isotactic and syndiotactic polypropelene structures is shown in Fig. 6.

The same type of homogeneous catalyst, which at low temperature homopolymerizes propylene to syndiotactic polymer, was used at higher temperatures (for example, 0°C) for the production of copolymers having a random

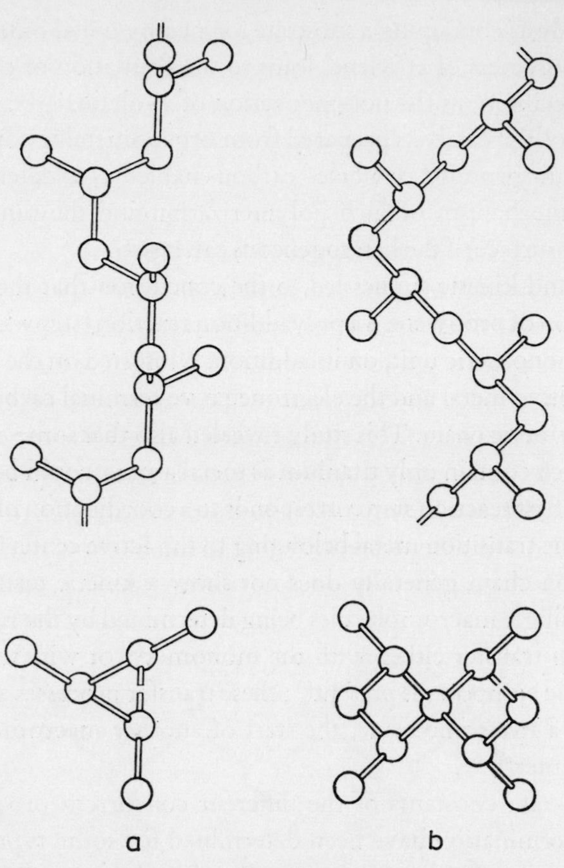

Fig. 6. Comparison between the side and end views of the chain structure of isotactic (a) and syndiotactic (b) polypropylenes (stable modifications) in the crystalline state.

distribution of propylene with ethylene[66]. These polymers, which are linear, are completely amorphous when the ethylene content decreases below 75 percent. They have a very flexible chain, due to the frequent CH_2-CH_2 bonds, while the relatively small number of $CH-CH_3$ groups is enough to hinder crystallization of the polymethylenic chain segments. These copolymers can be easily vulcanized through the use of peroxides; on the other hand the terpolymers, which contain not only ethylene and propylene but also small amounts (from 2 to 3 percent, by weight) of monomeric units, originated from the random copolymerizations of suitable diolefins[67] (or of cyclic compounds, such as cyclooctadiene, which can be prepared easily by dimerization of butadiene, following the method proposed by Wilke), can be vulcanized easily by the conventional methods used for the vulcanization of low-unsaturation rubber. They yield elastomers that are very interesting also from the practical point of view, because they can be obtained from low-priced materials and also because of their physical properties and resistance to aging.

2. *Ditactic polymers*

Polymers of 1-methyl-2-deuteroethylene. The study on the polymerization of differently deuterated propylenes, undertaken by us in order to arrive at more certain and univocal attributions of certain bands to the infrared spectrum of isotactic polypropylene, led us to the discovery of new interesting types of stereoisomerism in polymers of 1-methyl-2-deutero-ethylene, and generally in the case of polymers of 1,2-disubstituted ethylenes[68].

In fact, propylenes deuterated in the methylenic group can lead to monomer units having different steric structure depending on the relative orientation of the CH_3 and D substituents. Starting from these deuterated monomers showing phenomena of geometric isomerism, two types of polymers were ob-

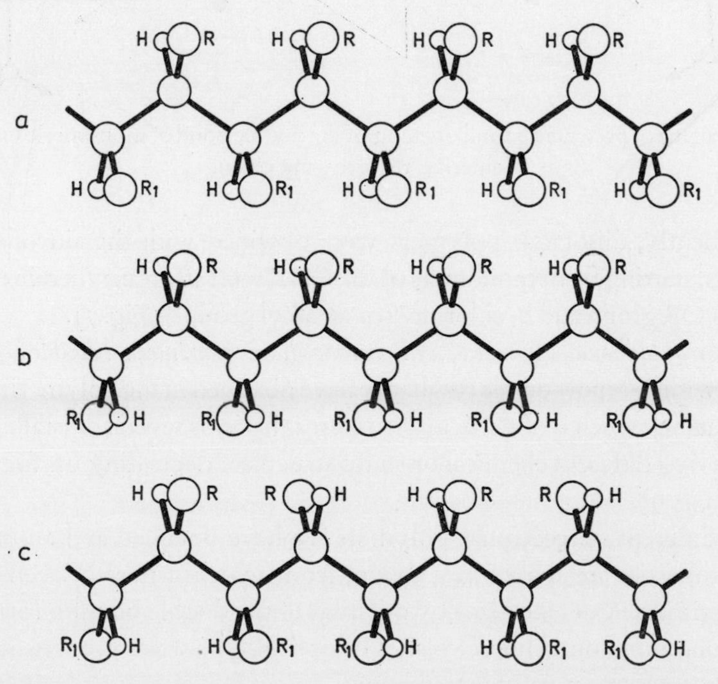

Fig. 7. Models of the chains of head-to-tail ditactic polymers supposed arbitrarily stretched on a plane, having, respectively, *threo*-diisotactic (a), *erythro*-diisotactic (b), and disyndiotactic (c) succession of the monomeric units.

tained. They exhibited the same X-ray spectra but different infrared spectra[69]. This means that such polymers possess the same helix structure as normal isotactic polypropylene, but that the relative orientation of D and CH_3 groups can lead to a new type of stereoisomerism. In general, starting from a monomer of the CH*A*=CH*B* type, three types of stereoregular isomers can be expected (see Fig. 7).

The type of stereoisomer obtainable by stereoregular polymerization depends on the mode of presentation and type of opening of the double bond of each monomer molecule on entering the growing chain (Fig. 8).

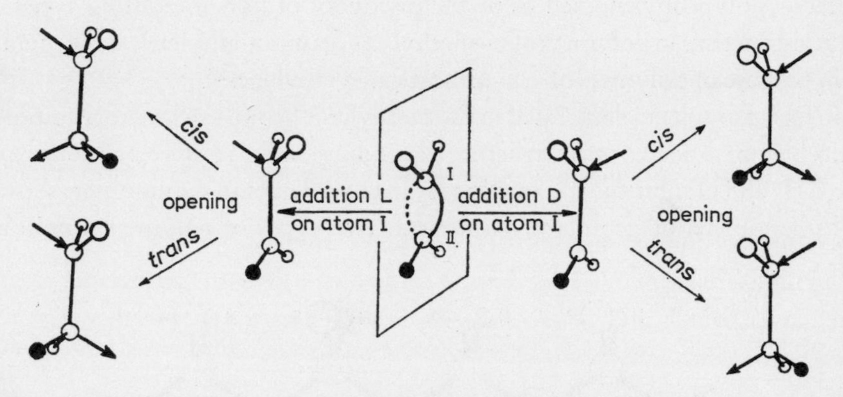

Fig. 8. Scheme of presentation and opening of the double bond of monomeric units when entering the growing chain.

Subsequently, diisotactic polymers were obtained with the aid of cationic catalysts, starting from monomers of the CH*A*= CH*B* type, wherein *A* designates an OR group and *B*, chlorine[70] or an alkyl group[71] (Fig. 7).

Stereoregular homopolymers of hydrocarbons having an internal double bond. First of all, I wish to report on the results we have obtained in the polymerization of cyclobutene, which is of particular interest as it yields several crystalline polymers having different chemical or steric structure, depending on the catalyst used[72] (Fig. 9).

The different stereoregular polymers we have obtained and a number of their properties are shown in Table 2, from which it may be seen that the polymerization can take place by opening of the double bond to form cyclic monomer units containing two sites of optical type stereoisomerism, so that crystalline polymers are of ditactic type.

In view of the fact that under suitable conditions it is possible to obtain two crystalline polymers containing enchained rings that show different physical

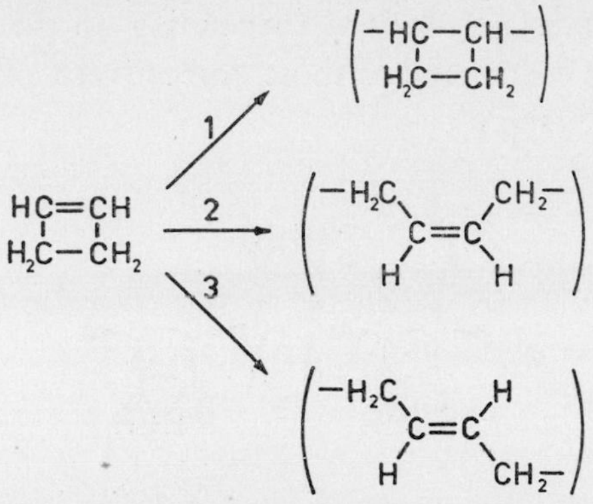

Fig. 9. Types of polymerization of cyclobutene: 1, cyclobutylenamer; 2, *cis*-1,4-polybutadiene; 3, *trans*-1,4-polybutadiene.

properties, we have ascribed the differences in their properties to the different steric structure and have attributed an *erythro*-diisotactic structure to one of them and an *erythro*-disyndiotactic structure to the other[73] (Fig. 10).

In the presence of other catalysts the ring opens to form unsaturated monomer units, which may show isomerism of geometric type. In this case, too,

Table 2

Stereospecific anionic coordinated polymerization of cyclobutene.

Catalytic system	*Prevailing chemical structure of polymer*	*Stereoregularity of polymer*	*Properties of crystalline polymers*		
			Density	*Melting temperature (°C)*	*Solubility in solvents*
$VCl_4 + Al(n\text{-}C_6H_{13})_3$	2-Polycyclobutylenamer	Presumably *erythro*-diisotactic	1.06	≃ 200	Insoluble in all the solvents below 150°C
V(acetylacetonate)$_3$ + $AlCl(C_2H_5)_2$	2-Polycyclobutylenamer	Presumably *erythro*-disyndiotactic	1.035	≃ 150	Soluble in tetralin at 150°C
$TiCl_4 + Al(C_2H_5)_3$	Polybutadiene	Prevailingly *cis*-1,4	Properties corresponding to those of 1,4-polybutadiene described in the literature		
$TiCl_3 + Al(C_2H_5)_3$	Polybutadiene	Prevailingly *trans*-1,4			

two different products are obtained (depending on the catalyst used), the properties of which correspond to those, respectively, of *cis*-1,4- and *trans*-1,4-polybutadiene[72] (Fig. 9).

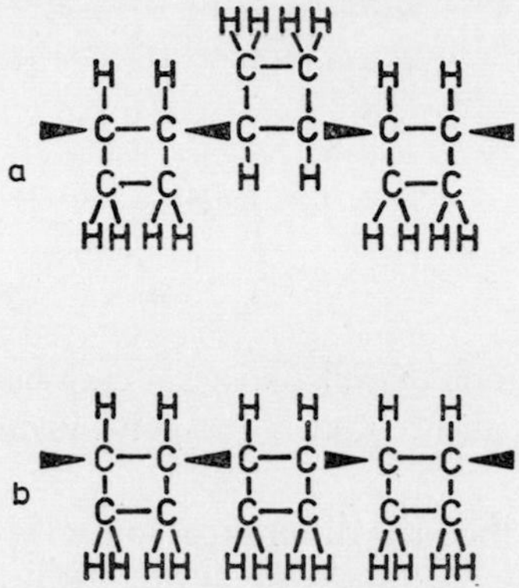

Fig. 10. Schematic drawing of the structures of *erythro*-diisotactic (a) and *erythro*-disyndiotactic (b) cyclobutylenamer.

Ditactic polymers are also obtained from certain monomers containing internal unsaturation, which are unable to homopolymerize but, as mentioned above, can copolymerize with ethylene, yielding crystalline, alternating copolymers of *erythro*-diisotactic structure. Among these monomers are *cis*-2-butene[35], cyclopentene[36], and cycloheptene[37]; *trans*-2-butene and cyclohexene behave in a different way and do not give crystalline copolymers.

Unlike the ditactic polymers of deuterated propylene, the ditactic polymers obtained by alternate copolymerization can exist in two disyndiotactic forms.

It is to be noted that the copolymerization of *cis*-2-butene is stereospecific only in the presence of heterogeneous catalysts of the type used in polymerizing α-olefins to isotactic polymers, while the copolymerization of cyclopentene and cycloheptene is also stereospecific when homogeneous catalysts are used. We have recently[74] proposed an interpretation of these facts based essentially on steric criteria.

3. Stereoregular polymers of conjugated diolefins

Stereoisomerism phenomena in the field of diolefins, and in particular of conjugated diolefins, are more complex than phenomena occurring in the case of monoolefinic monomers. In fact, besides the stereoisomerism phenomena observed in these last (isomerism due to asymmetric carbon atoms), isomerism phenomena of geometric type may also be present, depending on the *cis*- or *trans*-configuration of the residual double bonds present in the monomeric units.

Butadiene polymers. The simplest conjugated diolefin, 1,3-butadiene, can in fact yield two types of polymers, according to whether the polymerization takes place by opening of the vinyl bond (to form 1,2-enchained polymers)

$$\begin{array}{l} -CH-CH_2- \\ | \\ CH \\ \| \\ CH_2 \end{array}$$

or by opening of both conjugated double bonds (to form 1,4-enchained polymers)

$$-CH_2-CH=CH-CH_2-$$

In the first case, the same stereoisomerism phenomena observed in other vinyl polymers (for example, isotactic, syndiotactic, and atactic polymers) can be expected.

In the second case, each monomeric unit still contains a double bond in the 2-3 position, which can assume *cis*- or *trans*-configuration. Thus, four types of stereoregular polymers could be foreseen «*a priori*» and precisely: *trans*-1,4-, *cis*-1,4-, isotactic-1,2-, and syndiotactic-1,2-polybutadienes. All four these stereoisomers were prepared at my Institute with the aid of different stereospecific catalysts[75,76] with a high degree of steric purity (up to above 98 percent), as shown by infrared analysis[77].

X-Ray examination had made it possible for us not only to establish the steric structure of the different polymers but also to determine the conformation of the chains in the crystals and, for three of them, also a detailed lattice structure[21,78]. Fig. 11 shows the conformations of the chains of the various stereoisomers, while in Table 3 a number of physical characteristics of the single polymers are reported.

As mentioned above, stereoregularity in the field of butadiene polymers is not necessarily connected with the use of heterogeneous catalysts, and, in fact, all four regular stereoisomers can be obtained with the aid of homogeneous catalysts.

In the case of *cis*-1,4-polybutadiene, the highest steric purity is obtained by the use of homogeneous catalysts[76]. Of the four polybutadiene stereoisomers,

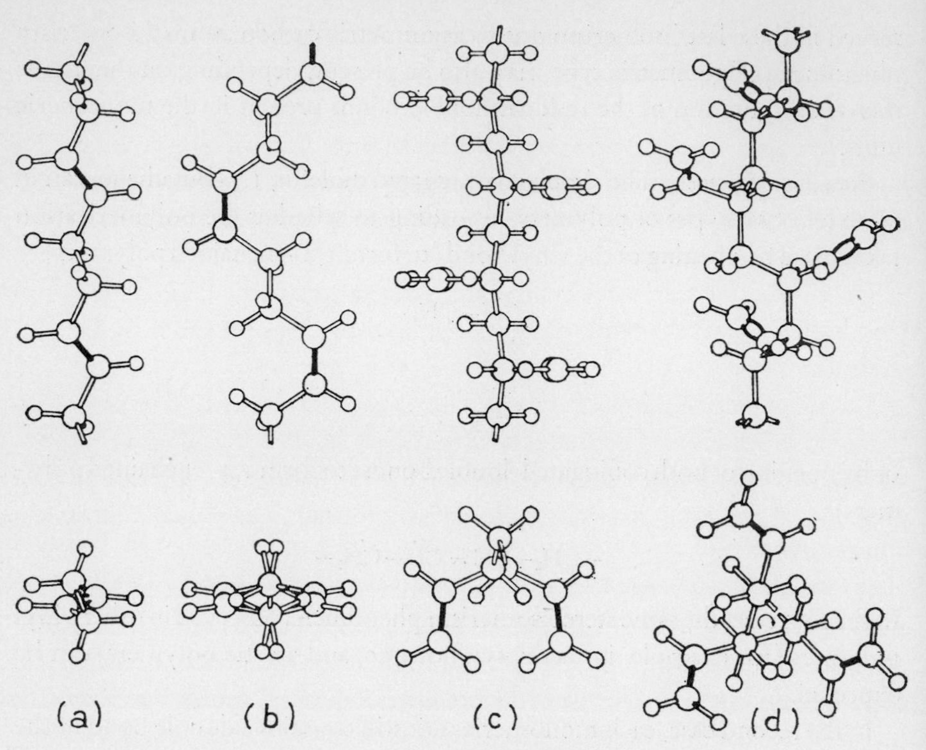

Fig. 11. Side and end views of the chain conformations of the four stereoisomers of polybutadiene: (a) *trans*-1,4; (b) *cis*-1,4; (c) syndiotactic-1,2; (d) isotactic-1.2.

Table 3

Some physical properties of the four stereoregular polymers of butadiene.

Polymer (infrared analysis)	*Melting point (°C)*	*Identity period (Å)*	*Density (g/ml)*
trans-1,4 (99–100%)	146[a]	4.85 (mod. I)	0.97
		4.65 (mod. II)	.93
cis-1,4 (98–99%)	2	8.6	1.01
Isotactic-1,2 (99% 1,2 units)	126	6.5	0.96
Syndiotactic-1,2 (99% 1,2 units)	156	5.14	.96

[a] *trans*-1,4-Polybutadiene exists in two crystalline modifications: one (mod. I) is stable below 75°C, the other (mod. II) is stable between about 75°C and the melting point of the polymer.

the *cis*-1,4 stereoisomer is of particular interest also from a practical viewpoint. Its preparation and properties have been investigated by a large number of workers[79].

Isoprene polymers. The two polyisoprene geometrical isomers were already known in nature: natural rubber (*cis*-1,4 polymer) and gutta-percha and balata (*trans*-1,4 polymers). Both were obtained by synthesis through stereospecific polymerization.

The *cis*-1,4 polymer was obtained for the first time in the United States by Goodrich's workers[80], while the *trans*-1,4 polymer was prepared by us[81] at the beginning of 1955.

The other stereoisomers, having 1,2- or 3,4-enchainment, have not been prepared as yet in such a degree of steric purity as to yield crystalline products. In fact, the only known polymer having 3,4-enchainment, obtained in the presence of the same catalysts yielding syndiotactic 1,2-polybutadiene, is amorphous.

1,3-Pentadiene polymers. Unlike butadiene polymers, the stereoregular polymers of 1,3-pentadiene obtained so far contain at least one asymmetric carbon atom in the monomer unit. Furthermore, for some of them it is possible to expect geometric isomers, due to the presence of internal double bonds which may have *cis*- or *trans*-configuration, so that all the polymers will show two centers of steric isomerism. And in fact polymers having 3,4-enchainment, containing two asymmetric carbon atoms, show two sites of optical isomerism; all the others exhibit one site of optical isomerism and one of geometric isomerism (1,2 and 1,4 units).

On the assumption that only polymers showing stereoregularity in both possible sites (ditactic polymers) will be crystalline, 11 crystalline pentadiene polymers can be expected:

(*1*) Polymers having 3,4-enchainment (Fig. 12a): (*i*) *erythro*-diisotactic polymer, (*ii*) *threo*-diisotactic polymer, (*iii*) Syndiotactic polymer.

(*2*) Polymers having 1,2-enchainment (Fig. 12b): (*iv*, *v*) Isotactic polymers containing, respectively, one *cis*- or *trans*-double bond in the side-chain, (*vi*, *vii*) Syndiotactic polymers containing, respectively, one *cis*- or *trans*-double bond in the side-chain.

(*3*) Polymers having 1,4-enchainment (Fig. 12c): (*viii*, *ix*) *cis*-1,4-isotactic and syndiotactic polymers, respectively, (*x*, *xi*) *trans*-1,4-isotactic and syndiotactic polymers, respectively.

Of these stereoisomers the only three so far known were prepared in my Institute: *trans*-1,4-isotactic[82], *cis*-1,4-isotactic[83], and *cis*-1,4-syndiotactic

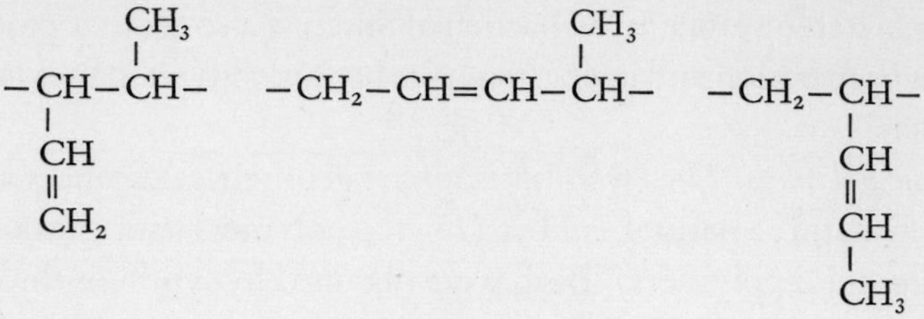

CH_3 $\vert$ $-CH-CH-$ $\vert$ CH $\Vert$ CH_2	CH_3 $\vert$ $-CH_2-CH=CH-CH-$	$-CH_2-CH-$ $\vert$ CH $\Vert$ CH $\vert$ CH_3
3,4 (a)	1,4 (b)	1,2 (c)

Fig. 12. Structure of 1,3-pentadiene polymers.

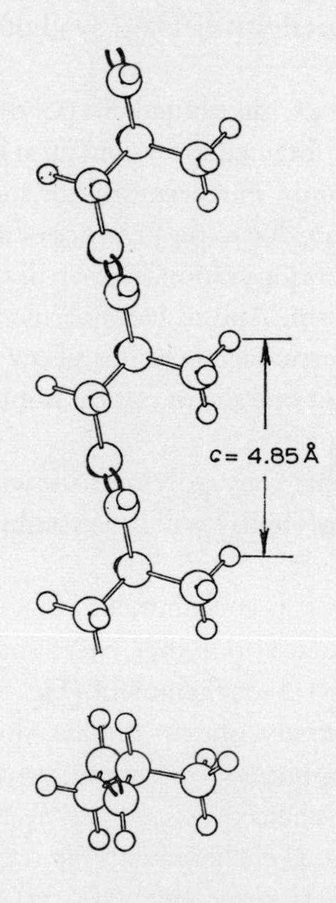

Fig. 13. Side and end views of the macromolecule of isotactic *trans*-poly(1-methylbuta-1,3-diene) (that is, *trans*-1,4-polypentadiene) in the crystalline state.

Table 4

Some physical properties of the three stereoregular isomers of 1,3-polypentadiene known so far.

Polymer	*Infrared analysis*	*Identity period (Å)*	*Melting (point °C)*	*Density (g/ml)*
Isotactic *trans*-1,4	*trans*-1,4 (98–99%)	4.85	96	0.98
Isotactic *cis*-1,4	*cis*-1,4 (85%)	8.1	44	0.97
Syndiotactic *cis*-1,4	*cis*-1,4 (90%)	8.5	53	1.01

polymer[84]. In Table 4 a number of physical properties characteristic of these isomers are reported; Figs. 13 and 14 show the conformation of the chains in the crystals.

As could be expected, the best elastic properties in vulcanized polymers are observed for *cis*-1,4-polymers, owing to their melting point, which is slightly below the melting point of natural rubber.

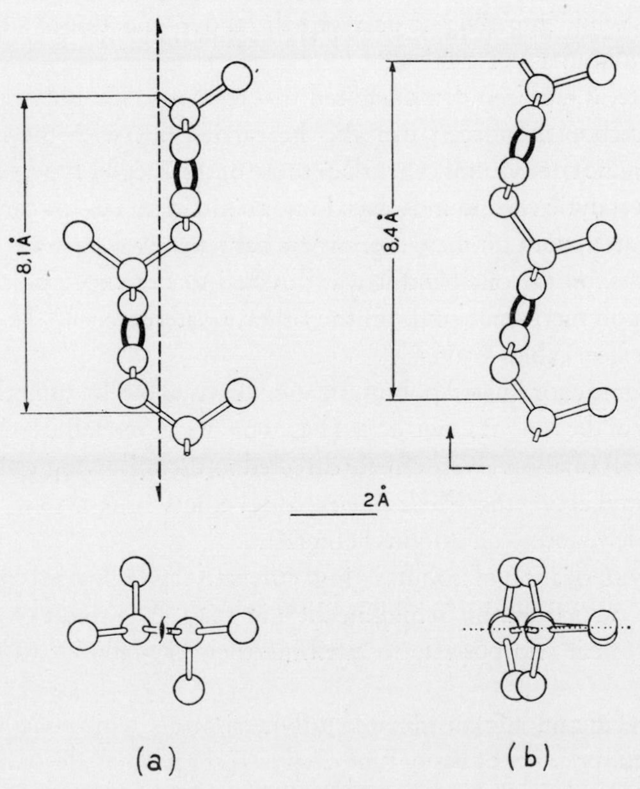

Fig. 14. Side and end views of the macromolecule of isotactic *cis*-1,4-polypentadiene (a) and syndiotactic *cis*-1,4-polypentadiene (b).

III. *Stereospecificity in Polymerization of Nonhydrocarbon Monomers*

Unlike the polymerization of unsatured hydrocarbons, and particularly of α-olefins, the polymerization of monomers containing functional groups, in the presence of catalysts based on organometallic compounds, has not been investigated until recently. This is due to the fact that the functional groups contained in such monomers can react with organometallic catalysts through reactions that are well known in the field of classical organic chemistry, such as Grignard reactions, Michael's reaction, or splitting of an ether bond.

Initially it was feared that these reactions might involve both deactivation of the catalytic agent and total or partial alteration of the said monomers.

In 1956 we demonstrated for the first time in the case of acrylonitrile[85] and its homologs that, by suitably selecting the transition metal compounds and organometallic compounds forming the catalytic complex, it is possible to bring about stereospecific, anionic coordinated polymerization of these monomers while impeding or delaying the above-mentioned side reactions between monomer and catalyst.

Therefore, it has been demonstrated that stereospecific polymerization of nonhydrocarbon monomers can also be carried out with the use of pure organometallic compounds other than those of the Ziegler type, or even with the aid of catalytic compounds that do not contain metal-to-carbon bonds.

The research work on these monomers has taken two separate but parallel paths; that is, on the one hand it was directed to stereospecific cationic coordinated polymerization and, on the other, to stereospecific anionic polymerization (see Tables 5 and 6).

The cationic coordinated polymerizations carried out by us in the presence of catalysts of the type of Lewis acids (based on organometallic compounds or Friedel-Craft catalysts) were chiefly directed to the following classes of monomers: vinyl alkyl ethers[86,87], alkenyl alkyl ethers[70], alkoxy-styrenes[88], vinylcarbazole[89], and β-chlorovinyl ethers[71].

The polymerization of isobutyl vinyl ethers to crystalline polymers had already been carried out by Schildknecht[49] in 1949. As a result of our further research work it was possible to attribute their crystallinity to an isotactic structure[86].

Stereospecific anionic coordinated polymerization, which is in general carried out in the presence of basic-type catalysts (organometallic or metal amidic compounds, alcoholates) was chiefly investigated in connection with the following classes of monomers: higher homologs of acrylonitrile[90], vinyl-

Table 5
Nonhydrocarbon monomers polymerized in a stereospecific way by coordinated cationic catalysis in the homogeneous phase.

Monomer	*Type of catalyst*	*Type of stereo-specificity in the polymer*
Vinylalkyl ether	$Al(C_2H_5)Cl_2$	Isotactic
trans-Alkenylalkyl ether	$Al(C_2H_5)Cl_2$	*threo*-Diisotactic
cis-β-Chlorovinylalkyl ether	$Al(C_2H_5)Cl_2$	*erythro*-Diisotactic
trans-β-Chlorovinylalkyl ether	$Al(C_2H_5)Cl_2$	*threo*-Diisotactic
o-Methoxystyrene	$Al(C_2H_5)Cl_2$	Isotactic
N-Vinylcarbazole	$Al(C_2H_5)Cl_2$	
N-Vinyldiphenylamine	$Al(C_2H_5)Cl_2$	
Benzofuran	$AlCl_3$	

Table 6
Nonhydrocarbon monomers polymerized in the stereospecific way by coordinated anionic catalysis in the homogeneous phase.

Monomer	*Type of catalyst*	*Type of stereospecificity in the polymer*
Vinylpyridine	$Mg(C_6H_5)Br$	Isotactic
Acrylonitrile	$Cr(Acac)_3+Zn(C_2H_5)_2$	Syndiotactic
α-Substituted acrylonitrile	$Mg(C_2H_5)_2$	
Sorbates	Butyl-Li	*erythro*-Diiso-*trans*-tactic
Acrylates	Mg amides	Isotactic
Aliphatic aldehydes	$Al(C_2H_5)_3$	Isotactic

pyridine[91], sorbates[92], acrylates[93], and aliphatic aldehydes[94].

Unlike the α-olefin polymerization, which requires the presence of a catalyst containing a crystalline substrate in order that it may proceed in a stereospecific isotactic manner, the polymerization of nonhydrocarbon monomers containing functional groups or atoms having free electron pairs (such as, for example, ethereal, carbonylic, or carboxylic oxygen; aminic, amidic, or nitrilic nitrogen) can proceed in a stereospecific way also in the absence of a solid substrate–that is, in a homogeneous phase. Here the stereospecificity–which in this case is also connected with a constant orientation and constant mode of presentation, on polymerizing, of the monomer units with respect to the growing chain and to the catalytic agent–is due to the coordination of an electron pair in the monomer with the metal of the catalytic agent by means of a dative bond[47,95]. As the olefinic double bond too is necessarily bound to the active center, such monomers appear to be doubly linked to the complex formed by the catalytic agent and the terminal group of the growing chain. A predetermined steric orientation is thus made possible.

Likewise, both the diolefins containing two olefin groups bound to the catalyst complex and certain aromatic α-olefins, wherein the second anchoring point is provided by the aromatic group π-linked to a catalyst containing a highly electropositive atom with a very small radius (lithium)[96], can be polymerized stereospecifically even in the homogeneous phase.

The coordination of the monomer with the catalytic agent, which is the indispensable step preceding any stereospecific polymerization both in the homogeneous and in the heterogeneous phase, has been particularly well exemplified by the stereospecific polymerization of 2-vinylpyridine in the presence of organometallic compounds of magnesium[97].

In fact, the presence of Lewis bases in the polymerization of this monomer exerts a determining influence on its behavior in the polymerizations. Compounds having a higher degree of basicity than vinylpyridine itself (for example, pyridine) form stable coordination compounds with the catalyst, thus impeding the coordination of the monomer; in this way, not only does the catalytic activity appear very much reduced, but also the stereospecificity disappears and the polymer obtained is atactic. Compounds having a lower degree of basicity than the monomer (aliphatic ethers) compete with the monomer only in so far as the association with the catalyst is concerned. Accordingly this does not result in the disappearance of the catalyst reactivity, but only in its reduction along with the degree of stereospecificity of the reaction.

The asymmetric synthesis of optically active high polymers, starting from monomers showing no centers of optical-type asymmetry, constituted a particular, more advanced case of isotactic stereospecific polymerization.

IV. *Asymmetric Synthesis of High Polymers and Interpretation of Stereospecific Polymerization*

In fact, whereas in the normal stereospecific polymerization to isotactic polymers a succession of monomer units with a given configuration takes place in each single macromolecule so that enantiomorphous macromolecules in equal amounts are present in crude polymers, in the case of asymmetric synthesis one of the two enantiomorphous isomers of the monomer unit is contained in higher amounts.

It should be noted that isotactic high polymers of α-olefins or of other simple vinyl monomers cannot show detectable optical activity, since an ideal isotactic polymer of infinite length does not contain asymmetric carbon atoms, and in isotactic polymers having finite length[98] the optical activity, due to a difference in the terminal groups, can be detected only in oligomers and decreases with increase in the molecular weight. This is due to the fact that the asymmetry of each asymmetric carbon atom is to be ascribed not to the chemical difference of contiguous groups linked to the said carbon atom but to a difference in length of the chain segments linked to it[99].

In fact, in the case of poly-α-olefins, optically active polymers were obtained by polymerization only from monomers having an asymmetric carbon atom[100].

On the basis of our investigations it has been possible to obtain optically active polymers from monomers containing no centers of optical asymmetry only when, during the polymerization, monomer units are incorporated so as to develop new asymmetric centers. The asymmetry of the new centers arises from a difference in the chemical constitution of the groups contiguous to the carbon atoms themselves[101-104].

Such a result was obtained by means of stereospecific polymerization processes, operating under conditions that allow asymmetric induction to favor the formation of one of the two enantiomorphous structures of the monomer unit.

The methods that have led us to the asymmetric synthesis of polymers of substituted diolefins and of certain heterocyclic, unsaturated compounds are of two types.

(*1*) The first is the use of normal stereospecific catalysts wherein at least one group bound to the organometallic compound used in the catalyst preparation, which will be the terminal group of the macromolecules, is optically active[101]. In this case the asymmetric induction is probably due to the particular configuration of the terminal group of the growing chain bound to the catalyst.

(*2*) A second method is based on the use of conventional stereospecific catalysts prepared without using optically active alkyls, provided they are complexed with optically active Lewis bases, such as β-phenylalanine[102], or with the use of an optically active transition metal compound[104] (Table 7).

In the first case, as the polymerization proceeds, the optical activity de-

Table 7

Asymmetric synthesis of polymers.

Monomer	*Catalyst*	*Structure of polymer*	$[\alpha]_D$
Methyl sorbate	(S) Iso-amyl lithium	*erythro*-Diiso-*trans*-tactic	− 7.9
Butyl sorbate	Butyl lithium-(−) MEE[a]	*erythro*-Diiso-*trans*-tactic	+ 8.4
Benzofuran	$AlCl_3$-(+)PHE[b]	Not determined	+69
trans-1,3-Pentadiene	(+)$Al(iC_5H_{11})_3 + VCl_3$	Iso-*trans*-tactic	− 1.05
trans-1,3-Pentadiene	$Al(C_2H_5)_3$-(−)$Ti(OC_{10}H_{19})_4$[c]	Iso-*cis*-tactic	−22

[a] MEE, Menthyl ethyl ether.
[b] PHE, β-phenylalanine.
[c] $Ti(OC_{10}H_{19})_4$, titanium tetramentholate.

creases, as could be expected in view of the fact that any accidental inversion of configuration exerts an action not confined to one monomer unit only, but tending to extend to subsequent units.

In the second case, on the other hand, the induction is due to the asymmetry of the optically active counterion[105], which maintains its steric structure also in the case where the asymmetric polymerization gives low optical yields.

These results can be extended to the interpretation of stereospecific catalysis of vinyl monomers. They suggest that a higher stereospecificity can be expected when using catalysts, the active centers of which are *per se* asymmetric, than when symmetric catalysts are used, in which the stereospecificity derives from asymmetric induction brought about by the configuration assumed by the last polymerized unit.

Even before the discovery of the asymmetric synthesis of high polymers, we attributed[106] the stereospecificity of certain heterogeneous catalysts, prepared by reaction of solid titanium halides, to the fact that the active centers contain surface atoms of a transition metal having coordination number 6. In fact it is known that, in such a case, when at least two of the coordinated groups show a different chemical nature with respect to the others, enantiomorphous structures of the surface complexes can exist.

The high stereospecificity of such catalysts is probably due to the fact that the initial complex maintains its asymmetry even when linked to the growing chain.

An interesting aspect of the asymmetric polymerization of benzofuran

Table 8

Autocatalytic effect in the asymmetric polymerization of benzofuran in the presence of $AlCl_3$-(+)β-phenylalanine.

Run	*Weight*[a]	[α]	$\Delta W[\alpha]/\Delta W$[b]
A	1.48	31.0	
A	2.30	46.7	75
A	2.72	52.7	88
B	1.22	50.2	
B	4.60	76.3	86
C	0.68	51.5	
C	2.30	69.3	77

[a] Weight (W) in grams of polymer per millimole of phenylalanine.

[b] $\Delta W[\alpha]/\Delta W$, Optical activity of the polymer formed between the two subsequent drawings.

consists in an autocatalytic effect observed in the first reaction period. In fact it was noticed that the optical activity of the polymers increases as the polymerization proceeds[107] (Table 8).

To clarify this phenomenon further, polymerization runs have been performed in the presence of optically active polybenzofuran previously obtained.

Although the sign of the optical activity always corresponds to that of the β-phenylalanine complexed with the counter ion, nevertheless the presence of preformed polymer, obtained in the same polymerization or added to the catalytic system at the beginning of the polymerization, causes an increase in the optical activity of the polymer newly formed.

Such an observation may have an interest that goes beyond the interpretation of stereospecific polymerization; in fact it can suggest suitable patterns characteristic of certain biological processes in which the formation of asymmetric molecules or groups of a given type is connected with the preexistence of optically active macromolecules.

1. H. Staudinger, *Die hochmolekularen organischen Verbindungen*, Springer, Berlin, 1932.
2. A. A. Morton, E. E. Magat and R. L. Letsinger, *J. Am. Chem. Soc.*, 69 (1947) 950.
3. G. R. Levi and G. Natta, *Atti Accad. Nazl. Lincei, Rend.*, [6] 2 (1925) 1; G. Natta, *ibid.*, [6] 2 (1925) 495; G. Natta and A. Rejna, *ibid.*, [6] 4 (1926) 48; G. Natta, *Nuovo Cimento*, 3 (1926) 114; G. Natta and E. Casazza, *Atti Accad. Nazl. Lincei, Rend.*, [6] 5 (1927) 803; G. Natta, *ibid.*, [6] 5 (1927) 1003; G. Natta, *Gazz. Chim. Ital.*, 58 (1928) 344; G. Natta and L. Passerini, *ibid.*, 58 (1928) 472; G. Natta and M. Strada, *ibid.*, 58 (1928) 419; G. Natta, *ibid.*, 58 (1928) 619, 870; G. Natta and M. Strada, *Atti Accad. Nazl. Lincei, Rend.*, [6] 7 (1928) 1024; G. Natta and L. Passerini, *Gazz. Chim. Ital.*, 58 (1928) 597, 59 (1929) 280; G. Natta and L. Passerini, *Atti Accad. Nazl. Lincei, Rend.*, [6] 9 (1929) 557; G. Natta and L. Passerini, *Gazz. Chim. Ital.*, 59 (1929) 129; G. Bruni and G. Natta, *Rec. Trav. Chim.*, 48 (1929) 860; G. Natta and L. Passerini, *Gazz. Chim. Ital.*, 59 (1929) 620; G. Natta, *Atti III Congr. Nazl. Chim. Pura e Appl., Firenze*, 1929, p. 347; G. Natta and L. Passerini, *ibid.*, p. 365; G. Natta, *Atti Accad. Nazl. Lincei, Rend.*, [6] 11 (1930) 679; G. Natta and A. Nasini, *Nature*, 125 (1930) 457; G. Natta, *ibid.*, 126 (1930) 97, 127 (1931) 129, 235.
4. G. Natta, M. Baccaredda and R. Rigamonti, *Gazz. Chim. Ital.*, 65 (1935) 182; G. Natta, M. Baccaredda and R. Rigamonti, *Monatsh. Chem.*, 66 (1935) 64; G. Natta, M. Baccaredda and R. Rigamonti, *Sitzunger Akad. Wiss. (Wien)*, 144 (1935) 196; G. Natta and M. Baccaredda, *Atti Accad. Nazl. Lincei, Rend.*, [6] 23 (1936) 444; G. Natta and R. Rigamonti, *ibid.*, [6] 24 (1936) 381.
5. G. Natta, *Giorn. Chim. Ind. ed Appl.*, 12 (1930) 13; G. Natta, *Österr. Chemiker-Ztg.*, 40 (1937) 162; G. Natta, P. Pino, G. Mazzanti and I. Pasquon, *Chim. Ind.* (*Milan*), 35 (1953) 705.
6. G. Natta and M. Strada, *Giorn. Chim. Ind. ed Appl.*, 12 (1930) 169, 13 (1931) 317; G. Natta and R. Rigamonti, *ibid.*, 14 (1932) 217.
7. G. Natta, *Chim. Ind.* (*Milan*), 24 (1942) 43; G. Natta and G. F. Mattei, *ibid.*, 24 (1942) 271; G. Natta and G. Negri, *Dechema Monograph.*, 21 (1952) 258.
8. G. Natta, P. Pino and R. Ercoli, *J. Am. Chem. Soc.*, 74 (1952) 4496.
9. G. Natta and E. Mantica, *Gazz. Chim. Ital.*, 81 (1951) 164.
10. K. Ziegler, *Angew. Chem.*, 64 (1952) 323.
11. G. Natta, P. Pino and M. Farina, *Ric. Sci. Suppl.*, 25 (1955) 120.
12. K. Ziegler, E. Holzkamp, H. Breil and H. Martin, *Angew. Chem.*, 67 (1955) 541.
13. G. Natta, P. Pino and G. Mazzanti, *Brit. Pat.*, 810,023; *U.S. Pat.* 3,112,300 and 3,112,301 (Italian priority, 8 June 1954).
14. G. Natta, *Atti Accad. Nazl. Lincei, Mem.*, [8] 4 (1955) 61; G. Natta, *J. Polymer Sci.*, 16 (1955) 143; G. Natta, P. Pino, P. Corradini, F. Danusso, E. Mantica, G. Mazzanti and G. Moraglio, *J. Am. Chem. Soc.*, 77 (1955) 1708; G. Natta, P. Pino and G. Mazzanti, *Chim. Ind. (Milan)*, 37 (1955) 927.
15. G. Natta, P. Pino and G. Mazzanti, *Gazz. Chim. Ital.*, 87 (1957) 528.
16. G. Natta and P. Corradini, *Atti Accad. Nazl. Lincei, Mem.*, [8] 4 (1955) 73.
17. G. Natta and P. Corradini, *Atti Accad. Nazl. Lincei, Rend.*, [8] 18 (1955) 19.
18. G. Natta and R. Rigamonti, *Atti Accad. Nazl. Lincei, Rend.*, [6] 24 (1936) 381.

19. G. Natta, L. Porri, G. Zanini and L. Fiore, *Chim. Ind. (Milan)*, 41 (1959) 526.
20. G. Natta, L. Porri, P. Corradini and D. Morero, *Atti Accad. Nazl. Lincei, Rend.*, [8] 20 (1956) 560.
21. G. Natta and P. Corradini, *Atti Accad. Nazl. Lincei, Rend.*, [8] 19 (1955) 229; G. Natta and P. Corradini, *J. Polymer Sci.*, 20 (1956) 251; G. Natta and L. Porri, *Belgian Pat.* 549,544.
22. M. L. Huggins, G. Natta, V. Desreux and H. Mark, *J. Polymer Sci.*, 56 (1962) 153.
23. L. Pauling and R. B. Corey, *Proc. Natl. Acad. Sci. (U.S.)*, 37 (1951) 205.
24. C. W. Bunn, *Proc. Roy. Soc. (London), Ser. A*, 180 (1942) 67.
25. G. Natta, P. Pino and G. Mazzanti, *Italian Pat.* 526, 101, *Brit. Pat.* 828,791; G. Natta, P. Corradini, I. W. Bassi and L. Porri, *Atti Accad. Nazl. Lincei, Rend.*, [8] 24 (1958) 121.
26. G. Natta and P. Corradini, *Nuovo Cimento, Suppl.*, [10] 15 (1960) 9; P. Corradini, *Atti Accad. Nazl. Lincei, Rend.*, [8] 28 (1960) 632.
27. G. Perego and I. W. Bassi, *Makromol. Chem.*, 61 (1963) 198.
28. I. W. Bassi, personal communication.
29. G. Natta, P. Corradini and I. W. Bassi, *Nuovo Cimento, Suppl.*, [10] 15 (1960) 68.
30. G. Natta, P. Corradini and I. W. Bassi, *Nuovo Cimento, Suppl.*, [10] 15 (1960) 52.
31. G. Natta, P. Corradini and I. W. Bassi, *Atti Accad. Nazl. Lincei, Rend.*, [8] 23 (1957) 363.
32. G. Natta, P. Corradini and I. W. Bassi, *Nuovo Cimento, Suppl.*, [10] 15 (1960) 83.
33. I. W. Bassi, *Atti Accad. Nazl. Lincei, Rend.*, [8] 29 (1960) 193.
34. G. Dall'Asta and I. W. Bassi, *Chim. Ind. (Milan)*, 43 (1961) 999.
35. G. Natta, G. Dall'Asta, G. Mazzanti, I. Pasquon, A. Valvassori and A. Zambelli, *J. Am. Chem. Soc.*, 83 (1961) 3343; G. Natta, G. Dall'Asta, G. Mazzanti and F. Ciampelli, *Kolloid-Z.*, 182 (1962) 50; P. Corradini and P. Ganis, *Makromol. Chem.*, 62 (1963) 97.
36. G. Natta, G. Dall'Asta, G. Mazzanti, I. Pasquon, A. Valvassori and A. Zambelli, *Makromol. Chem.*, 54 (1962) 95.
37. G. Natta, G. Dall'Asta and G. Mazzanti, *Chim. Ind. (Milan)*, 44 (1962) 1212.
38. G. Natta, G. Mazzanti, G. F. Pregaglia and M. Binaghi, *J. Am. Chem. Soc.*, 82 (1960) 5511.
39. G. Natta, G. Mazzanti, G. F. Pregaglia and G. Pozzi, *J. Polymer Sci.*, 58 (1962) 1201.
40. D. Sianesi, G. Pajaro and F. Danusso, *Chim. Ind. (Milan)*, 41 (1959) 1176; G. Natta, *Makromol. Chem.*, 35 (1960) 93.
41. G. Natta, L. Porri, A. Carbonaro and G. Lugli, *Makromol. Chem.*, 53 (1962) 52.
42. G. Natta, *J. Polymer Sci.*, 34 (1959) 531; G. Natta and I. Pasquon, *Advan. Catalysis*, 9 (1959) 1.
43. G. Natta, *Angew. Chem.*, 68 (1956) 393; *Chim. Ind. (Milan)*, 38 (1956) 751.
44. G. Natta, *Ric. Sci., Suppl.*, 28 (1958) 1.
45. G. Natta, F. Danusso and D. Sianesi, *Makromol. Chem.*, 30 (1959) 238; F. Danusso, *Chim. Ind. (Milan)*, 44 (1962) 611.
46. G. Natta, *Experientia, Suppl.*, 7 (1957) 21; *Materie Plastiche*, 21 (1958) 3.
47. G. Natta and G. Mazzanti, *Tetrahedron*, 8 (1960) 86.

48. D.J.Cram and K.R.Kopecky, *J.Am.Chem.Soc.*, 81 (1959) 2748; D.J.Cram and D.R.Wilson, *ibid.*, 85 (1963) 1249; M.Szwarc, *Chem.Ind.* (*London*), (1958) 1589; G.E.Ham, *J.Polymer Sci.*, 40 (1959) 569, 46 (1960) 475.
49. C.E.Schildknecht, S.T.Gross, H.R.Davidson, J.M.Lambert and A.O.Zoss, *Ind. Eng.Chem.*, 40 (1948) 2104.
50. G.Natta, L.Porri, A.Carbonaro and G.Stoppa, *Makromol.Chem.*, 77 (1964) 114; G.Natta, L.Porri and A.Carbonaro, *ibid.*, 77 (1964) 126.
51. G.Natta, P.Corradini and G.Allegra, *Atti Accad.Nazl.Lincei, Rend.*, [8] 26 (1959) 155.
52. G.Natta, P.Corradini and G.Allegra, *J.Polymer Sci.*, 51 (1961) 399; G.Allegra, *Nuovo Cimento*, [10] 23 (1962) 502.
53. G.Natta, *Actes II Congr.Intern.de Catalyse, Paris, 1960*, 1961, p.39; *Chim.Ind.* (*Milan*), 42 (1960) 1207.
54. G.Natta, I.Pasquon, A.Zambelli and G.Gatti, *J.Polymer Sci.*, 51 (1961) 387.
55. G.Natta, P.Corradini, I.W.Bassi and L.Porri, *Atti Accad.Nazl.Lincei, Rend.*, [8] 24 (1958) 121.
56. G.Natta, P.Pino, E.Mantica, F.Danusso, G.Mazzanti and M.Peraldo, *Chim.Ind.* (*Milan*), 38 (1956) 124.
57. G.Natta, G.Pajaro, I.Pasquon and V.Stellacci, *Atti Accad.Nazl. Lincei, Rend.*, [8] 24 (1958) 479.
58. G.Natta, P.Pino, G.Mazzanti and R.Lanzo, *Chim.Ind.* (*Milan*), 39 (1957) 1032.
59. G.Natta, I.Pasquon and E.Giachetti, *Angew.Chem.*, 69 (1957) 213.
60. G.Natta, I.Pasquon, E.Giachetti and F.Scalari, *Chim.Ind.* (*Milan*), 40 (1958) 103.
61. G.Natta, I.Pasquon and E.Giachetti, *Chim.Ind.* (*Milan*), 40 (1958) 97; G.Natta, I.Pasquon, E.Giachetti and G.Pajaro, *ibid.*, 40 (1958) 267.
62. G.Natta and I.Pasquon, *Advan. Catalysis*, 11 (1959) 1.
63. G.Natta and I.Pasquon, *Volume Corso Estivo Chimica Macromolecole, Varenna, 1961*, C.N.R., Rome, 1963, p.75.
64. G.Natta, I.Pasquon and A.Zambelli, *J.Am.Chem.Soc.*, 84 (1962) 1488.
65. G.Natta, I.Pasquon, P.Corradini, M.Peraldo, M.Pegoraro and A.Zambelli, *Atti Accad.Nazl.Lincei, Rend.*, [8] 28 (1960) 539.
66. G.Natta, G.Mazzanti, A.Valvassori, G.Sartori and D.Fiumani, *J.Polymer Sci.*, 51 (1961) 411.
67. G.Natta, *Rubber Plastics Age*, 38 (1957) 495; G.Natta and G.Crespi, *Rubber Age* (*N.Y.*), 87 (1960) 459; G.Natta, G.Crespi and M.Bruzzone, *Kautschuk Gummi*, 14 (1961) 54WT; G.Natta, G.Crespi, E.di Giulio, G.Ballini and M.Bruzzone, *Rubber Plastics Age*, 42 (1961) 53; G.Natta, G.Crespi, G.Mazzanti, A.Valvassori, G.Sartori and P.Scaglione, *Rubber Age* (*N.Y.*), 89 (1961) 636; G.Natta, G.Crespi and G.Mazzanti, *Proc.Rubber Technol.Conf., 4th, London, 1962*; G.Crespi and E.di Giulio, *Rev.Gen.Caoutchouc*, 40 (1963) 99; G.Natta, G.Crespi, G.Mazzanti, A. Valvassori and G.Sartori, *Chim.Ind.* (*Milan*), 45 (1963) 651.
68. G.Natta, M.Farina and M.Peraldo, *Atti Accad.Nazl.Lincei, Rend.*, [8] 25 (1958) 424.
69. M.Peraldo and M.Farina, *Chim.Ind.* (*Milan*), 42 (1960) 1349.

70. G. Natta, M. Peraldo, M. Farina and G. Bressan, *Makromol. Chem.*, 55 (1962) 139.
71. G. Natta, M. Farina, M. Peraldo, P. Corradini, G. Bressan and P. Ganis, *Atti Accad. Nazl. Lincei, Rend.*, [8] 28 (1960) 442.
72. G. Dall'Asta, G. Mazzanti, G. Natta and L. Porri, *Makromol. Chem.*, 56 (1962) 224.
73. G. Natta, G. Dall'Asta, G. Mazzanti and G. Motroni, *Makromol. Chem.*, 69 (1963) 163.
74. G. Dall'Asta and G. Mazzanti, *Makromol. Chem.*, 61 (1963) 178.
75. G. Natta, L. Porri, P. Corradini and D. Morero, *Chim. Ind. (Milan)*, 40 (1958) 362; G. Natta, L. Porri and A. Mazzei, *ibid.*, 41 (1959) 116; G. Natta, L. Porri and A. Carbonaro, *Atti Accad. Nazl. Lincei, Rend.*, [8] 31 (1961) 189; G. Natta, L. Porri and L. Fiore, *Gazz. Chim. Ital.*, 89 (1959) 761; G. Natta, L. Porri, G. Zanini and L. Fiore, *Chim. Ind. (Milan)*, 41 (1959) 526; G. Natta, L. Porri, G. Zanini and A. Palvarini, *ibid.*, 41 (1959) 1163.
76. G. Natta, L. Porri, A. Mazzei and D. Morero, *Chim. Ind. (Milan)*, 41 (1959) 398; G. Natta, *Rubber Plastics Age*, 38 (1957) 495; G. Natta, *Chim. Ind. (Milan)*, 39 (1957) 653; G. Natta, *Rev. Gen. Caoutchouc*, 40 (1963) 785.
77. D. Morero, A. Santambrogio, L. Porri and F. Ciampelli, *Chim. Ind. (Milan)*, 41 (1959) 758.
78. G. Natta, P. Corradini and L. Porri, *Atti Accad. Nazl. Lincei, Rend.*, [8] 20 (1956) 728; G. Natta and P. Corradini, *Angew. Chem.*, 68 (1956) 615; G. Natta, P. Corradini and I. W. Bassi, *Atti Accad. Nazl. Lincei, Rend.*, [8] 23 (1957) 363; G. Natta and P. Corradini, *Nuovo Cimento, Suppl.*, [10] 15 (1960) 122.
79. See, for example, *Belgian Pat.* 551, 851 (1956), Phillips Company U.S.; *Belgian Pat.* 573, 680 (1958) and 575, 507 (1959), Montecatini, Milan, Italy; see G. Natta, G. Crespi, G. Guzzetta, S. Leghissa and F. Sabbioni, *Rubber Plastics Age*, 42 (1961) 402; G. Crespi and U. Flisi, *Makromol. Chem.*, 60 (1963) 191.
80. S. E. Horne *et al.*, *Ind. Eng. Chem.*, 48 (1956) 784.
81. G. Natta, L. Porri and G. Mazzanti, *Belgian Pat.*, 545, 952 (Italian priority, March 1955).
82. G. Natta, L. Porri, P. Corradini, G. Zanini and F. Ciampelli, *Atti Accad. Nazl. Lincei, Rend.*, [8] 29 (1960) 257; *J. Polymer Sci.*, 51 (1961) 463.
83. G. Natta, L. Porri, G. Stoppa, G. Allegra and F. Ciampelli, *J. Polymer Sci.*, 1B (1963) 67.
84. G. Natta, L. Porri, A. Carbonaro, F. Ciampelli and G. Allegra, *Makromol. Chem.*, 51 (1962) 229.
85. G. Natta and G. Dall'Asta, *Italian Pat.*, 570, 434 (1956).
86. G. Natta, I. W. Bassi and P. Corradini, *Makromol. Chem.*, 18–19 (1955) 455.
87. G. Natta, G. Dall'Asta, G. Mazzanti, U. Giannini and S. Cesca, *Angew. Chem.*, 71 (1959) 205.
88. G. Natta, G. Dall'Asta, G. Mazzanti and A. Casale, *Makromol. Chem.*, 58 (1962) 217.
89. G. Natta, G. Mazzanti, G. Dall'Asta and A. Casale, *Italian Pat.*, 652, 763 (1960).
90. G. Natta, G. Mazzanti and G. Dall'Asta, *Italian Pat.*, 643, 282 (1960); G. Natta, G. Dall'Asta and G. Mazzanti, *Italian Pat.* 648, 564 (1961).
91. G. Natta, G. Mazzanti, G. Dall'Asta and P. Longi, *Makromol. Chem.*, 37 (1960) 160.

92. G. Natta, M. Farina, P. Corradini, M. Peraldo, M. Donati and P. Ganis, *Chim. Ind. (Milan)*, 42 (1960) 1360.
93. G. Natta, G. Mazzanti, P. Longi and F. Bernardini, *Chim. Ind. (Milan)*, 42 (1960) 457.
94. G. Natta, G. Mazzanti and P. Corradini, *Atti Accad. Nazl. Lincei, Rend.*, [8] 28 (1960) 8; G. Natta, P. Corradini and I. W. Bassi, *ibid.*, [8] 28 (1960), 284; G. Natta, G. Mazzanti, P. Corradini and I. W. Bassi, *Makromol. Chem.*, 37 (1960) 156; G. Natta, P. Corradini and I. W. Bassi, *J. Polymer Sci.*, 51 (1961) 505.
95. I. W. Bassi, G. Dall'Asta, U. Campigli and E. Strepparola, *Makromol. Chem.*, 60 (1963) 202.
96. D. Braun, W. Betz and W. Kern, *Makromol. Chem.*, 28 (1958) 66.
97. G. Natta, G. Mazzanti, P. Longi, G. Dall'Asta and F. Bernardini, *J. Polymer Sci.*, 51 (1961) 487.
98. Actually an isotactic chain of finite length does include asymmetric carbon atoms, but each one is neutralized by another at an equal distance from the center of the main chain:

$$\begin{array}{cccc} CH_3- & CH_3 & \times\quad CH_3- & CH_3 \\ | & | & | & | \\ H & H & H & H \end{array}$$

Hence the fully isotactic polypropylene is a particular case of a *meso*-configuration and must be optically inactive.
99. G. Natta, P. Pino and G. Mazzanti, *Gazz. Chim. Ital.*, 87 (1957) 528.
100. P. Pino, G. P. Lorenzi and L. Lardicci, *Chim. Ind. (Milan)*, 43 (1960) 711; P. Pino and G. P. Lorenzi, *J. Am. Chem. Soc.*, 82 (1960) 4745; W. J. Baileg and E. T. Yates, *J. Org. Chem.*, 25 (1960) 1800.
101. G. Natta, M. Farina, M. Peraldo and M. Donati, *Chim. Ind. (Milan)*, 42 (1960) 1363; G. Natta, M. Farina and M. Donati, *Makromol. Chem.*, 43 (1961) 251.
102. G. Natta, M. Farina, M. Peraldo and G. Bressan, *Chim. Ind. (Milan)*, 43 (1961) 161; *Makromol. Chem.*, 43 (1961) 68.
103. G. Natta, L. Porri, A. Carbonaro and G. Lugli, *Chim. Ind. (Milan)*, 43 (1961) 529.
104. G. Natta, L. Porri and S. Valente, *Makromol. Chem.*, 67 (1963) 225.
105. M. Farina and G. Bressan, *Makromol. Chem.*, 61 (1963) 79.
106. G. Natta, *Ric. Sci., Suppl.*, 28 (1958) 1.
107. G. Natta, G. Bressan and M. Farina, *Atti Accad. Nazl. Lincei, Rend.*, [8] 34 (1963) 475; M. Farina, G. Natta and G. Bressan, *Symp. Macromolecular Chemistry I. U. P. A. C., Paris, 1963*; *J. Polymer Sci.*, C4, (1964) 141.

Biography

Giulio Natta was born at Imperia on February 26, 1903. He graduated in Chemical Engineering at the Polytechnic of Milan in 1924 and passed the examinations entitling him to teach there in 1927. In 1933 he was established on the staff of Pavia University as a full professor and at the same time was appointed director of the Institute of General Chemistry at that University, where he stayed till 1935, that is until he was appointed full professor in physical chemistry at the University of Rome. From 1936 to 1938 he was full professor and director of the Institute of Industrial Chemistry at the Polytechnic of Turin. He has been full professor and director of the Department of Industrial Chemistry at the Milan Polytechnic since 1938.

Now a world famous scientist, Prof. Natta began his career with a study of solids by means of X-rays and electron diffraction. He then used the same methods for studying catalysts and the structure of some high organic polymers (the latter from 1934). His kinetic research on methanol synthesis, on selective hydrogenation of unsaturated organic compounds and on oxo-synthesis led to an understanding of the mechanism of these reactions and to an improvement in the selectivity of catalysts.

In 1938 Prof. Natta began to study the production of synthetic rubber in Italy; he took part in research work on butadiene and was the first to accomplish physical separation of butadiene from 1-butadiene by a new method of extractive distillation.

In 1938 he began to investigate the polymerisation of olefins and the kinetics of subsequent concurrent reactions. In 1953, with financial aid from a large Italian chemical company, Montecatini, Prof. Natta extended the research conducted by Ziegler on organometallic catalysts to the stereospecific polymerization, thus discovering new classes of polymers with a sterically ordered structure, *viz.* isotactic, syndiotactic and di-isotactic polymers and linear non-branched olefinic polymers and copolymers with an atactic (or sterically non-ordered) structure. These studies, which were developed for industrial application in Montecatini's laboratories, led to the realisation of a thermoplastic material, isotactic polypropylene, which Montecatini were the first to produce on an industrial scale, in 1957, in their Ferrara plant. This product has been marketed successfully as a plastic material, by the name of Moplen, as a synthetic fibre, by the name of Meraklon, as a monofilament by the name of Merakrin and as packing film, by the name of Moplefan.

By X-ray investigations, Prof. Natta has also succeeded in determining the exact arrangement of chains in the lattice of the new crystalline polymers he has discovered.

No less important is his later research which led to the synthesis of completely new elastomers, in two different ways: by polymerization of butadiene into *cis*-1,4 polymers with a very high degree of steric purity and by copolymerization of ethylene with other α-olefins (propylene), originating extremely interesting materials such as saturated synthetic rubbers. The vulcanisation of these rubbers was made possible by the usual methods used for natural rubber, with the introduction of unsaturated monomeric units (terpolymers containing ethylene and propylene). The processes for the asymmetric synthesis, which allow the production of optically active macromolecules from optically inactive monomers, are of great scientific importance, due to their similarity to the natural biological processes. Other interesting results obtained by Natta in the field of macromolecular chemistry concern the synthesis of crystalline alternating copolymers of different couples of monomers and the synthesis of various sterically ordered polymers of non-hydrocarbon monomers.

Prof. Natta's scientific and technical activity is documented in over 700 published papers, of which about 500 concern stereoregular polymers, and by a large number of patents in many different countries. In 1961 he was made an honorary life member of the New York Academy of Sciences of which he had been a fellow since 1958. In 1955 he became a «national member» of the Accademia dei Lincei; he is also a member of the Istituto Lombardo di Scienze e Lettere and of the Accademia delle Scienze of Turin. He was made honorary member of the Austrian (1960), Belgian (that awarded him the STAS medal) (1962), and Swiss (1963) Chemical Societies. Professor Natta received a gold medal from the town of Milan (1960), from the President of the Italian Republic (1961, reserved to those who gained merits in the field of school, culture and art), the first international gold medal of the synthetic rubber industry (1961); a gold medal from the Milan district (1962) and from the Society of Plastic Engineers (New York, 1963), the Perrin medal from the French Chemical Physical Society, and the Lavoisier medal from the Chemical Society of France (both in 1963), the Perkin gold medal of the English

Society of Dyers and Colourists (1963), the John Scott award from the Board of Directors of the City Trust of Philadelphia, and the Medal «Leonardus Vincius Florentinus Doctor Ingenieurs» of FIDIIS, Paris (1971). The Turin University gave him an honorary degree in pure chemistry, and in 1963 Prof. Natta received an honorary degree from Mainz University.

Prof. Natta is a honorary member of the Industrial Chemical Society of Paris (1966) and of the Chemical Society of London (1970); an honorary member of the Rotary Club; associated foreign member of the Académie des Sciences de l'Institut de France (1964); member of the National Academy of XL, Rome (1964); joined member of the International Academy of Astronautics, Paris (1965); foreign member of the Academy of Sciences of Moscow, U.S.S.R. (1966); honorary president of the Italian Section of the Society of Plastics Engineers (SPE). He holds the following awards and honorary degrees: gold medal of the Union of Italian Chemists (1964); gold medal «Lomonosov» of the Moscow Academy of Sciences (1969); the «Carl-Dietrich-Harries-Plakette» of the Deutsche Kautschuk Gesellschaft, Frankfurt/Main (1971); honorary degrees from the University of Genoa (1964), the Polytechnic Institute of Brooklyn, New York (1964), the Catholic University of Louvain, Belgium (1965), and in 1971 from ESPI, University of Paris.

Chemistry 1964

DOROTHY CROWFOOT HODGKIN

«for her determinations by X-ray techniques of the structures of important biochemical substances»

Chemistry 1964

Presentation Speech by Professor G. Hägg, Member of the Royal Swedish Academy of Sciences

Your Majesties, Your Royal Highnesses, Ladies and Gentlemen.

Exactly 50 years ago, a Nobel Prize was awarded which we have much reason to be reminded of today. Max von Laue was awarded the 1914 Nobel Prize for physics for, according to the citation, «his discovery of the diffraction of X-rays by crystals». It is this phenomenon which has formed the basis of the work for which Mrs. Dorothy Crowfoot Hodgkin has been awarded the Nobel Prize for chemistry this year.

Very soon after von Laue's discovery, the two English scientists Bragg, father and son, began to apply X-ray diffraction in order to determine how the atoms of a compound are situated in relation to each other in a crystal. In other words, they tried to find out what is usually known as the «structure» of the compound. Their successes in this field resulted in their being jointly awarded the 1915 Nobel Prize for physics.

Knowledge of a compound's structure is absolutely essential in order to interpret its properties and reactions and to decide how it might be synthetized from simpler compounds. To begin with, only very simple structural problems could be solved by X-ray diffraction, and these problems were taken almost entirely from the field of inorganic chemistry. Organic compounds, compounds containing carbon, usually have more complicated structures, and these presented too many difficulties at this stage. However, even then considerable possibilities existed for determining how the atoms of an organic compound are bonded to each other, by purely chemical methods. These methods were based largely upon the knowledge obtained from the latter half of the nineteenth century concerning the geometry of the bonds directed from a carbon atom. Large molecules were broken down into components whose structures were already known, and when some idea had been obtained of how these components were joined together in the large molecule this could often be confirmed by synthetizing the molecule.

Gradually, however, such large and complicated molecules were reached that these «classical» methods no longer yielded a result. This was particularly so in the case of the structures of many of the molecules which form part of living organisms and participate in the vital processes. In these instances it was necessary to obtain help from the field of physics, and in the first place use was made of X-ray diffraction by crystals of the compound concerned. During the period following the discovery of X-ray diffraction, this method of structure determination had been developed to such a degree that by the 1940's it began to be possible to use it for solving the structures of organic compounds which were insoluble by classical methods.

However, even today structure determination by X-ray methods does not yield a direct route from the experimental data to the structure. In complicated cases the scientist only obtains a result after considerable mental effort, in which chemical knowledge, imagination and intuition play a significant part. In addition, the experimental data often have to be processed using different mathematical treatments, which must be varied according to the circumstances. Add to this the fact that the more complicated the structure, the greater becomes the volume of experimental data which must be amassed and processed. For relatively simply built compounds it was possible to carry out the calculations with pencil and paper. Nowadays it is nearly always necessary to use electronic computers, and their arrival has made an enormous difference to the possibility of carrying out structure determinations. However, it is not usually possible to just feed in the experimental data, and get out the figures which give the final structure; the scientist's ability to handle the data is still of vital importance. It is in this respect that Mrs. Hodgkin has shown such exceptional skill.

Mrs. Hodgkin has carried out a large number of structure determinations, primarily of substances which are of importance biochemically and medically, but two of these substances deserve especial mention. These are penicillin and vitamin B_{12}, whose structures have become completely and definitely known through her efforts.

The use of penicillin in medicine began to be tested about the beginning of the second world war, and its exceptional antibiotic properties meant that the demand increased enormously. It was therefore obviously desirable to find out whether penicillin itself or other related compounds having a similar effect could be prepared by chemical methods. For this purpose it was essential to determine the composition and structure of penicillin, and a large number of chemists and X-ray crystallographers in both England and the U.S.A. were put on to this problem. Mrs. Hodgkin was to play a leading part in the X-ray crystallographic work, and it was chiefly her efforts which brought it to a satisfactory conclusion. The work was begun in 1942 and the structure was elucidated after four years' intensive work. This was marked by close cooperation between organic chemists, X-ray crystallographers and scientists in other branches of physical chemistry and physics. A number of X-ray crystallographic methods were also used here for the first time.

Mrs. Hodgkin's determination of the structure of penicillin bears evidence of exceptional skill and great perseverance. The difficulties were considerable, but this was not because the molecule was particularly large. However, it possessed some unknown features, which meant that the chemical properties did not give sufficient guidance.

In 1948 Mrs. Hodgkin began her attempts to determine the structure of vitamin B_{12}, which had been isolated in the same year. This vitamin can be synthetized by certain bacteria and fungi, of which some play an active part in the digestive processes of animals. The production of B_{12} is most pronounced in the ruminants, who seem to require this vitamin in particularly large amounts. In most of the other higher animals, for example in man, the production of B_{12} is small, and their food must therefore contain sufficient quantities of ready-made B_{12}. Lack of B_{12} in the diet, or a reduced ability to absorb this vitamin *via* the walls of the alimentary canal, leads in man to the fatal blood condition of pernicious anaemia. The illness can always be arrested by injections of B_{12} which is only needed in very small quantities. It is still not clear how B_{12} functions in the metabolic processes, but in order to begin to come to grips with this problem it is essential to know the structure in detail.

In 1956, after eight years' work, Mrs. Hodgkin and her collaborators had clarified the B_{12} structure. Never before had it been possible to determine the exact structure of so large a molecule, and the result has been seen as a triumph for X-ray crystallographic techniques. It was also, however, a triumph for Mrs. Hodgkin. It is certain that the goal would never have been reached at this stage without her skill and exceptional intuition.

There is reason to hope that the detailed knowledge of the B_{12} structure, revealed as a result of this work, will make it possible both to understand how this vitamin assists in the body's metabolism and to synthetize it. For the time being it has to be produced *via* bacterial fermentation.

Professor Hodgkin. You have for many years directed your efforts towards the determination of crystal structures by means of X-ray diffraction techniques. You have solved a large number of structural problems, the majority of great importance in biochemistry and medicine, but there are two landmarks which stand out. The first is the determination of the structure of penicillin, which has been described as a magnificent start to a new era of crystallography. The second, the determination of the structure of vitamin B_{12}, has been considered the crowning triumph of X-ray crystallographic analysis, both in respect of the chemical and biological importance of the results and the vast complexity of the structure.

Scientists working in many different fields, in X-ray crystallography, in chemistry, and in medicine admire the great determination and skill, involving what can only be described as gifted intuition, which has always been the mark of your work.

In recognition of your services to science the Royal Swedish Academy of Sciences decided to award you this year's Nobel Prize for Chemistry. To me has been granted the privilege of conveying to you the most hearty congratulations of the Academy and of requesting you to receive your prize from the hands of his Majesty the King.

DOROTHY CROWFOOT HODGKIN

The X-ray analysis of complicated molecules

Nobel Lecture, December 11, 1964

I first met the subject of X-ray diffraction of crystals in the pages of the book W.H. Bragg wrote for school children in 1925, ‹Concerning the Nature of Things›. In this he wrote: «Broadly speaking, the discovery of X-rays has increased the keenness of our vision over ten thousand times and we can now ‹see› the individual atoms and molecules.» I also first learnt at the same time about biochemistry which provided me with the molecules it seemed most desirable to ‹see›. At Oxford, seriously studying chemistry, with Robinson and Hinshelwood among my professors, I became captivated by the edifices chemists had raised through experiment and imagination - but still I had a lurking question. Would it not be better if one could really ‹see› whether molecules as complicated as the sterols, or strychnine were just as experiment suggested? The process of ‹seeing› with X-rays was clearly more difficult to apply to such systems than my early reading of Bragg had suggested; it was with some hesitation that I began my first piece of research work with H.M. Powell on thallium dialkyl halides, substances remote from, yet curiously connected with, my later subjects for research.

A series of lucky accidents (a chance meeting in a train between an old friend of mine, Dr. A.F. Joseph and Professor Lowry was one) took me to Cambridge to work with J.D. Bernal in 1932. There our scientific world ceased to know any boundaries. In a sub-department of Mineralogy, changed during my stay into one of Physics, we explored the crystallography of a wide variety of natural products, the structure of liquids and particularly water, Rochelle salt, isomorphous replacement and phase determination, metal crystals and pepsin crystals, and speculated about muscular contraction. Our closest friends were biologists and biochemists. I left Cambridge with great reluctance to try to settle down academically and to try to solve at least one or two of the many problems we had raised.

I do not need here to give a detailed account of the theoretical background of structure analysis by the X-ray diffraction of crystals since this was done long ago by W.L. Bragg[1] and again two years ago, very beautifully, by Perutz and Kendrew[2]. The experimental data we have to employ are the

X-ray diffraction spectra from the crystal to be studied, usually recorded photographically and their intensities estimated by eye. These spectra correspond with a series of harmonic terms which can be recombined to give us a representation of the X-ray scattering material in the crystal, the electron density. The calculation involves the summation of a Fourier series in which the terms have the amplitudes and phases of the observed spectra; both depend on the positions of the atoms in the crystal, but only the amplitudes are easily measurable. As Perutz and Kendrew explained, the introduction of additional heavy atoms into a crystal under investigation at sites which can be found, may make it possible to calculate phase angles directly from the observed amplitudes of the spectra given by the isomorphous crystals. One is then in the position that, from a sufficient number of measurements, one can calculate directly the electron density and see the whole structure spread out before one's eyes. However, the feat involved in the calculations described two years ago was prodigious – tens of thousands of reflections for five or six crystals were measured to provide the electron-density distribution in myoglobin and haemoglobin. More often, and with most crystals, the conditions for direct electron-density calculation are not initially met and one's progress towards the final answer is stepwise; if some of the atoms can be placed, particularly the heavier atoms in the crystal, calculations, necessarily imperfect, of the electron density can be started from which new regions in the crystal may be identified; the calculation is then repeated until the whole atomic distribution is clear. At the outset of my research career, two essential tools became available, the Patterson synthesis and Beevers and Lipson strips. Patterson showed that a first Fourier synthesis calculated directly from the raw data, without phase information, represented the inter-atomic vector distribution in the crystal structure[3]. This was capable, in simple structures, of showing the whole atomic arrangement and, in more complicated ones, at least of indicating the positions of heavy atoms. Beevers and Lipson strips[4] provided the means for a poor crystallographer to start calculating – each strip represents the wave-like distribution corresponding with a single term. I still have the letter Beevers and Lipson wrote offering me a box for £5 – I bought it.

Our early attempts at structure analysis now seem to be very primitive. The crystal structures of cholesteryl chloride and bromide proved not sufficiently isomorphous to solve by direct-phase determination. We moved over to cholesteryl iodide, where the heavier atom was both easier to place in the crystal from the Patterson synthesis (Fig. 1) and contributed more to the scattering[5]. Harry Carlisle showed it was possible to place the atoms in three

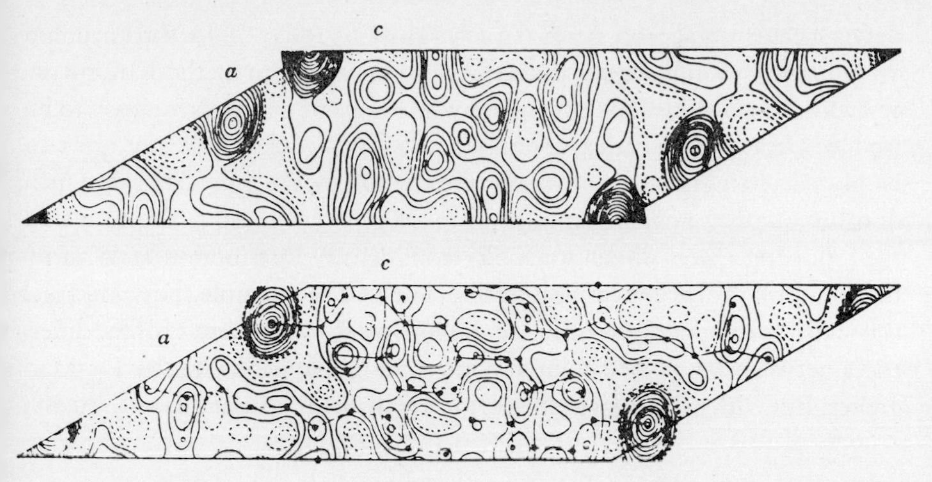

Fig. 1. The initial stages of the X-ray analysis of cholesteryl iodide. Above Patterson projection along b. The heavy peaks indicate I — I vectors. Below, electron-density projection calculated with terms given the phase angles of the iodine contributions. The outlines of two sterol molecules related by a two-fold screw axis are drawn in.

dimensions by calculating sections and lines in the three-dimensional electron-density distribution with phases derived at first from the iodine contributions alone; it took him months to make calculations on Beevers–Lipson strips which now would take fewer hours. The atomic arrangement found completely confirmed the sterol formula as revised by Rosenheim and King and Wieland and Dane, following Bernal's first X-ray measurements[6]. We sought for a compound of more unknown structure.

We were encouraged to try our operations on penicillin by Chain and Abraham before ever the antibiotic itself was crystallised; I grew crystals for X-ray analysis from 3 mg of the sodium salt flown over during the war from the Squibb Research Institute to Sir Henry Dale; the crystals were grown under the watchful eyes of Kathleen Lonsdale, who brought them to me from London. Later, we also grew crystals of potassium and rubidium benzylpenicillin, hoping again for an isomorphous series. But first the sodium salt was not isomorphous with the other two, then the potassium and rubidium ions were in such positions in the structure that they did not contribute to many of the reflections. The solution followed from a comparison of very imperfect maps calculated for the two series. But the methods by which these maps were obtained were more a consequence of the ingenuity of my collaborators, Charles Bunn and Barbara Low, combined with our low computing power, than general processes for structure solving today[7]. This little structure would now be handled quite differently, by heavy-atom methods, using one crystalline form alone. For example, by biosynthetic methods it is easy to introduce a heavy atom such as bromine into the molecule; the heavy atom can be placed unambiguously in three dimensions by the calculated Patterson distribution; the remaining atomic positions appear with no difficulty at all in the following three-dimensional electron-density distribution and, on refinement, the atoms appear beautifully clearly. The example shown in Fig. 2 is actually bromophenoxymethylpenicillin[8], prepared for a study of the differences between benzylpenicillin and the acid-stable penicillins by Dr. Margreiter. But with molecules of this sort it is really not necessary to introduce an

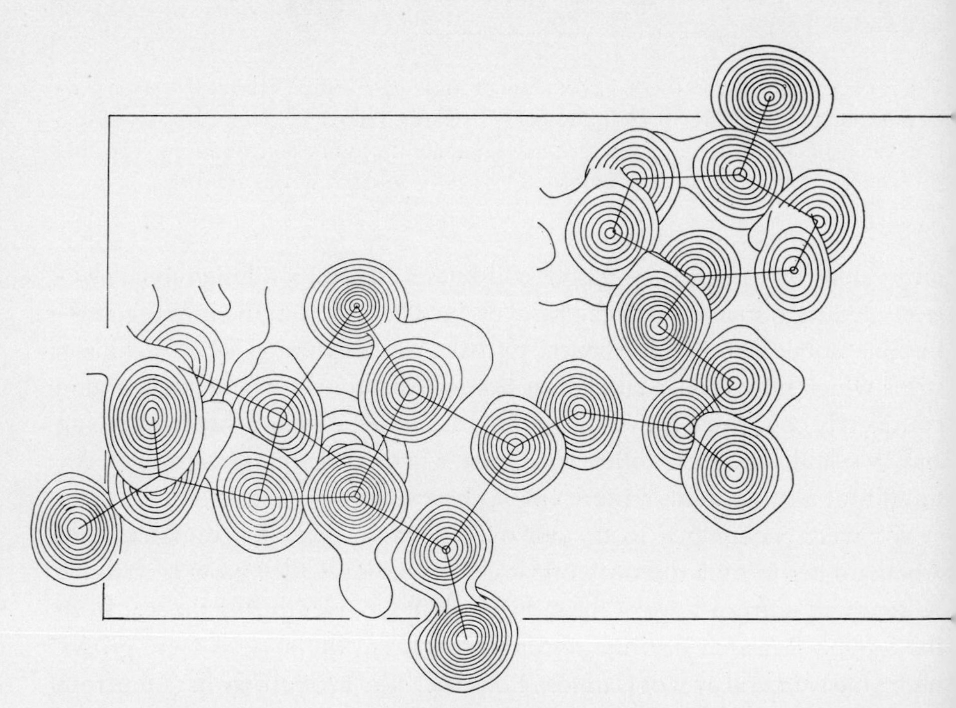

Fig. 2. *p*-Bromophenoxymethylpenicillin. Electron density near the atomic centres shown projected on the *c* plane. The contours are drawn at intervals of 1 e/A³ (K. J. Watson).

extra heavy atom at all. The sulphur atom present is itself relatively heavy enough to operate for structure finding purposes in sodium benzylpenicillin only a little less effectively then bromine. As Maslen and Abrahamsson showed, in relation to penicillin V itself[9] and cephalosporin C (ref. 10), with only a little more trouble one can place the sulphur atom unambiguously in the crystal structure and use the vector distribution relative to this to find the remaining atoms. Fig. 3 shows the electron density in the crystal of cephalosporin Cc, a very interesting antibiotic prepared by Abraham and Newton. Here it is easy to see the chemical structure, a four-membered β-lactam ring attached to a six-membered sulphur-containing ring–derived most probably like the corresponding penicillin, from δ-(α-aminoadipoyl)-cysteinyl valine but in an oxidised form. The natural antibiotic, cephalosporin C, loses on hydrolysis an acetyl group to give the more stable Cc molecule with the carboxyl group combined in a five-membered lactone ring. It crystallises with one molecule of acetic acid of crystallisation, fitting in between the long chains in the crystal[11].

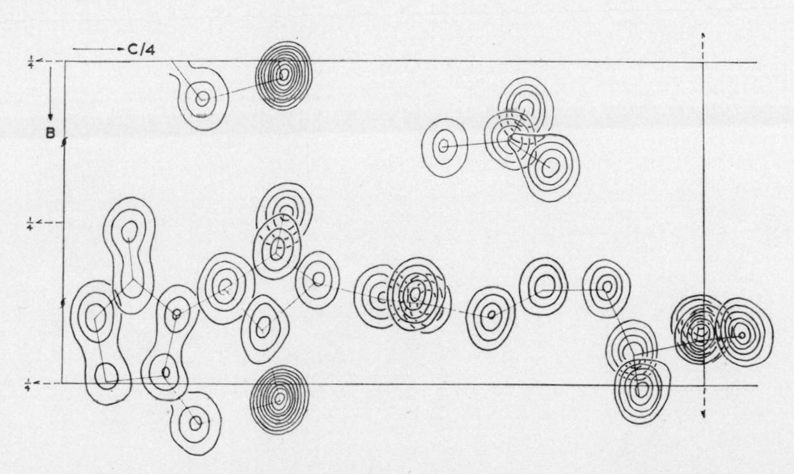

Fig. 3. Cephalosporin Cc. Electron density near the atomic centres shown projected along *a*. Contours at intervals of 2 e/A^3 (R. Diamond).

The X-ray scattering effect of sulphur is roughly one sixth of that of the rest of the molecule of cephalosporin C, approximately the same as that of cobalt in relation to vitamin B_{12}, cyanocobalamin. At the time that Dr. Lester Smith brought us his first red crystals of this, the antipernicious anaemia factor, shortly after its first isolation by Dr. Folkers and his colleagues, we knew nothing at all about the molecule. Two X-ray photographs, taken overnight, showed that it had a molecular weight of the order of 1500. It is of such complexity that even its analytical formula follows best from its X-ray analysis. Formally, the process of structure determination followed the course outlined earlier – 3D Patterson, 3D Fourier, atom sorting in rounds of calculation – the outline hardly gives an accurate impression of the stages of confused half knowledge through which we passed. Again an important part in the analysis was played by our having a view of the corrin nucleus surrounding the cobalt atom in quite different crystals, the cyanocobalamin crystals and also some, or perhaps I should say one, crystal of a hexacarboxylic acid derived by degradation from them by Cannon, Johnson and Todd[12]. And we were greatly helped by friends with computers; on a particularly happy day Kenneth Trueblood, on a casual summer visit to Oxford, walked into the laboratory and offered to carry out any additional calculations we needed on a fast computer in California, free and for nothing and with beautiful accuracy. Extracts from the calculations he carried out still provide some of the best examples of the processes I have been describing, of the gradual appearance of precise peaks marking atomic positions through stages in the electron-density calculations[13]. Examples are given in Fig. 4.

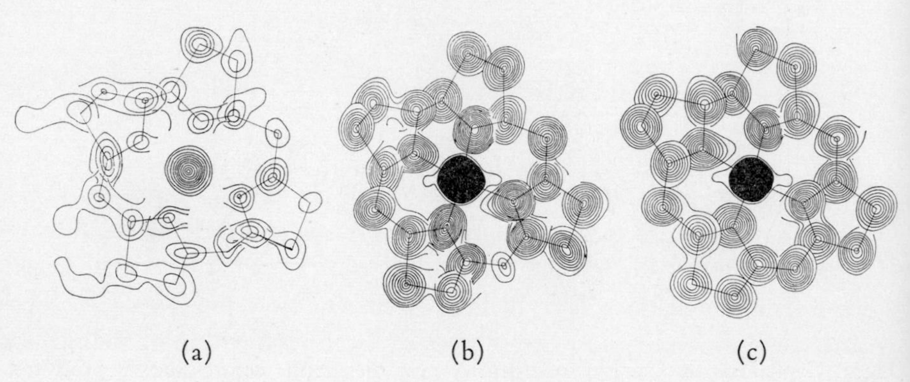

Fig. 4. Electron-density peaks over the corrin nucleus in the hexacarboxylic acid at different calculation stages. Terms phased on contributions calculated for (a) cobalt, (b) with the nucleus atoms less C-10, (c) cobalt with all the nucleus atoms. Contours at 1 e/A^3, except over Co.

Today our best evidence for the structure of the nucleus in B_{12} comes from the X-ray analysis of cobyric acid, Factor V 1a, orange-red crystals isolated in very small quantities from sewage sludge by Bernhauer, Wagner and Wahl (ref. 14). The X-ray analysis was achieved by what still seems to me a remarkable operation. The crystals are monoclinic, $P2_1$, with two molecules in the unit cell, and X-ray photographs, taken of them with copper $K\alpha$-radiation

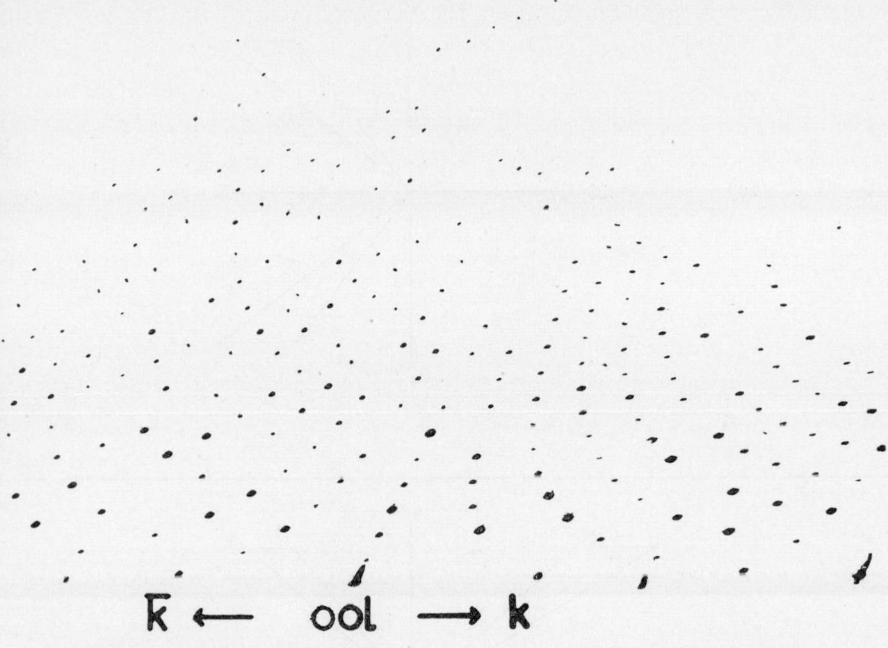

Fig. 5. *0kl* reflections from Factor V 1a. Note dissymmetry across line marked ← →.

show very markedly the effects of anomalous dispersion – compare Fig. 5, $F_{hkl} \neq F_{\overline{hkl}}$. The effects are due to a small phase change introduced by the scattering at the cobalt atom which has an absorption edge near the wave length of copper $K\alpha$-radiation. They make it possible to use yet another method of phase angle determination first suggested by Bijvoet, Ramachandran and others[15,16], and illustrated in Fig. 6. By measuring the intensities of both F_{hkl} and $F_{\overline{hkl}}$ reflections, Dale and Venkatesan were able to assign rather accurate phase angles to 1994 reflections – about half the total observed. The calculation requires a knowledge of the cobalt atom position, easily found from a Patterson synthesis. The first three dimensional electron density map calculated with just these 1994 terms showed the whole molecule and crystal structure clearly defined; the chemical formula of cobyric acid (I) could have been written with very little hesitation from this map alone although, strictly, it shows only part of each atom (Fig. 7a). With further rounds of calculation the full electron density is introduced; even many of the hydrogen atom positions then appear as individual peaks as in Fig. 7b. It is clear that the molecule is present in the form of an aquo- or hydroxy-cyanide, where the cyanide group

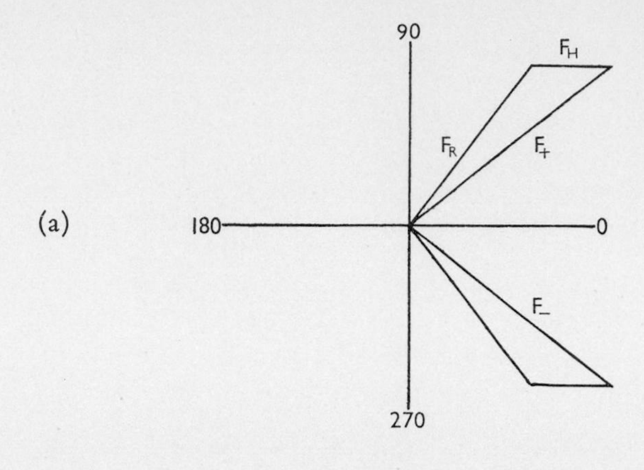

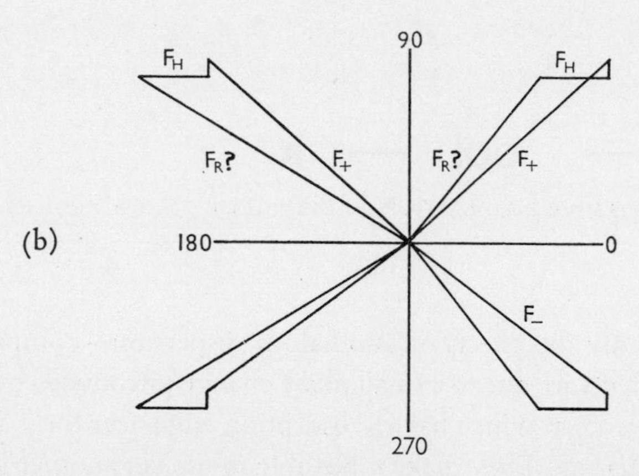

Fig. 6. Phase determination in a monoclinic crystal such as cobyric acid. (a) In a pair of isomorphous crystals. Here where F values for the substituted and unsubstituted crystals and for the heavy atoms (F, F_R and F_H respectively) can all be found there is an ambiguity in the phase-angle determination about the line 0–180 in the Argand diagram. (b) In a single crystal containing an anomalous scatterer, F_{hkl} and $F_{\overline{hkl}}$, F+ and F− can be distinguished, but in general F_R is unknown, and the phase angle is ambiguous about the line 90–270°. In the study of cobyric acid, the phase angle nearer F_H was chosen.

replaces the nucleotide in cyanocobalamin and the water molecule the cyanide group[17].

Many details of the crystal structure of factor V 1a are interesting in relation to its chemistry (Fig. 8). Thus the amide groups on the periphery of the molecule are all hydrogen-bonded to those of neighbouring molecules either

NH_2–CO–CH_2–CH_2 CH_3 CH_3 CH_2–CO–NH_2
C C7
NH_2–CO–CH_2 CH–C C B CH–CH_2–CH_2–CO–NH_2
C A N HOH N=C 8
CH_3 C N → Co+ ← CH
CH_3 CH–N C N–C CH_3
NH_2–CO–CH_2–CH D N C C
C C CH CH_3
$^-$OOC–CH_2–CH_2 CH_3 CH_3 CH_2–CH_2–CO–NH_2

Formula I.

directly or through water molecules, except one, that on ring B. This is turned inwards to the hydrogen bond with the water molecule on the cobalt atom; the forces involved are sufficient to distort the position of the β-carbon atom, C-7, to which it is attached. At the same time, the amide nitrogen atom is brought close to C-8 with which it could readily react to form the lactam

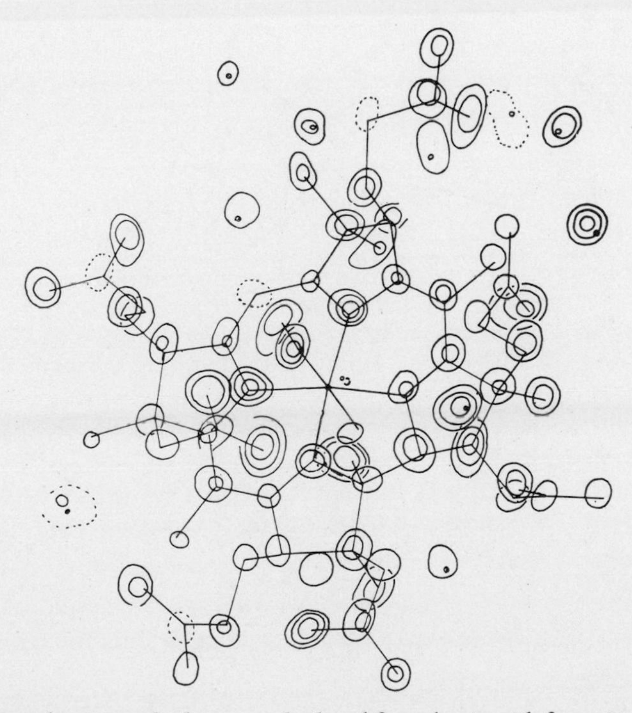

Fig. 7a. Electron density peaks from $\varrho 1$ calculated for cobyric acid, factor V 1a. The contours are at intervals of 1 e/A^3, starting with the 2 electron contour. The 1-electron contour is dotted where necessary to define an atomic position.

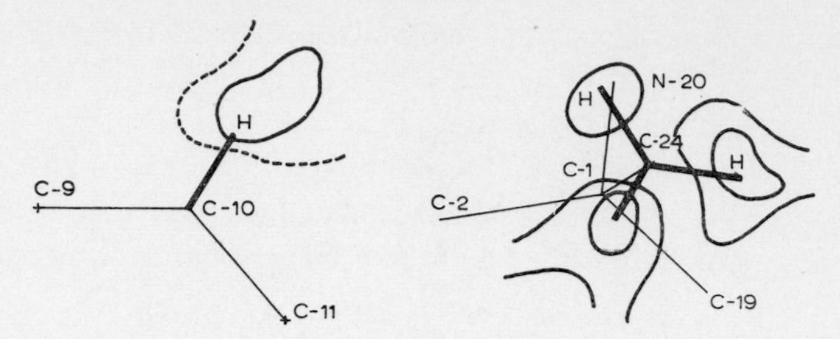

Fig. 7b. Part of difference map $\Delta\varrho 6$, calculated for cobyric acid showing density due to hydrogen atoms attached to C-24 and C-10.

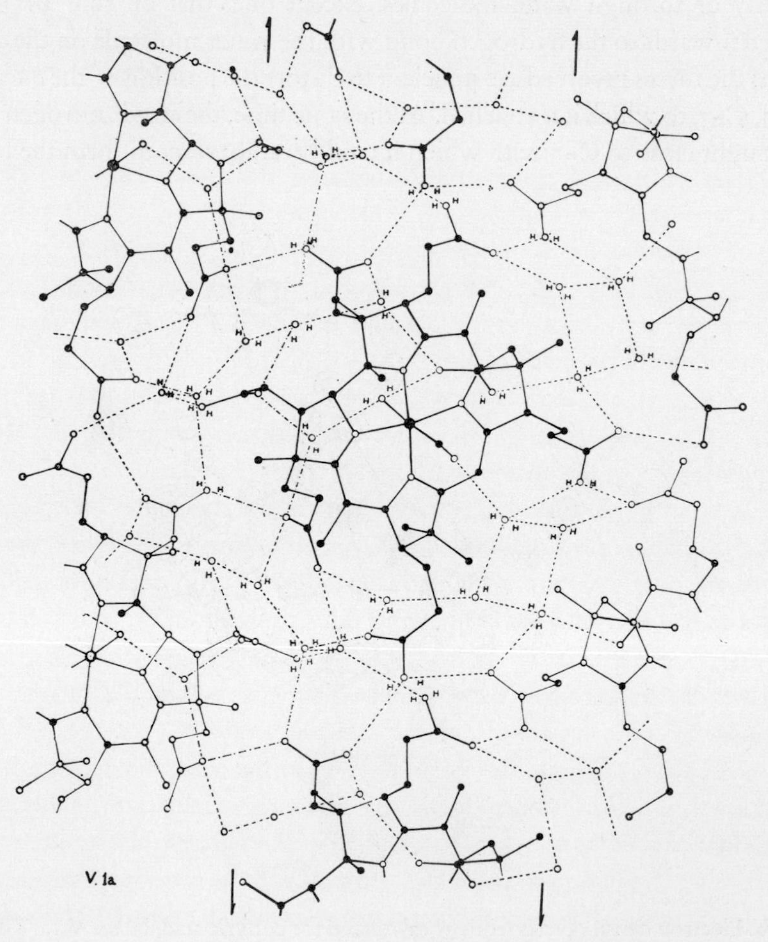

Fig. 8. The crystal structure of cobyric acid projected along *a*. Molecules centred at x are shown in strong lines, those at $\bar{x}$ in thin lines. Hydrogen bonds are dotted.

ring observed in the hexa acid, mentioned earlier. That the oxygen atom attached to cobalt is part of a water molecule, not a hydroxyl group, is suggested by the fact that these orange-red crystals separate at acid pH (*cf.* the haemoglobins and myoglobin). In our crystal this oxygen atom makes a second contact through a water molecule with the carboxyl group attached to a neighbouring molecule. It would, in fact, be very easy to change the system

```
H–O–H·····O–H·····O–C–O
  |       |       |
          H
```

to

```
HO·····H–O·····HO–C=O
 |       |        |
         H
```

by a movement of two hydrogen atoms within the crystal; it would be interesting to see if this movement occurs, as in Rochelle salt, under the influence of an electric field.

The interatomic distances in the inner ring of factor V 1a, which has been called the corrin ring, conform very closely with the distances proposed[12] for a structure containing six resonating double bonds, so closely as to leave almost no doubt of the correct formulation of its chemical structure (*cf.* Fig. 9). It was all the same, quite a moment in my life when J. D. Dunitz showed last summer at the Royal Society very closely similar figures derived by the X-ray analysis of the nickel corrin derivative synthesized early this year by Eschenmoser and his coworkers[18]. The molecule as a whole is much smaller (Fig. 10) but the identity of the nucleus with that in cobyric acid is certain.

Very recently, Dunitz and Meyer have refined the nickel corrin structure through several more stages; their latest interatomic distances shown in Fig. 11 are now so close to the earlier proposed theoretical figures that one begins to feel that the small remaining deviations are likely to be real, *e.g.* in C–N 9–21 and 11–22. There is a tendency in the natural series also for these bonds to be longer than the distances given by the simple theory at first proposed.

One of the features of the corrin nucleus in the natural compounds is that even the inner ring is not quite planar. The same is true of the synthetic nucleus. In nickel corrin, the distortion is small but very regular and tetrahedral in character in relation to the nickel atom; alternate nitrogen atoms, bonded to it, lie above and below the least squares plane passing through the nickel and four nitrogen atoms. The non-planarity of the system as a whole no doubt derives from the stereochemistry of the five-membered ring C-1, C-19,

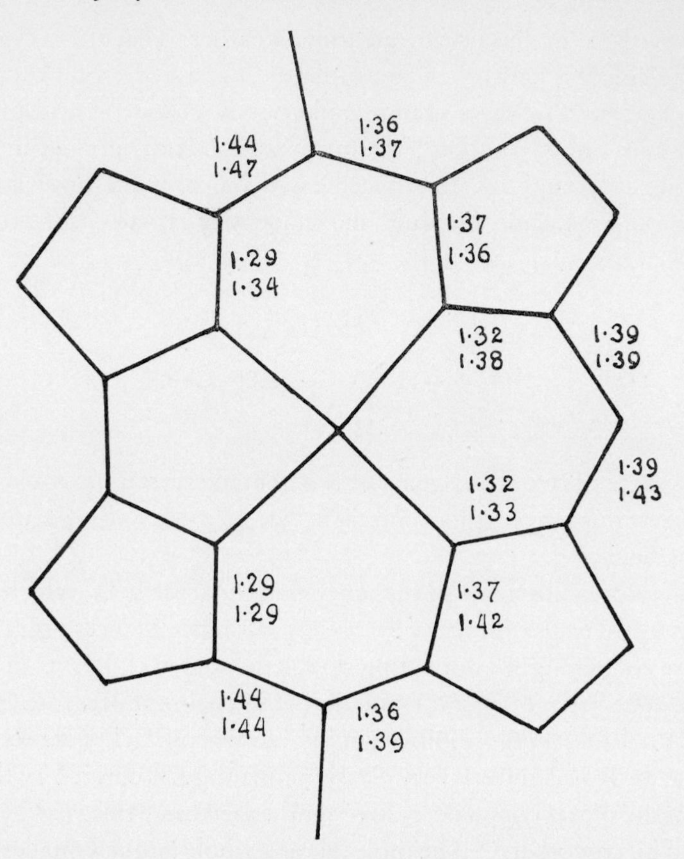

Fig. 9. Interatomic distances (below) found from $\varrho 6$ for cobyric acid compared with those suggested for a system containing six resonating double bonds.

N-23, Ni, N-20. In the natural corrins, the deviations are rather different in compounds with or without the nucleotide. In the unsubstituted compounds, and particularly in cobyric acid, the deviations are in the same sense as in the nickel corrin derivative and as in this molecule, C-5 and C-15, which carry C-35 and C-53 respectively, are on opposite sides of the plane containing the cobalt and four inner nitrogen atoms. In the nucleotide-containing compounds these two atoms are on the same side of the inner plane. In both series, the most marked deviations occur in the region of C-35 which is in a very overcrowded situation.

Cobyric acid is the natural precursor of the most remarkable molecule of our series, Co-5′-deoxyadenosyl-cobalamin, the coenzyme discovered by Barker, which ought, most properly to be called vitamin B_{12}. Crystals of

Fig. 10. Electron-density peaks over the atoms in the nickel corrin derivative (Dunitz and Meyer).

this compound were grown from water in capillary tubes by Dr. Galen Lenhert in 1960 from material supplied by Dr. Barker, and X-ray photographs were taken of them *in situ*, in their mother liquor. Again the intensities of the X-ray diffraction spectra were measured, Patterson and electron-density distributions were calculated, atoms belonging to the best known parts of the cobalamin molecule being placed first in the calculations. At this point, I should pause to say that a great advantage of X-ray analysis as a method of chemical structure analysis is its power to show some totally unexpected and surprising structure with, at the same time, complete certainty. Fig. 12 illustrates the structure we found – first, the electron-density map calculated over the region of uncertain structure, with the known part of the molecule placed – then the whole atomic arrangement that is derivable from the completed map. Clearly, in this molecule, cobalt was shown to be attached direct to the 5′-carbon atom of the adenosyl residue as in formula II[19]. There followed directly an explanation of the observed great instability of the molecule to light and cyanide ions – instability which had led to the failure of earlier investigations to recognise its existence and to isolate, in its place, cyanocobalamin.

The detailed geometry of the coenzyme molecule as a whole is fascinating in its complexity. The peculiar form of the corrin ring with the direct link

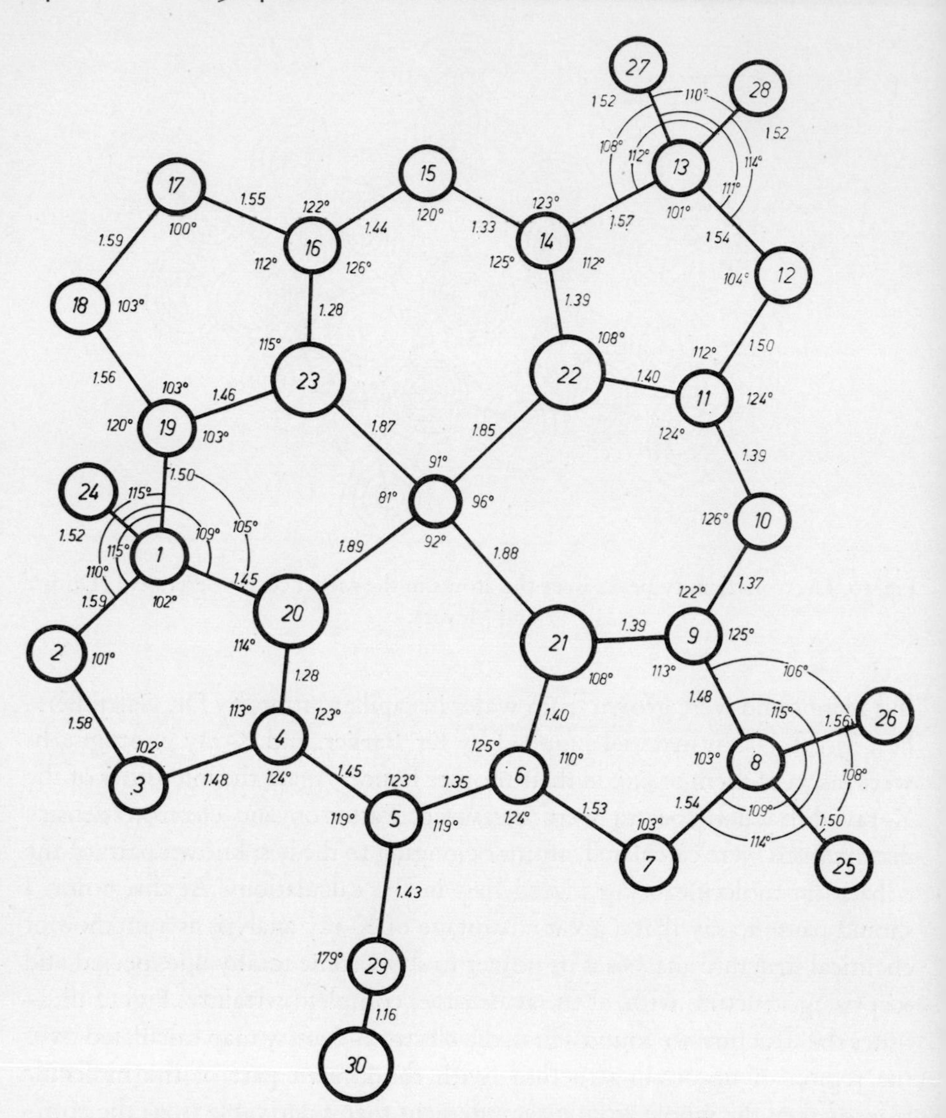

Fig. 11. Interatomic distances measured in the nickel corrin derivative (Dunitz and Meyer).

between rings A and D and the position of the methyl group at C-24 make the two sides of the corrin ring system very different stereochemically and the differences are reinforced by the positions of the methyl group and acetamide and propionamide residues attached at the carbon atoms (*cf.* Fig. 13). Approach to the cobalt atom from the lower side of the molecule is hindered by

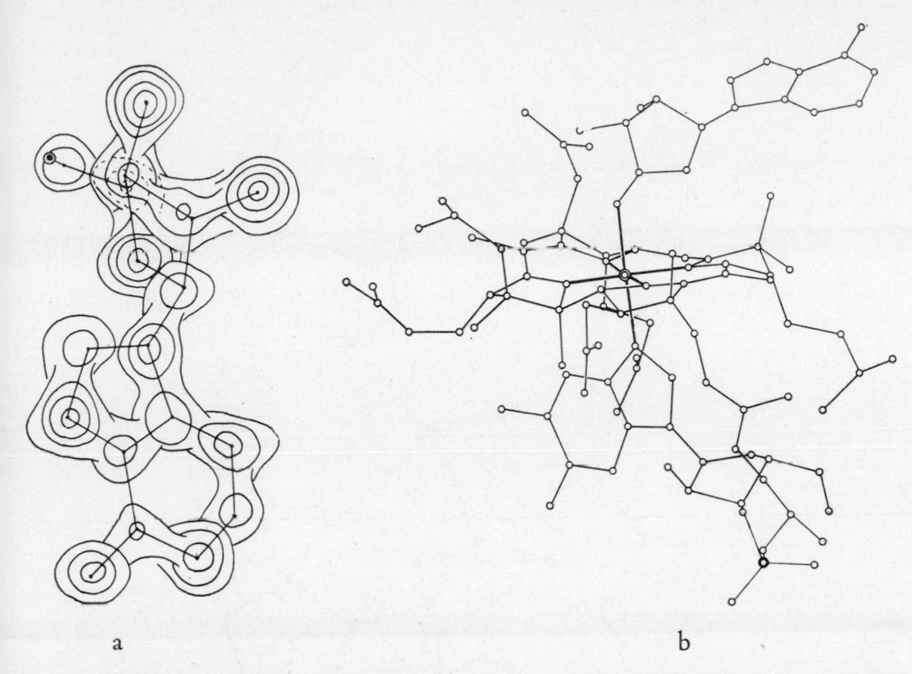

Fig. 12. (a) Electron-density peaks over the adenosyl residue calculated from X-ray data on dimethylbenzimidazole cobamide coenzyme. The calculation illustrated was one in which phase angles were computed from the positions of all the atoms in the molecule except those shown. Although only part of the electron density appears, the relative weights of the atoms are in agreement with their chemical structure. (b) The atomic positions found for the coenzyme molecule projected along the crystallographic *b* axis.

the methyl group C-24 and all groups attached here are rather loosely bound and easily displaced. At the upper site, on the other hand, attachment of a wide variety of ligands is possible; once in position they are enclosed by non-polar groups, the methyl and methylene groups projecting normal to the plane of the ring. Here they may be positively protected from immediate reaction with the surrounding solvent, for use when required in different biochemical transformations. In terms of this structure, one can begin to understand some of the uses to which the cobalamin nucleus is put in nature, for example, its part in the transfer of methyl groups. In the laboratory, methylcobalamin can be made by a series of reactions involving the reduction of aquocobalamin to a compound, probably cobalamin hydride, which easily exchanges with methyl compounds such as diazomethane or methyl sulphate; in nature, reduction also seems necessary for methyl transfer; the experiments of D.D.

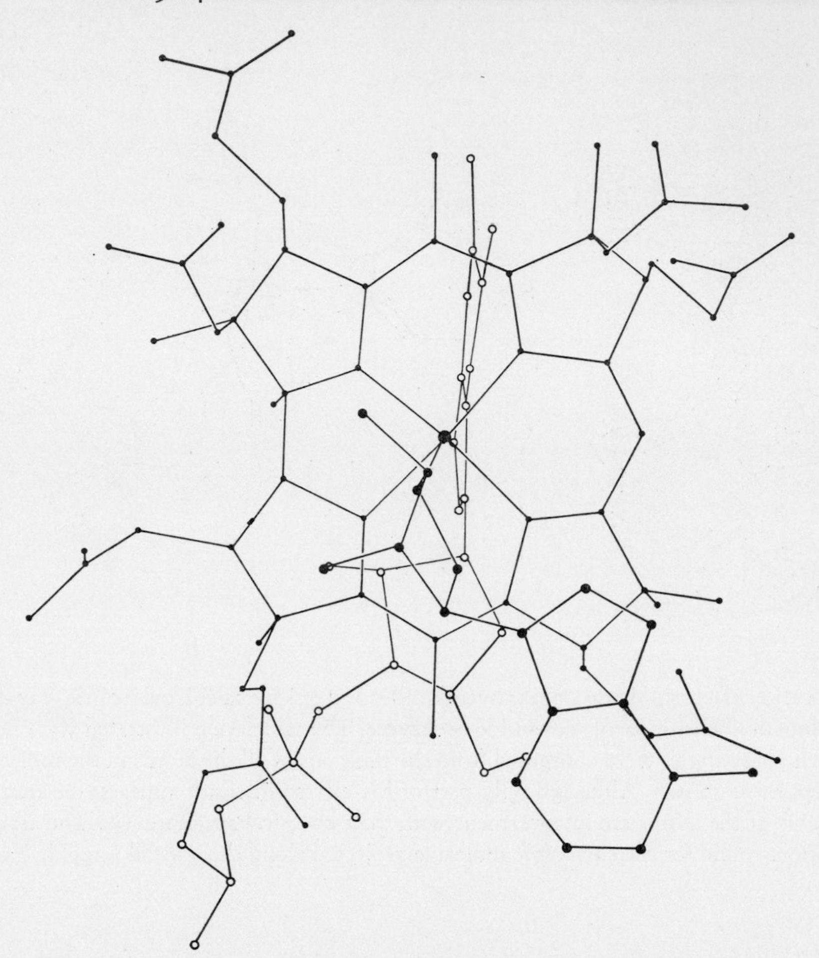

Fig. 13a. The positions found for the atoms in the coenzyme molecule projected (a) onto the least square squares plane passing through cobalt and the four inner nitrogen atoms, (b) and (c) across this plane.

Woods and his colleagues strongly suggest that methylcobalamin is the actual intermediate in one of the pathways by which methyl groups are transferred from methyltetrahydrofolate to homocysteine to form methionine[20]. Other effects of the B_{12} coenzymes are still not easy to explain in detail, particularly the part they play in the isomerisation reactions which led to their discovery. How so complex a system can effect the simple and fundamental migration process that changes methylmalonate to succinate, for example, is still a problem. One can see features of the system that might be important, the variable

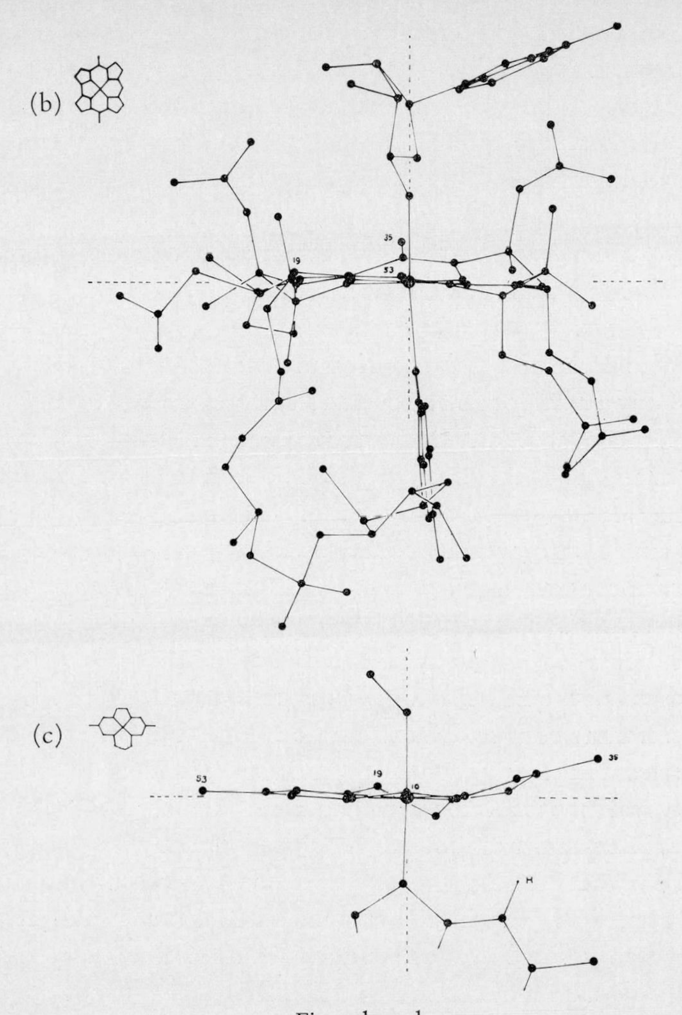

Fig. 13b and c.

valency of cobalt and the possibility of changes within the corrin nucleus, the fact that the cobalt to carbon bond readily breaks to give rise to free radicals[21]. But one would like to be able to observe the molecule in action, held in the meshes of the necessary enzyme, and this may be both possible and necessary for the detection of unstable intermediates in the reaction path.

There have been some practical results following from our knowledge of the structure of the molecules so far described. Penicillin V has been synthesised by Sheehan and Henery-Logan[22]. The coenzyme, a few years ago obtainable with difficulty in milligram quantities, can now be prepared very easily from cyanocobalamin – Dr. Lester Smith has made crystals 0.5 cm across. Soon, and no doubt very beautifully, cyanocobalamin itself will be synthesised by R. B. Woodward. But microorganisms are even more efficient than chemists at synthesis of molecules of this magnitude and will most likely continue to provide the main supplies of these compounds to be used in medicine. What we most hope to gain from knowledge of the structure and synthesis of these molecules is a complete understanding of their biogenesis and the part they play in metabolism. This should enable natural processes to be controlled when they go astray.

I should not like to leave an impression that all structural problems can be settled by X-ray analysis or that all crystal structures are easy to solve. I seem to have spent much more of my life not solving structures than solving them. I will illustrate some of the difficulties to be overcome by considering our efforts to achieve the X-ray analysis of insulin.

Insulin is a molecule of weight about 6000, larger than any so far described, though small if considered as a protein and compared with myoglobin and haemoglobin. Although its complete chemical structure is now known from Sanger's researches[23], it is quite unclear what exactly the molecule does that makes it so necessary to life. Our hope, following the kind of reasoning outlined above, is that a complete knowledge of the molecular geometry, how the peptide chains fit together within the molecule and the molecules within crystals, may make it possible for us to understand and control its behaviour. It crystallises in a number of different modifications and their very different degrees of complexity seem significant. In acid insulin salts the molecule appears to be dimeric, the two insulin molecules in one dimer being related to one another by two-fold axes of symmetry[24]. Crystallographic two-fold axes also relate insulin molecules in a cubic metal-free form of insulin first observed by Abel[25] in 1927. These crystals have so far only been obtained as extremely small rhombic dodecahedra; their habit and symmetry suggest that here six insulin dimers may be grouped in a larger aggregate, with symmetry 23, reminiscent of structures proposed for the smaller viruses[26]. But it is difficult to get sufficient X-ray diffraction effects from these crystals to check the hypothesis. In all insulin crystals and in solutions which contain adequate zinc, or some other similar bivalent metal ions, a definite aggregate of six insulin molecules appears; having, indeed, the molecular weight 36000 first recorded by Svedberg[27] in 1935. This insulin hexamer corresponds with the unit cell in the rhombohedral crystals first investigated and is the asymmetric unit in the monoclinic form which appears in the presence of phenol. The proportion of

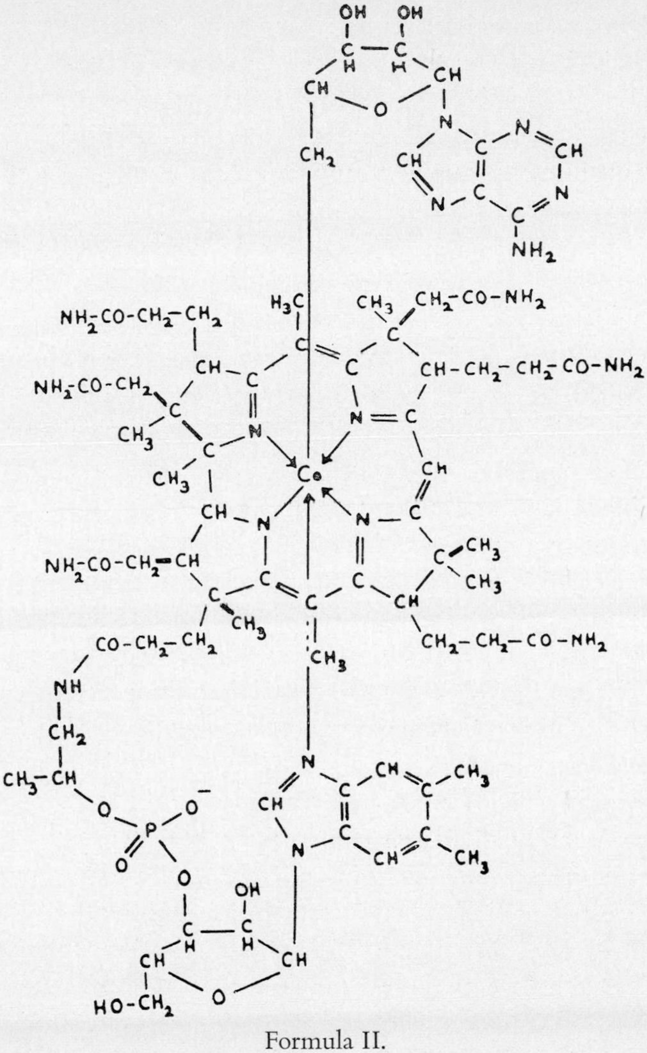

Formula II.

(a) (b) (c)

Fig. 14. Crystals of insulin (a) cubic; (b) monoclinic and (c) rhombohedral (taken from Schlichtkrull, *Insulin Crystals*, 1958).

zinc to insulin molecules is 2:6 and the symmetry relations strongly suggest that the zinc is situated on the three-fold axes around which are arranged insulin molecules related to one another both by two- and three-fold axes. The symmetry relations have been explored by the use of the functions described by Rossman and Blow[28]. In the presence of halide, a slightly different packing is adopted with 4 Zn: 6 insulin molecules. All these crystals give beautiful X-ray diffraction effects from which it ought to be possible to solve the structure to atomic resolution. Here the zinc present is much too light, about 0.01 in scattering effect, to be used for phase determination as we used cobalt in B_{12}. We need to introduce additional heavy atoms into the crystal structure. This is not difficult to do in quantity. But it seems very difficult to limit the crystal uptake to the one per insulin molecule which should be easy to place by X-ray methods, either by chemical reaction or by cocrystallisation, the methods adopted by Kendrew and Perutz. We are driven either to try to solve an initially complex problem, concerned with the heavy atom distribution in the crystals, or to do more chemistry in the hope of binding one heavy atom alone to each insulin molecule. In practice, I suppose, we shall attempt both. There are encouraging features of our present experiments that lure us on – some of our heavy atom containing crystals show very marked anomalous absorption effects, from which some initial phasing evidence can be obtained. But the electron-density maps we have so far calculated are far too imperfect and difficult to interpret for me to present them today*.

It will be clear from all that I have said so far that my research owes a debt I cannot adequately pay to the work of others, my colleagues who have provided many of the ideas I have used and many interesting examples of similar analyses, my collaborators, without whose brains and hands and eyes very little would have been done. I should also like to remember here today many whose friendship and encouragement I have greatly enjoyed. I will name three particularly, W. T. Astbury, I. Fankuchen and K. Linderström-Lang, because they would themselves so much have enjoyed this occasion.

* Subsequent examination suggests these maps are not so imperfect as I supposed and are interpretable.

1. W.L.Bragg, *Nobel Lectures, Chemistry, 1901–1921*, Elsevier, Amsterdam, 1967, p. 370.
2. M.F.Perutz and J.C.Kendrew, *Nobel Lectures, Chemistry, 1942–1962*, Elsevier, Amsterdam, 1964 pp.653,676.
3. A.L.Patterson, *Z.Krist., A*, 90 (1935) 517.
4. C.A.Beevers and H.Lipson, *Proc.Phys.Soc. (London)*, 48 (1936) 772.
5. C.H.Carlisle and D.Crowfoot, *Proc.Roy.Soc. (London), Ser.A*, 184 (1945) 64.
6. J.D.Bernal, *Nature*, 129 (1932) 277.
7. D.Crowfoot, C.W.Bunn, B.W.Rogers-Low and A.Turner Jones, *The Chemistry of Penicillin*, Princeton University Press, 1949, p.310.
8. K.J.Watson, to be published.
9. S.Abrahamson, D.C.Hodgkin and E.N.Maslen, *Biochem.J.*, 86 (1963) 514.
10. D.C.Hodgkin and E.N.Maslen, *Biochem.J.*, 79 (1961) 393.
11. R.D.Diamond, D.Phil.Thesis, *A Crystallographic Study of the Structure of Some Antibiotics*, Oxford, 1963.
12. D.C.Hodgkin, J.Kamper, J.Lindsey, M.Mackay, J.Pickworth, J.H.Robertson, C.B.Shoemaker, J.G.White, R.J.Prosen and K.N.Trueblood, *Proc.Roy.Soc. (London), Ser.A*, 242 (1957) 228.
13. D.C.Hodgkin, J.Pickworth, J.H.Robertson, R.J.Prosen, R.A.Sparks and K.N. Trueblood, *Proc.Roy.Soc. (London), Ser.A*, 251 (1959) 306.
14. K.Bernhauer, F.Wagner and D.Wahl, *Biochem.Z.*, 334 (1961) 279.
15. A.F.Peerdeman and J.M.Bijvoet, *Acta Cryst.*, 9 (1956) 1012.
16. G.N.Ramachandran and S.Raman, *Current Sci. (India)*, 25 (1956) 348; S.Raman, *Proc.Indian Acad.Sci.*, 47 (1958) 1.
17. D.Dale, D.C.Hodgkin and K.Venkatesan, *Crystallography and Crystal Perfection*, Academic Press, London, 1963, p.237.
18. E.Bertele, H.Boos, J.D.Dunitz, F.Elsinger, A.Eschenmoser, I.Felner, H.P.Gribi, H.Gschwend, E.F.Meyer, M.Pesaro and R.Scheffold, *Angew.Chem.*, 76 (1964) 393; *Angew.Chem., Intern.Ed.*, 3 (1964) 490.
19. P.G.Lenhert and D.C.Hodgkin, *Nature*, 192 (1961) 937.
20. M.A.Foster, M.J.Dilworth and D.D.Woods, *Nature*, 201 (1964) 39.
21. For a review see K.Bernhauer, O.Muller and F.Wagner, *Angew.Chem.*, 3 (1964) 200.
22. J.C.Sheehan and K.R.Henery-Logan, *J.Amer.Chem.Soc.*, 81 (1959) 3089.
23. A.P.Ryle, F.Sanger, L.F.Smith and R.Kitai, *Biochem.J.*, 60 (1955) 541.
24. B.W.Low and J.R.Einstein, *Nature*, 186 (1960) 470.
25. J.J.Abel, E.M.K.Geiling, C.A.Rouiller, F.K.Bell and O.Wintersteiner, *J.Pharmacol.Exptl.Therap.*, 31 (1927) 65.
26. M.M.Harding, D.C.Hodgkin, A.F.Kennedy, A.O'Connor and P.D.J.Weitzmann, in preparation.
27. K.Marcker, *Acta Chem.Scand.*, 14 (1960) 2071.
28. M.G.Rossman and D.M.Blow, *Acta Cryst.*, 15 (1962) 24; E.Coller, M.M.Harding, D.C.Hodgkin and M.G.Rossman, in preparation.

Biography

Dorothy Crowfoot was born in Cairo on May 12th, 1910 where her father, John Winter Crowfoot, was working in the Egyptian Education Service. He moved soon afterwards to the Sudan, where he later became both Director of Education and of Antiquities; Dorothy visited the Sudan as a girl in 1923, and acquired a strong affection for the country. After his retirement from the Sudan in 1926, her father gave most of his time to archaeology, working for some years as Director of the British School of Archaeology in Jerusalem and carrying out excavations on Mount Ophel, at Jerash, Bosra and Samaria.

Her mother, Grace Mary Crowfoot (born Hood) was actively involved in all her father's work, and became an authority in her own right on early weaving techniques. She was also a very good botanist and drew in her spare time the illustrations to the official Flora of the Sudan. Dorothy Crowfoot spent one season between school and university with her parents, excavating at Jerash and drawing mosaic pavements, and she enjoyed the experience so much, that she seriously considered giving up chemistry for archaeology.

She became interested in chemistry and in crystals at about the age of 10, and this interest was encouraged by Dr. A. F. Joseph, a friend of her parents in the Sudan, who gave her chemicals and helped her during her stay there to analyse ilmenite. Most of her childhood she spent with her sisters at Geldeston in Norfolk, from where she went by day to the Sir John Leman School, Beccles, from 1921–28. One other girl, Norah Pusey, and Dorothy Crowfoot were allowed to join the boys doing chemistry at school, with Miss Deeley as their teacher; by the end of her school career, she had decided to study chemistry and possibly biochemistry at university.

She went to Oxford and Somerville College from 1928–32 and became devoted to Margery Fry, then Principal of the College. For a brief time during her first year, she combined archaeology and chemistry, analysing glass tesserae from Jerash with E. G. J. Hartley. She attended the special course in crystallography and decided, following strong advice from F. M. Brewer, who was then her tutor, to do research in X-ray crystallography. This she began for part II Chemistry, working with H. M. Powell, as his first research student on thallium dialkyl halides, after a brief summer visit to Professor Victor Goldschmidt's laboratory in Heidelberg.

Her going to Cambridge from Oxford to work with J. D. Bernal followed from a chance meeting in a train between Dr. A. F. Joseph and Professor Lowry. Dorothy Crowfoot was very pleased with the idea; she had heard Bernal lecture on metals in Oxford and became, as a result, for a time, unexpectedly interested in metals; the fact that in 1932 he was turning towards sterols, settled her course.

She spent two happy years in Cambridge, making many friends and exploring with Bernal a variety of problems. She was financed by her aunt, Dorothy Hood, who had paid all her college bills, and by a £75 scholarship from Somerville. In 1933, Somerville, gave her a research fellowship, to be held for one year at Cambridge and the second at Oxford. She returned to Somerville and Oxford in 1934 and she has remained there, except for brief intervals, ever since. Most of her working life, she spent as Official Fellow and Tutor in Natural Science at Somerville, responsible mainly for teaching chemistry for the women's colleges. She became a University lecturer and demonstrator in 1946, University Reader in X-ray Crystallography in 1956 and Wolfson Research Professor of the Royal Society in 1960. She worked at first in the Department of Mineralogy and Crystallography where H. L. Bowman was professor. In 1944 the department was divided and Dr. Crowfoot continued in the subdepartment of Chemical Crystallography, with H. M. Powell as Reader under Professor C. N. Hinshelwood.

When she returned to Oxford in 1934, she started to collect money for X-ray apparatus with the help of Sir Robert Robinson. Later she received much research assistance from the Rockefeller and Nuffield Foundations. She continued the research that was begun at Cambridge with Bernal on the sterols and on other biologically interesting molecules, including insulin, at first with one or two research students only. They were housed until 1958 in scattered rooms in the University museum. Their researches on penicillin began in 1942 during the war, and on vitamin B_{12} in 1948. Her research group grew slowly and has always been a somewhat casual organisation of students and visitors from various universities, working principally on the X-ray analysis of natural products.

Dorothy Hodgkin took part in the meetings in 1946 which led to the foundation of the International Union of Crystallography and she has visited for scientific purposes many countries, including China, the U. S. A. and the U. S. S. R. She was elected a Fellow of the Royal Society in 1947, a foreign member of the Royal Netherlands Academy of Sciences in 1956, and of the American Academy of Arts and Sciences (Boston) in 1958.

In 1937 she married Thomas Hodgkin, son of one historian and grandson of two others, whose main field of interest has been the history and politics of Africa and the Arab world, and who is at present Director of the Institute of African Studies at the University of Ghana, where part of her own working life is also spent. They have three children and three grandchildren. Their elder son is a mathematician, now teaching for a year at the University of Algiers, before taking up a permanent post at the new University of Warwick. Their daughter (like many of her ancestors) is an historian – teaching at girls' secondary school in Zambia. Their younger son has spent a pre-University year in India before going to Newcastle to study Botany, and eventually Agriculture. So at the present moment they are a somewhat dispersed family.

Chemistry 1965

ROBERT BURNS WOODWARD

«for his outstanding achievements in the art of organic synthesis»

Chemistry 1965

Presentation Speech by Professor A. Fredga, member of the Nobel Committee for Chemistry of the Royal Swedish Academy of Sciences

Your Majesty, Your Royal Highnesses, Ladies and Gentlemen.

In our days, the chemistry of natural products attracts a very lively interest. New substances, more or less complicated, more or less useful, are constantly discovered and investigated. For the determination of the structure, the architecture of the molecule, we have to-day very powerful tools, often borrowed from Physical Chemistry. The organic chemists of the year 1900 would have been greatly amazed if they had heard of the methods now at hand. However, one cannot say that the work is easier; the steadily improving methods make it possible to attack more and more difficult problems and the ability of Nature to build up complicated substances has, as it seems, no limits.

In the course of the investigation of a complicated substance, the investigator is sooner or later confronted by the problem of synthesis, of the preparation of the substance by chemical methods. He can have various motives. Perhaps he wants to check the correctness of the structure he has found. Perhaps he wants to improve our knowledge of the reactions and the chemical properties of the molecule. If the substance is of practical importance, he may hope that the synthetic compound will be less expensive or more easily accessible than the natural product. It can also be desirable to modify some details in the molecular structure. An antibiotic substance of medical importance is often first isolated from a microorganism, perhaps a mould or a germ. There ought to exist a number of related compounds with similar effects; they may be more or less potent, some may perhaps have undesirable secondary effects. It is by no means certain, or even probable, that the compound produced by the microorganism – most likely as a weapon in the struggle for existence – is the very best from the medical point of view. If it is possible to synthesize the compound, it will also be possible to modify the details of the structure and to find the most effective remedies.

The synthesis of a complicated molecule is, however, a very difficult task; every group, every atom must be placed in its proper position and this should be taken in its most literal sense. It is sometimes said that organic synthesis is at the same time an exact science and a fine art. Here Nature is the uncontested master, but I dare say that the prize-winner of this year, Professor Woodward, is a good second.

Professor Woodward has a special liking for synthetic undertakings which are generally regarded as practically impossible. I shall here touch upon a number of his most famous achievements, some of the substances in question being well-known from the columns of the daily press. During World War II, Professor Woodward synthesized *quinine*, the well-known antimalarial. Later followed the steroids *cholesterol* and *cortisone*. The related substance *lanosterol* is perhaps less familiar but very important from the scientific point of view. The synthesis of the famous poison *strychnine* caused a great sensation some ten years ago. Still more remarkable is perhaps the synthesis of *reserpine*, an alkaloid of great medical importance. Several other examples from the chemistry of the alkaloids could be mentioned, substances with strange names and interesting properties: *lysergic acid*, *ergonovine*, *ellipticine*, *colchicine*.

In the field of antibiotics Professor Woodward has, among many other things, established the structure of *aureomycin* and *terramycin*. He has also cleared the way for synthetic work within this group of substances, the so-called tetracyclines.

A very notable piece of work is the synthesis of *chlorophyll*, the green plant pigment which absorbs and transforms the radiant energy of the sun, the existence of which is thus a necessary condition for organic life on Earth. This work has greatly increased our knowledge of the chlorophyll molecule.

Professor Woodward's activity has by no means been restricted to synthetic work. He has established the structure of many important compounds, for instance the peculiar fish poison *tetrodotoxin*, causing numerous fatalities in Japan, and he has made an original and promising approach to the synthesis of polypeptides. He has also developed very interesting ideas about synthetic activity in Nature, the genesis of complicated molecules within the living organism. These theories have been confirmed by experiments with labelled molecules.

Professor Woodward's research work covers vast and various fields in Organic Chemistry. A leading feature is that the problems have been extremely difficult and that they have been solved with brilliant mastery. He has attacked them with a maximum of theoretical knowledge, a never-failing practical judgement and, not least, a genial intuition. He has, in a conspicuous way, widened the limits for what is practically possible. As a stimulating example he has exerted a profound influence on the organic chemistry of today.

Professor Woodward, I have here tried to give a brief survey of your more famous achievements in Organic Chemistry. It is sometimes said that you have demonstrated that nothing is impossible in organic synthesis. This is perhaps a slight exaggeration. You have, however, in a spectacular way expanded and enlarged the domain of the possible. It is also said that you stand out like a wizard. We know that in times long passed, chemistry was classified as an occult science. Anyhow, you have certainly not gained your scientific reputation by magical means, but by the penetrative intensity of your chemical thinking and the rigorous expert planning of your experiments. In these respects you hold a unique position among organic chemists of to-day. In recognition of your services to Chemical Science, the Royal Academy has decided to confer upon you the Nobel Prize of this year for your outstanding achievements in the art of organic synthesis.

To me has been granted the privilege to convey to you the most hearty congratulations of the Academy and to invite you to receive your prize from the hands of His Majesty the King.

R. B. WOODWARD

Recent advances in the chemistry of natural products

Nobel Lecture, December 11, 1965

The Nobel Prize in Chemistry for 1965 has been awarded for contributions to the art of chemical synthesis. It gives me much pleasure to record here my gratification with the citation, which properly signalizes an exciting and significant aspect of synthetic activity. But that aspect is one which is more readily – and I dare say more effectively – exemplified and epitomized than it is articulated and summarized. Having here this morning the responsibility of delivering a lecture on a topic related to the work for which the Prize was awarded, I have chosen to present an account of an entirely new and hitherto unreported investigation which, I hope, will illuminate many facets of the spirit of contemporary work in chemical synthesis.

Cephalosporin C, a product of the metabolism of *Cephalosporium acremonium*, was isolated in 1955 by Newton and Abraham[1] in an investigation notable for its perspicacity as well as its painstaking attention to detail. The investigation of the structure of the metabolite was successfully concluded in 1961 through studies in which both chemical[2] and X-ray crystallographic[3] techniques were employed. The molecular array (I) thus laid bare strikes one

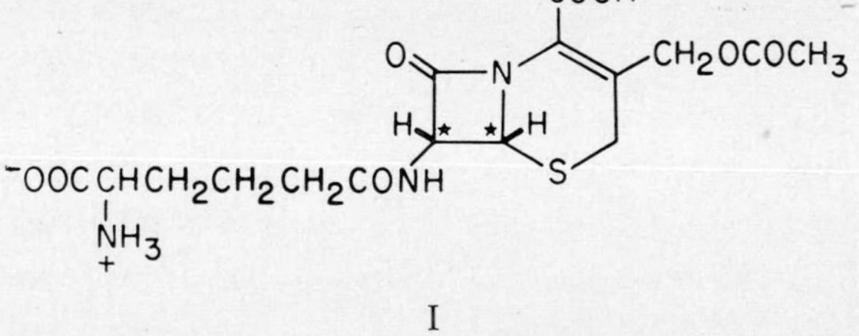

COOH
O
N
CH_2OCOCH_3
H
H
$^-OOCCHCH_2CH_2CH_2CONH$
S
$\overset{+}{N}H_3$

I

at once as having affinities with a hitherto well-known class of substances which has constituted one of the most challenging and recalcitrant synthetic objectives of our generation. I refer of course to the penicillins, of which penicillin G (II) – one of the earliest known and one which has been widely used in medicine – may serve as an example. There can be few organic chemists

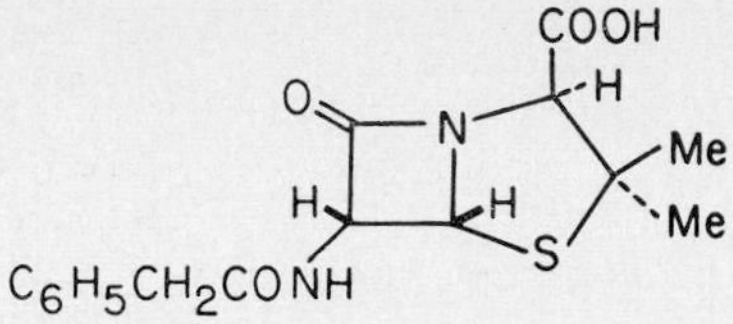

II

who do not know the fascinating history of the penicillins[4]. How, following up an early observation of Alexander Fleming, Chain and Florey isolated the first penicillin shortly after the outbreak of the Second World War. How the powerful practical desiderata of those trying times led to the establishment of a mammoth British/American program which had as its objectives the determination of the structure and the synthesis of the penicillins. How the chemical investigations, and especially the X-ray crystallographic studies of Dorothy Hodgkin conquered the structural problem, and how despite the best efforts of probably the largest number of chemists ever concentrated upon a single objective the synthetic problem had not been solved when the program was brought to a close at the end of the War. Many chemists continued to be fascinated by the problem, and some were still willing to gamble their skill against its obstinacy. In 1959, after more than a decade of intensive investigation, John Sheehan[5] succeeded in the development of methods by which penicillins could be prepared by total synthesis. That these methods have not come into practical use does not detract from this major achievement, but only emphasizes that the challenge presented to the synthetic chemist by the penicillins has not been exhausted.

A parenthesis is probably desirable at this point in order to allay some concern among those who have not been initiated in these matters. I have used the plural term «penicillins» because Nature provides several closely related substances, differing only in the acyl group attached to the nitrogen atom which is itself situated α to the lactam carbonyl group of the general structure (III). Furthermore, chemists have found ways of removing these acyl groups from the natural representatives of the class, and attaching entirely new and

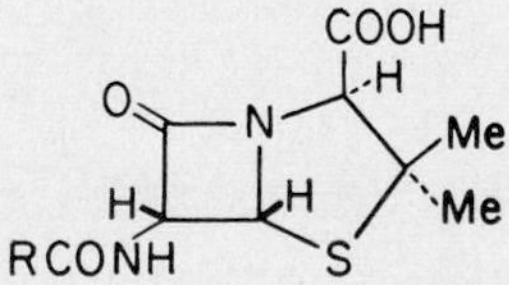

III

R	*Penicillin*	*R*	*Penicillin*
$C_6H_5CH_2$	G	$CH_3(CH_2)_6$	K
$CH_3CH_2CH{:}CHCH_2$	F	$C_6H_5OCH_2$	V
p-$HOC_6H_4CH_2$	X	$CH_2{:}CHCH_2SCH_2$	O

different groupings to the nitrogen atom thus freed. In this way, many hundreds of artificial penicillins have been prepared. The situation is similar in respect to cephalosporin C, in that a whole class of cephalosporins has been created by replacement of the ω-[D-(—)-α-aminoadipoyl] residue of the natural metabolite by numerous other acyl groups. In both classes – the penicillins and the cephalosporins – some of the derived substances possess properties which confer on them special utility in medicine. Thus, cephalosporin C itself possesses antimicrobial activity of a relatively low order of magnitude, which, however, early attracted special interest because it persisted against organisms which had become resistant to the penicillins. In some of the derived cephalosporins this especially interesting aspect of the antimicrobial activity is retained, while at the same time the level of activity is much heightened. Further, the activity extends over the range of Gram-negative and Gram-positive organisms. Consequently, some of these substances, of which cephalothin (IV) may serve as an example[6], have already achieved utility in medicine as broad-spectrum antibiotics of low toxicity, effective against penicillin-resistant organisms.

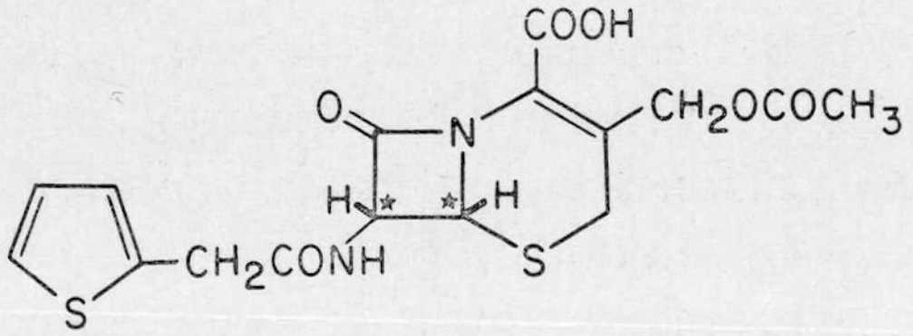

IV

In considering the development of a plan for the synthesis of any complicated substance, it is always desirable to look at the problem from an entirely fresh point of view. Nevertheless, in the case at hand, it was pertinent to examine whether the experience gained and the results achieved in synthetic studies on the penicillins might be usefully applicable to the structurally related cephalosporins. We rejected this possibility at the outset for several reasons. I have already alluded to the fact that the known penicillin syntheses, hard-won, brilliant achievements though they are, are lacking in practicality. Further, the problems which had had to be overcome in devising methods for penicillin synthesis had been quite difficult enough without adding to them the intricacies which would have been associated with the achievement of stereospecificity in the creation of asymmetric intermediates, and this aspect had been slighted. Finally, a special chemical point was of much importance. Thc β-lactam ring common to the penicillins and the cephalosporins is highly susceptible to hydrolytic cleavage. In the case, for example, of penicillin G (II), the product of this hydrolysis is penicilloic acid (V). The synthesis of

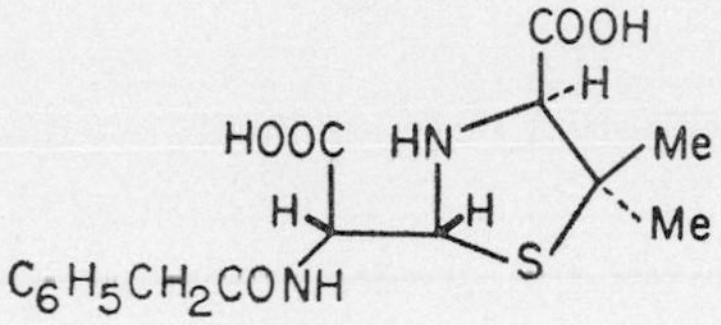

V

penicilloic acid and its analogs, at least by non-stereospecific methods, was a relatively simple problem, and by far the largest number of attempts to synthesize penicillins – and the only successful ones – involved the deceptively simple task of removing the elements of water from penicilloic acid analogs, with closure of the four-membered β-lactam ring. In the case of the cephalosporins the situation is strikingly different. Here the β-lactam ring is also easily cleaved, but the proximate product of the hydrolysis, which must have the structure (VI), is not a known substance. Its intricate and delicate con-

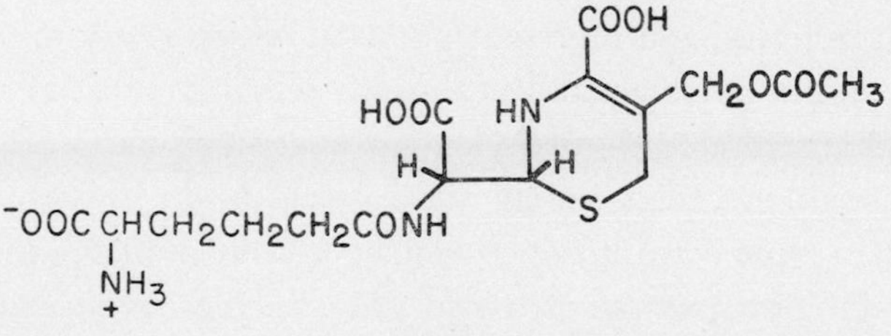

VI

stitution is such that it does not survive even the mild conditions of its generation from the corresponding lactam. Clearly then, it would be unwise to essay the synthesis of a cepahalosporin from such a hitherto unknown and obviously highly fugitive precursor.

Often in the course of synthetic work one or two key ideas set the style, development, and outcome of the investigation, while providing the flexibility essential for any long journey through unknown territory, beset with perils which at best can be only dimly foreseen. In planning our synthesis of cephalosporin the first of these definitive concepts was our choice of L(+)-cysteine (VII) as our starting material. This readily available substance pos-

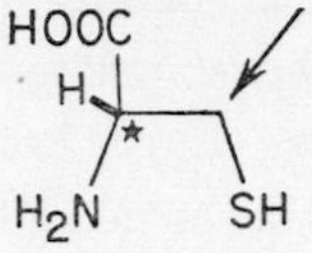

HOOC
H
*
H_2N SH

VII

sesses a two-carbon backbone to which are attached a carboxyl group, an α-nitrogen atom and a β-sulfur atom – in short, it presents in ready-made fashion a large portion of the crucial substituted β-lactam moiety of the cephalosporins. Furthermore, it is optically active, and the groups arranged about its one asymmetric carbon atom (starred in VII) are oriented in an absolute stereochemical sense precisely as are the similar groups in the objective; that is to say, as soon as the decision to use cysteine had been made, our stereochemical problem was in a sense already half solved, since the cephalosporin nucleus contains only one further asymmetric center (*cf.* stars in I and IV). On the other hand, advantageous as this choice obviously was in many ways, it was also clear that associated with it was a special problem which could by no means be viewed lightly. The cysteine molecule is a tightly assembled package of highly reactive groupings. The amino group, the sulfhydryl group, the carboxyl group, and the α-methine group each possess characteristic features of chemical reactivity, and represent points at which ready modification of the molecule might be expected. But the only remaining feature of the molecule, the simple saturated β-methylene group, represents a point at which there is little or no precedent for chemical attack. And yet, in the light of our plan we must in some way introduce a nitrogen atom at that point, preferably in a stereospecific manner (*cf.* arrow in VII). Further, even assuming that a method should be discovered for overcoming the defences of the molecule at that strong point, it was clear that we should be dealing with intermediates containing two electronegative atoms bound to the same carbon atom – a situation well known for its potentialities in conferring sensitivity and instability upon molecules so constituted. In sum, our initial decision placed us in the exhilarating position of having to make a discovery, and of being prepared to deal with substances of an especially precarious constitution.

Our first actual operations consisted quite naturally in so modifying the cysteine structure as to depress the reactivity of the amino, sulfhydryl, and carboxyl groups. Thus, the amino acid was first converted by reaction with acetone into the thiazolidine (VIII)[7], which in its turn reacted with *tert*-

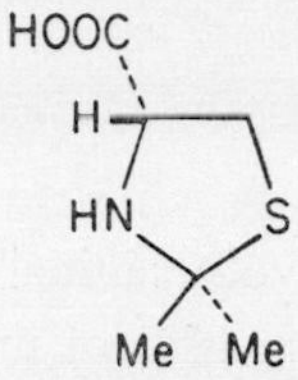

VIII 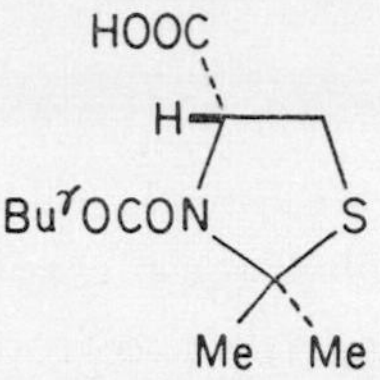 IX

butyloxycarbonyl chloride in the presence of pyridine to give the corresponding *N-tert*-butyloxycarbonyl compound (IX). Some special interest attaches to the fact that the acylation reaction undoubtedly takes place through internal delivery of the *N-tert*-butyloxycarbonyl group, which first becomes attached at the carboxyl site to give the mixed anhydride (X). The acylated thiazolidine was next converted into the methyl ester (XI) with diazomethane.

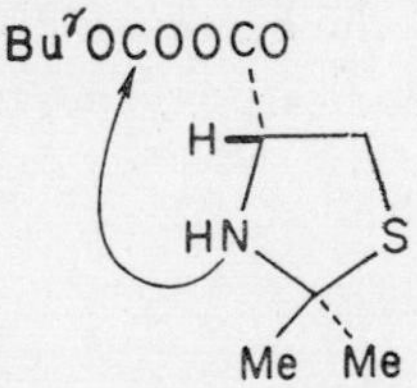

X 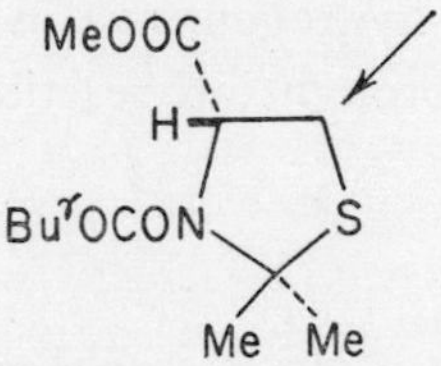 XI

These three simple changes had sufficed to convert the cysteine molecule into one whose methylene group (arrow in XI) might now enjoy a far better relative position in respect to reactivity as compared with that same grouping in the original cysteine. But they also served another function: by incorporating the methylene group in a ring, and thereby rendering rotation about the α,β carbon–carbon bond impossible we had set the stage for bringing about transformations at the methylene group in a stereospecific manner.

I shall not detail here the many weapons which were brought into play against that still expectedly recalcitrant methylene grouping. Suffice it to say that the protected ester (XI) reacted with excess dimethyl azodicarboxylate at 105° during forty-five hours to give the hydrazo diester (XII) in almost quantitative yield. It is of special interest that we have been able to assemble

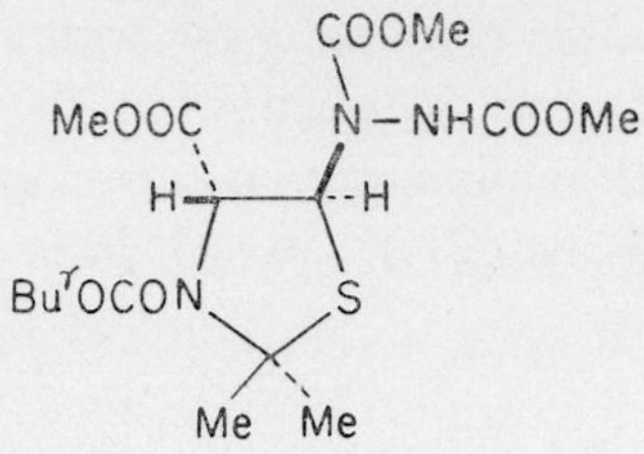

XII

evidence which suggests that this novel reaction involves initial attack of the sulfur atom upon the azo grouping, and that the formation of this bond may be concerted with migration of hydrogen from the methylene group to the second nitrogen atom (*cf.* XIII→XIV→XV). Thus, if substances containing

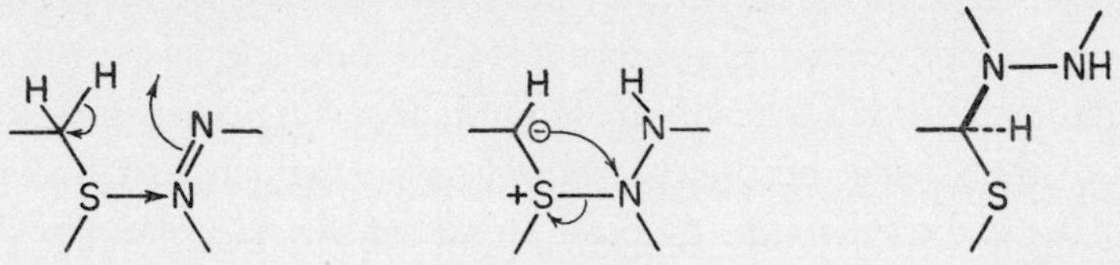

XIII XIV XV

free active hydrogen, such as the acid (IX), and the benzenesulfonyl amide (XVI) are brought into reaction with dimethyl azodicarboxylate, attack upon

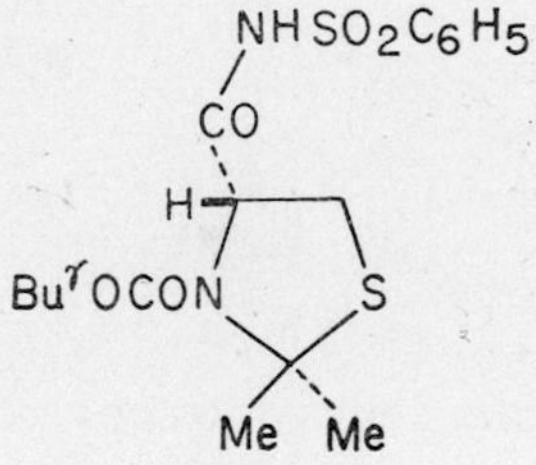

XVI

a methylene group is not observed, and the products (XVII and XVIII)

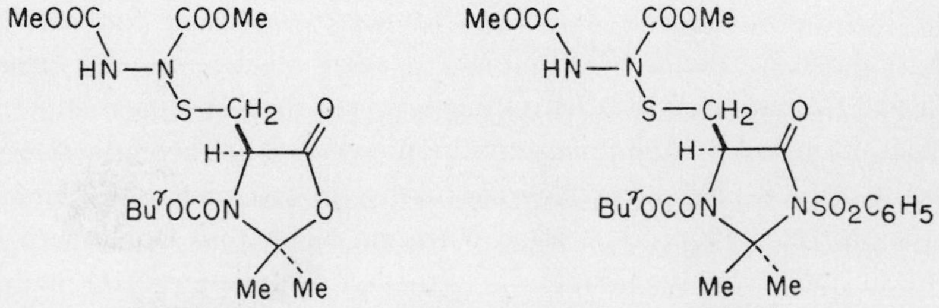

XVII XVIII

contain an actual sulfur–nitrogen bond (*cf.* XIX). Further, we have been unable to observe an intramolecular version of the reaction–for example, with the compound (XX)–a circumstance which we connect with the very unfavourable geometry, in this case, of a transition state in which hydrogen moves as sulfur attacks nitrogen. Another special point of interest is that the

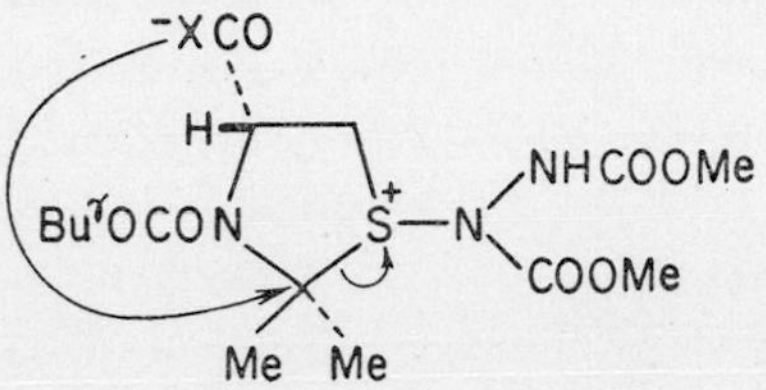

XIX

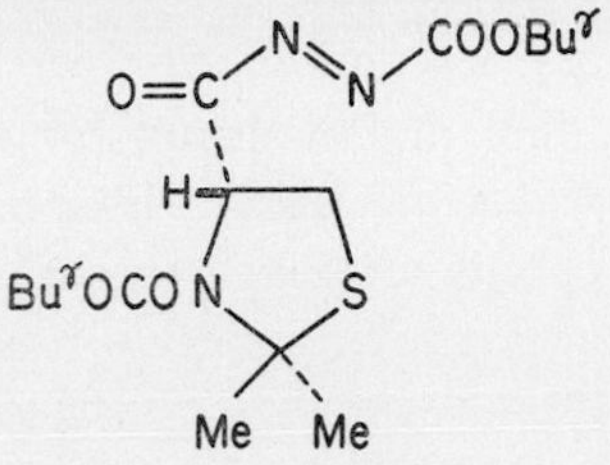

XX

conditions for the reaction, simple though they are, must be adhered to rigorously. At lower temperatures reaction is too slow to be useful, while if the temperature is raised only a small amount, the reaction is much less clean, and among the products is the *N*-methylated derivative (XXI)! In any event,

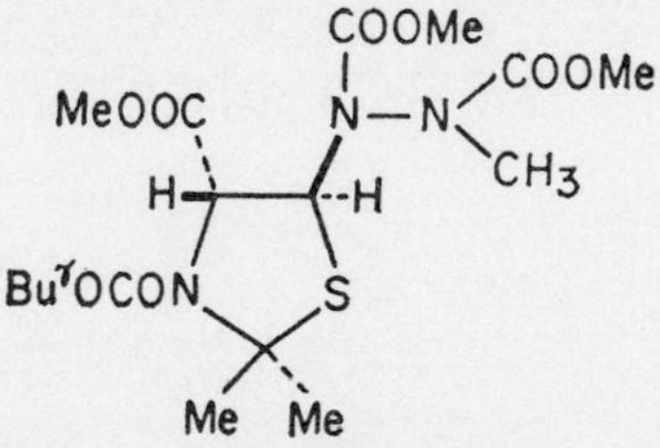

XXI

the new reaction was propitious in that we had achieved a substitution at the desired site, and in that the newly attached grouping was one which might be expected to exhibit selective reactivity. A very important point is that the reaction is stereospecific; the hydrazo diester grouping is introduced solely on one side of the ring. Clearly, this result is associated with the presence in (XI) of the carbomethoxyl grouping, whose bulk deprives the attacking moiety of the opportunity for attachment on the alternative side of the relatively rigid five-membered ring. Of course, the stereospecificity here exhibited was in a way of precisely the wrong kind, since what we required was the introduction of a nitrogen atom on the *same* side of the ring as the carbomethoxyl group. But this simply meant that we must now replace the newly introduced group with inversion of configuration at the β-carbon atom, a task which we might have taken in hand with little apprehension had it not been for the presence of the sulfur atom, whose attachment to the center at which inversion was required might render invertible intermediates non-existent or malefactory.

When the hydrazo diester (XII) was oxidized in boiling benzene for two hours, using somewhat more than two moles of lead tetraacetate, and the resulting reaction mixture was treated with excess anhydrous sodium acetate in boiling dry methanol for twenty-four hours, the *trans*-hydroxy ester (XXII) was produced. This sequence is not as simple as it might at first appear,

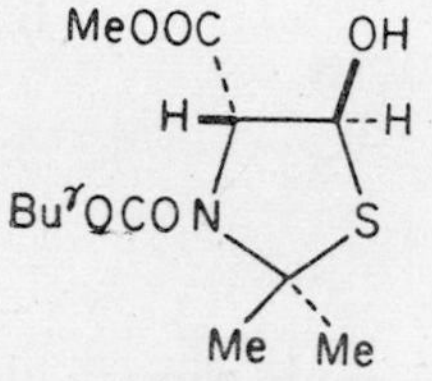

XXII

and we know some of the intermediary stages through which it proceeds. It is reasonable to suppose that the hydrazo compound, like all hydrazine derivatives, is susceptible to removal of two electrons by an oxidant. The resulting species (XXIII) must lose a carbomethoxyl group (starred) very readily to an

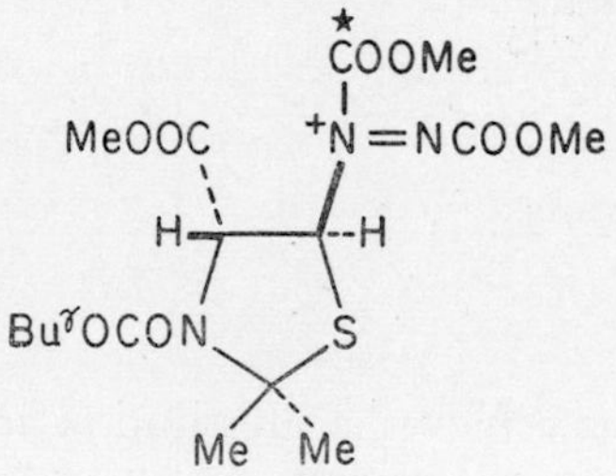

XXIII

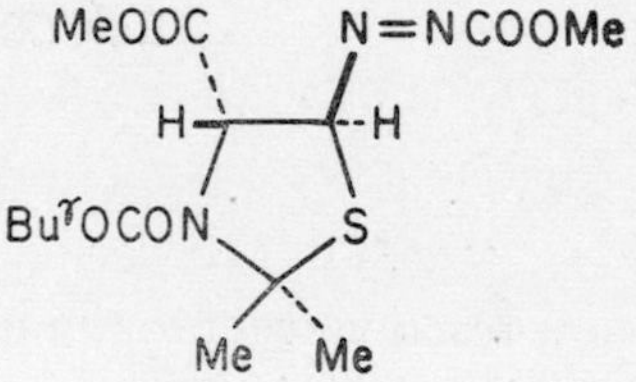

XXIV

available nucleophile. The product (XXIV) is of a type which would be expected to be transformed further by lead tetraacetate into an acetoxyazo compound (XXV); the spectroscopic evidence for the presence of such an

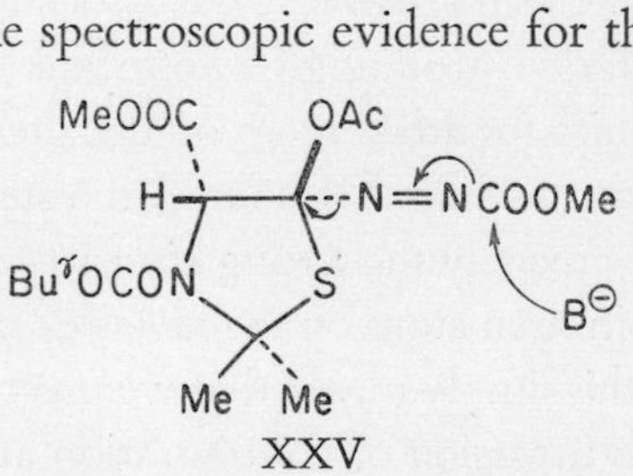

XXV

intermediate after the conclusion of the lead tetraacetate oxidation is convincing, when compared with the parallel characteristics of a similar compound (XXVI), isolated in the crystalline state and fully characterized, after

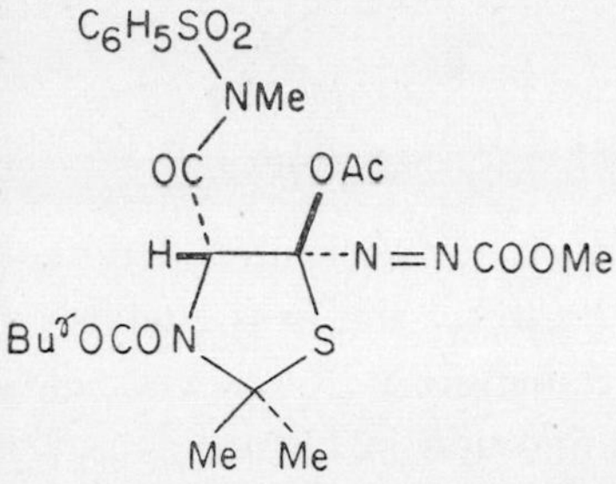

XXVI 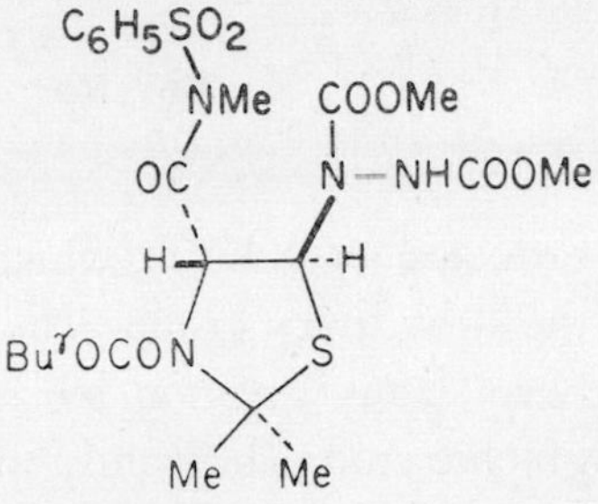 XXVII

oxidation of the sulfonamide (XXVII). The next change is the loss of a second carbomethoxyl group, again under nucleophilic attack (arrows in XXV), followed by loss of elementary nitrogen, and accession of a proton to the β-carbon atom. The final change is a simple base-catalyzed methanolysis of the acetoxy group. Special note should be made of the fact that these transformations involve a replacement at an asymmetric center. This replacement is stereoselective, in that by far the major product is the *trans*-acetoxy ester (XXVIII); no doubt the bulk of the adjacent carbomethoxyl group plays its role in forcing the acetoxy group into the more spacious location. None the less, a small amount of the *cis*-acetoxy ester (XXIX) is produced, but for reasons which will be developed shortly, this minor departure from stereospecificity is corrected almost at once.

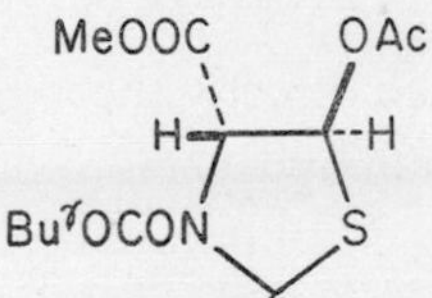

XXVIII 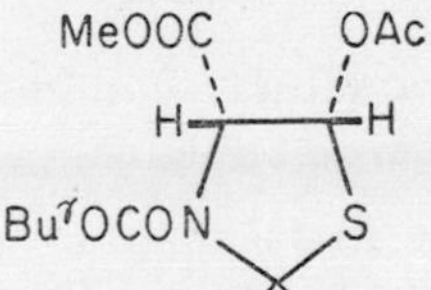 XXIX

The very existence of the *trans*-hydroxy ester (XXII) deserves special comment. It should first be noted that its structure was established with definitive rigor through its preparation by the action of diazomethane upon the corresponding acid (XXX). This acid was synthesized by a series of reactions similar to those outlined above, except that the carboxyl group of the thiazolidine (IX) was protected by the attachment of a β,β,β-trichloroethyl

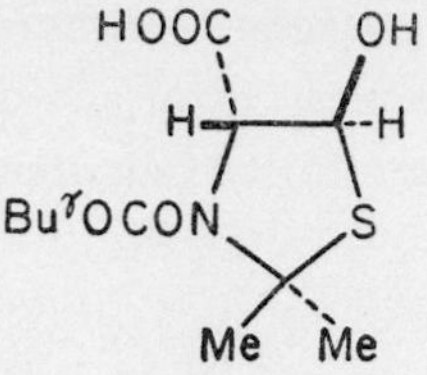

XXX 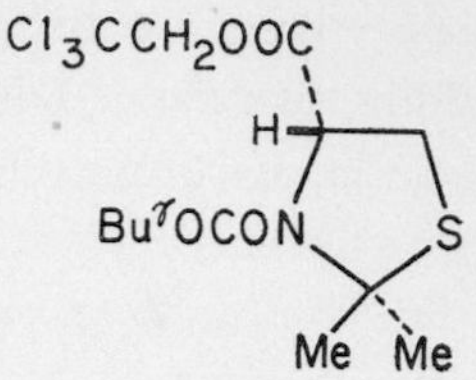 XXXI

grouping (*cf.* XXXI), which was removed reductively after introduction of the β-hydroxy group. The structure of the hydroxy acid was established beyond any question by a complete three-dimensional X-ray crystallographic study, brilliantly executed by Dr. Gougoutas in Cambridge. We have already alluded to the potentiality for instability inherent in the attachment of more than one electronegative atom to the same carbon atom, and it will be useful at this point to illustrate in some detail the factors which might have hurled us from the plateau on which we were now standing. Thus, the hydroxy ester (XXII) is an obvious candidate for participation in ring-chain tautomerism with an open-chain isomer (XXXII), which in its turn, possessing as it does a β-dicarbonyl system, could readily undergo essentially irreversible tautomerization to a stable β-hydroxyacrylic ester (XXXIII).

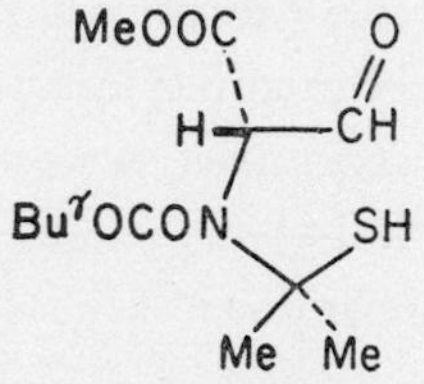

XXXII 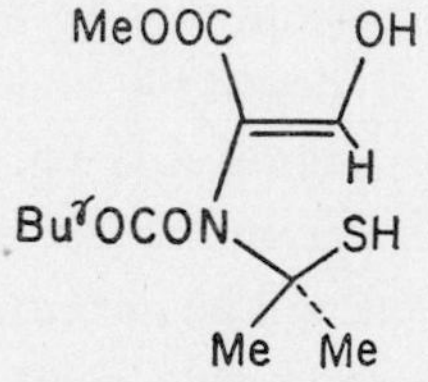 XXXIII

This same substance might alternatively be reached directly through a ready β-elimination of the sulfur atom. Either of these sulfhydryl tautomers might well lose thioacetone to give the corresponding *N*-monosubstituted urethane. Finally, it would not have been surprising if the newly introduced hydroxyl group, or indeed any of the newly introduced β-disposed group-

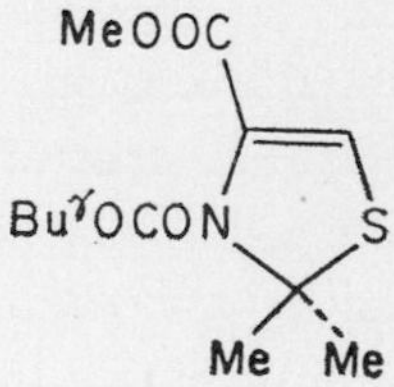

XXXIV

ings along the way, had been susceptible of ready elimination with the formation of the thiazoline (XXXIV). That the hydroxy ester is in the event a stable manipulable compound–that it does not too readily succumb to these potentialities for its self-destruction–was a major new result of our investigation so far. We can in fact say something more about these potentialities at this time. It was mentioned above that at one stage in the replacement of the hydrazo diester grouping by the hydroxyl group, a certain amount of *cis*-acetoxy ester (XXIX) is produced at an intermediary stage. Yet the sole product of the whole sequence is the *trans*-hydroxy ester (XXII). From our work with the trichloroethyl ester (XXXI), the *cis*- and *trans*-acetoxy acids (XXXV) and (XXXVI) have been isolated in the pure, crystalline state. Each

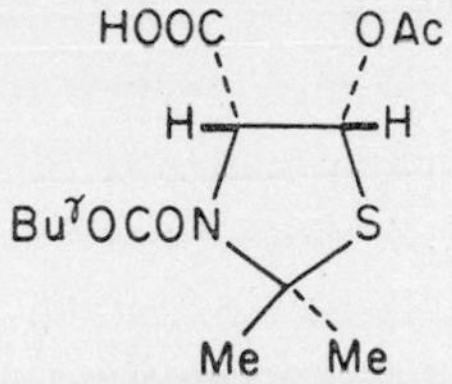

XXXV 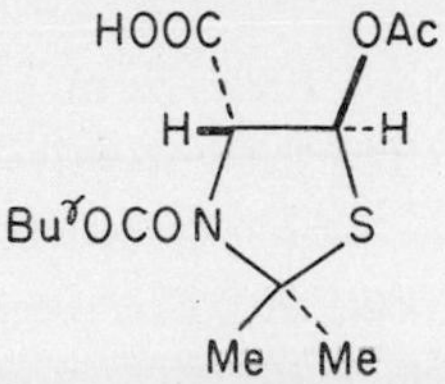 XXXVI

of these stereoisomeric substances is transformed on hydrolysis into the same *trans*-hydroxy compound (XXX)*. Thus, it is clear that the hydroxy compounds do participate in ring-chain tautomerism, but that reclosure of the open-chain aldehyde (*cf.* XXXII) is so much favored over enolization of the aldehyde group, or loss of thioacetone, that these latter changes, which would have been fatal to our prospects, do not obtrude.

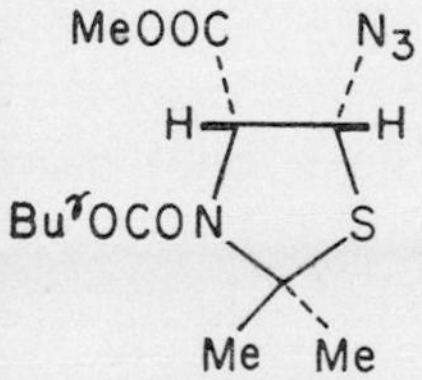

XXXVII 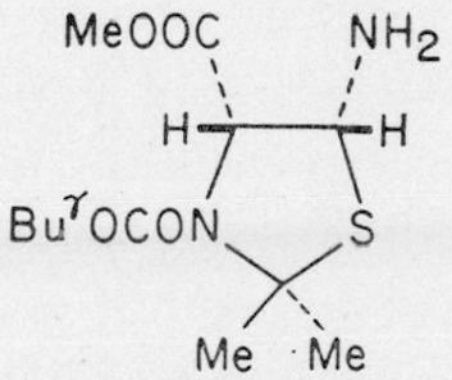 XXXVIII

The hydroxy ester (XXII) was now transformed by treatment in dimethylformamide with excess diisopropylethylamine and methanesulfonyl chloride, followed by concentrated aqueous sodium azide, to the *cis*-azido ester (XXXVII), which was in its turn reduced in methanol solution by aluminum

* Subsequent to the delivery of this lecture, we have demonstrated explicitly that the pure *cis*-acetoxy ester (XXIX) is converted in high yield to the *trans*-hydroxy ester (XXII) under our methanolysis conditions.

amalgam at $-15°$ during twenty-four hours to the *cis*-amino ester (XXXVIII). The structure of the latter was again confirmed by a complete X-ray crystallographic study carried out by Dr. Gougoutas. It is clear that this sequence of changes involves the intermediacy of the methanesulfonyl derivative (XXXIX), which does in the event undergo normal bimolecular nucleo-

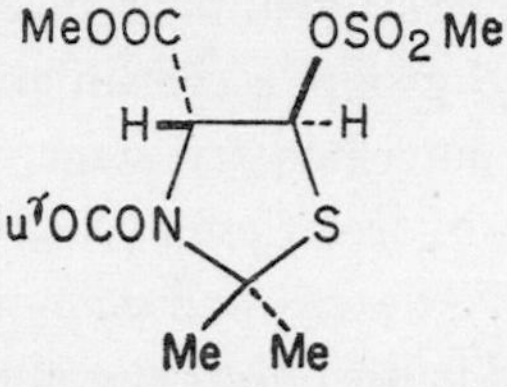

XXXIX 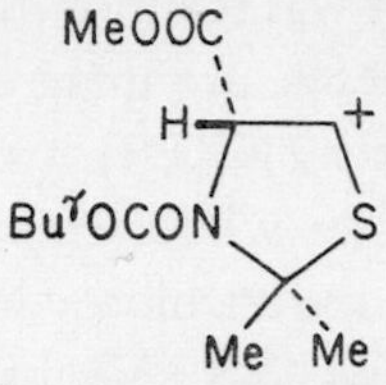 XL

philic displacement with inversion, in the classic mold, when attacked by azide ion. The dread possibility that the intermediary sulfonate might be too readily susceptible of ionization to a cation (XL), which would have led on to the thiazoline (XXXIV), or to stereochemically indiscriminate or undesired substitution at the β-carbon atom, was fortunately not manifest. No doubt avoidance of this danger is associated with the predictable relatively high energy of the electronic configuration (XLI).

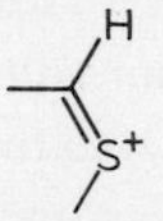

XLI

At this point we had succeeded in the major objective of introducing a properly oriented nitrogen atom into the β-position of the cysteine moiety. In short, the entire stereochemical problem presented by the cephalosporins had now been solved. It is appropriate here to introduce the second of the key ideas upon which our general plan was based. It was that we should attempt the preparation of the β-lactam (XLII), having it in mind that this substance,

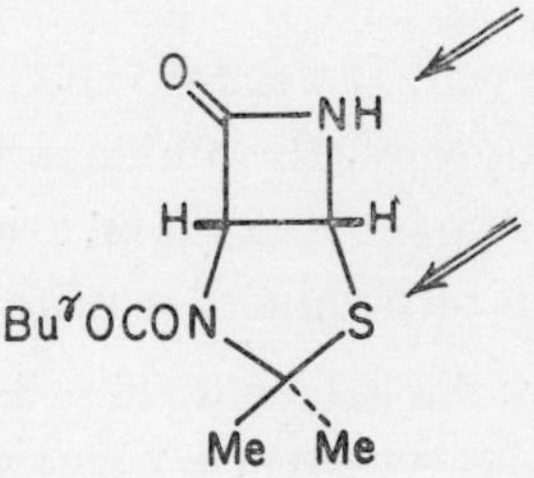

XLII

if procurable, would contain the basic structural elements common to the cephalosporins and the penicillins, and that it might serve as the source of a wide variety of known – and new – substances through the fusion of new rings at the presumably reactive nitrogen and sulfur atoms (arrows in XLII). The *cis*-amino ester (XXXVIII) now in hand differed from the desired lactam only by the elements of a molecule of methanol. The attachment of the amino and carbomethoxyl groups in (XXXVIII) to a relatively rigid ring system might be expected to favor formation of a new ring, and it was an interesting feature of the X-ray crystallographic study that the distance between the amino nitrogen atom and the carbonyl carbon atom was unusually low (2.82 Å). In all these circumstances, we felt that the stage had been well set, and we were gratified to find that when the *cis*-amino ester was treated with triisobutylaluminum in toluene it was in fact converted into the desired β-lactam (XLII). Again, the very existence of this substance, containing as it does potentialities for annihilation parallel to those discussed above in some detail for the hydroxy ester (XXII), further compounded by the considerable strain within the β-lactam ring, represents a major result of our investigation. In view of the importance of the intermediate its structure was established in detail and with complete rigor through yet a further three-dimensional X-ray crystallographic investigation by Dr. Gougoutas.

Our success with the remarkable series of substances I have described must tend to obscure the venturous spirit without which their investigation could not have been taken in hand. Lest it still be felt that our concern with the lability and versatility of our intermediates had been chimerical, it may be mentioned that the phosphinimine (XLIII), prepared from the azido ester (XXXVII) and tri-*n*-butylphosphine, gave on hydrolysis even under the mildest conditions, in addition to the *cis*-amino ester (XXXVIII), appreciable quantities of the *trans*-amino ester (XLIV) and the stable, non-cyclizable open-chain isomer (XLV)! Clearly, the formation of these substances involves subtly determined tautomeric changes closely parallel to those discussed in detail above in respect of the hydroxy ester (XXII).

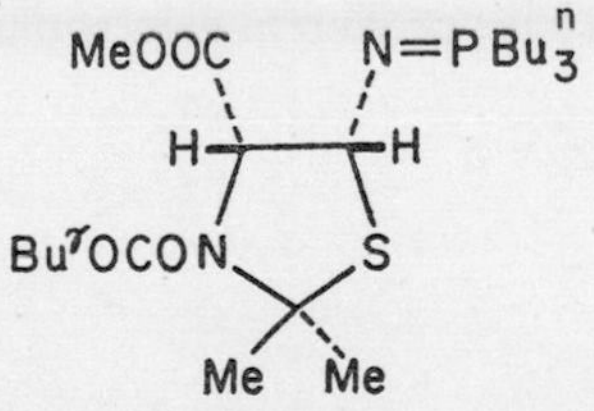

XLIII

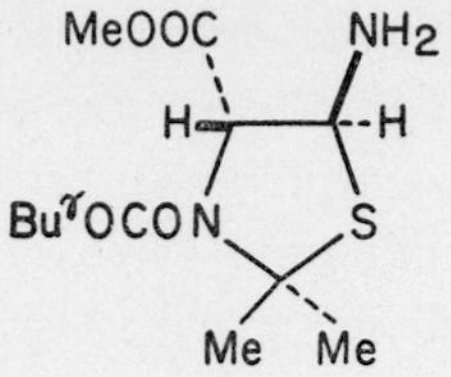

XLIV 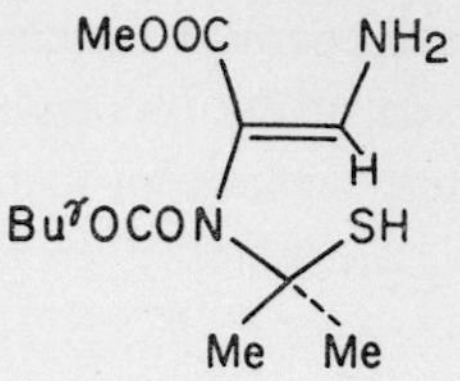 XLV

We were now ready to reduce to practice our presumption that the β-lactam (XLII) would be a versatile intermediate, capable of further development through fusion of further atomic groupings at the reactive nitrogen and sulfur atoms. In order to procure a suitable component for combination with the β-lactam, *d*-tartaric acid was converted into its di-β,β,β-trichloroethyl ester and the latter was oxidized, using sodium metaperiodate in aqueous methanol, to β,β,β-trichlorethyl glyoxylate (XLVI), isolated as the corresponding hydrate. This substance was condensed in aqueous solution with the sodium salt of malondialdehyde to give an aldol of the structure (XLVII). The aldol-condensation product in its turn lost a molecule of water when it was heated in normal octane, and the novel, highly reactive dialdehyde (XLVIII) was produced. This powerful electrophile was chosen in the hope that it might combine directly with a substance containing active hydrogen in a concerted cycloaddition process requiring no catalysis (*cf.* XLIX). The desire to avoid catalysts in reactions involving the β-lactam was of course a

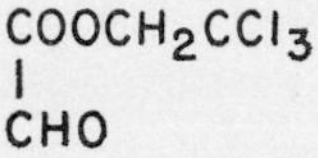

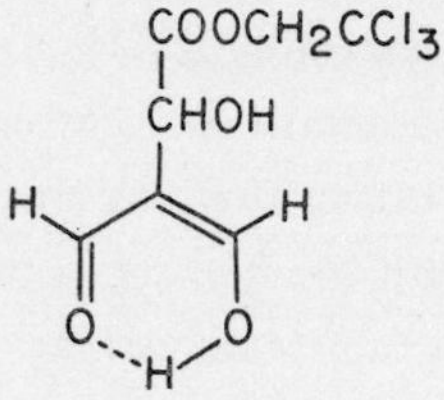

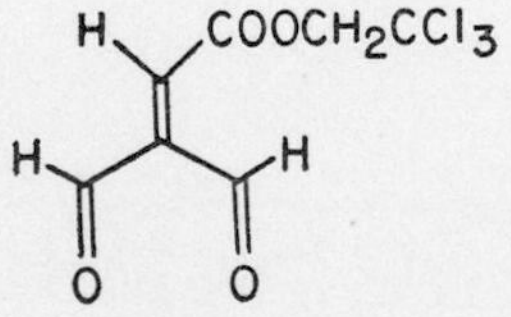

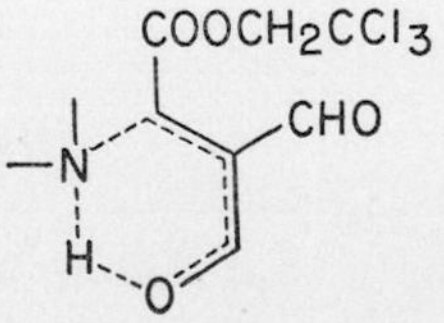

XLVI XLVII

XLVIII XLIX

consequence of our apprehension that such substances might well mobilize one or more of the capacities for self-destruction inherent in the intricate construction of our key intermediate.

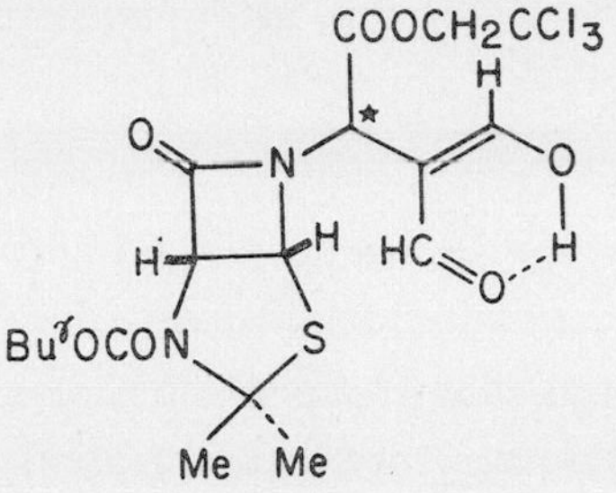

L

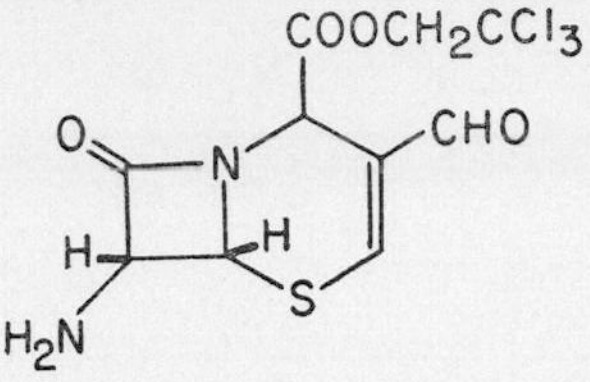

LI

In the event, when the β-lactam was heated with the dialdehyde in normal octane at 80° during sixteen hours, combination took place in the desired fashion, and the adduct (L) was produced. The latter in its turn, when allowed to stand in trifluoroacetic acid at room temperature during two and one-half hours, was transformed to the amino aldehyde (LI). The general nature of the processes involved in this latter change is clear. In particular, the crucial closure of the new six-membered ring is a consequence of the attack of the strongly electrophilic carbon atom of a protonated carbonyl group upon the nucleophilic sulfur atom (arrows in LII). The course of the ancillary changes

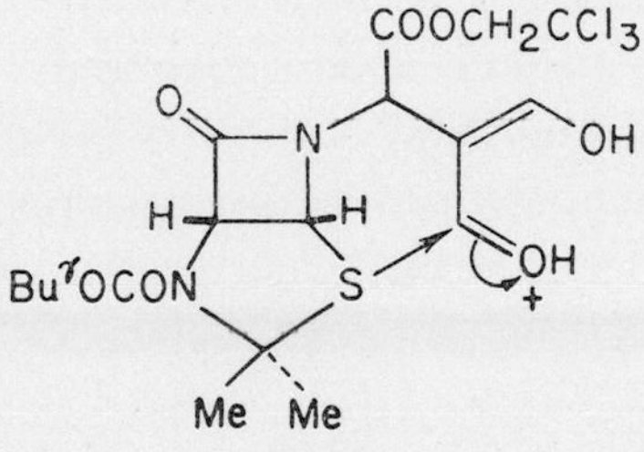

LII

need not be specified in detail. A number of more-or-less equivalent schemes may be considered, among which those portrayed in (LIII) and (LIV) should be included; a special point is that the amino group which is ultimately freed very probably appears at some time during reaction as the corresponding Schiff base (LV)–and we have found that Schiff bases are readily cleaved in trifluoroacetic acid solution. In any event, from the practical point of view it was most gratifying that the protecting groups–that is, the *N-tert*-butyl-

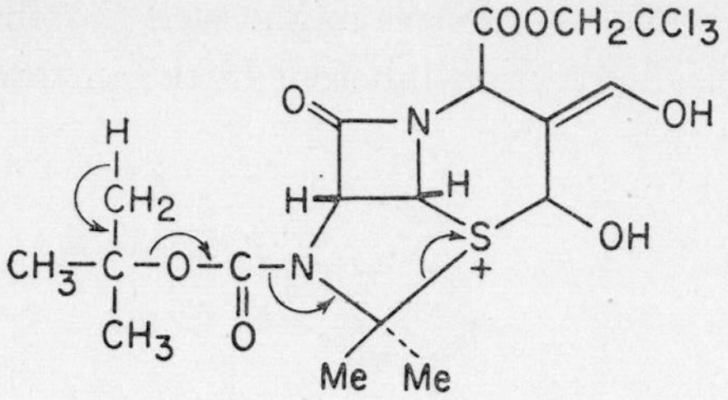

LIII

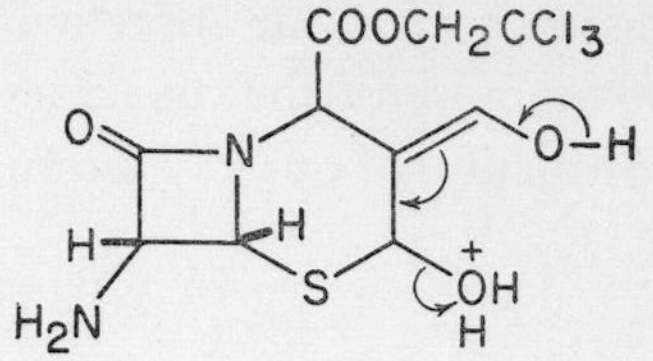

LIV

oxycarbonyl group and the bridging isopropylidene group which had so well served their several purposes and were now no longer wanted – were removed concomitantly with the crucial formation of the new six-membered ring.

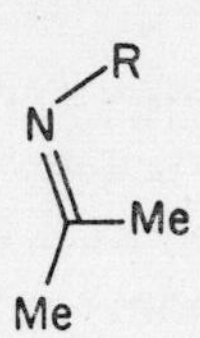

LV

Mention should be made at this point of a special stereochemical detail. The adduct (L) contains one asymmetric carbon atom (starred in L) in addition to those present in the β-lactam. The combination reaction gives both of the *a priori* possible products, which have been separated and carefully characterized. Although the matter is under active study, we cannot as yet make rigorous stereochemical assignments for the two isomers. In any event, the point is not an important one from the practical point of view, since as we shall see shortly, asymmetry at the center under discussion is expunged in subsequent operations.

The amino aldehyde (LI) was next acylated in benzene solution with thiophene-2-acetyl chloride in the presence of pyridine, and the resulting amide (LVI) was reduced, using diborane in tetrahydrofurane solution, to the

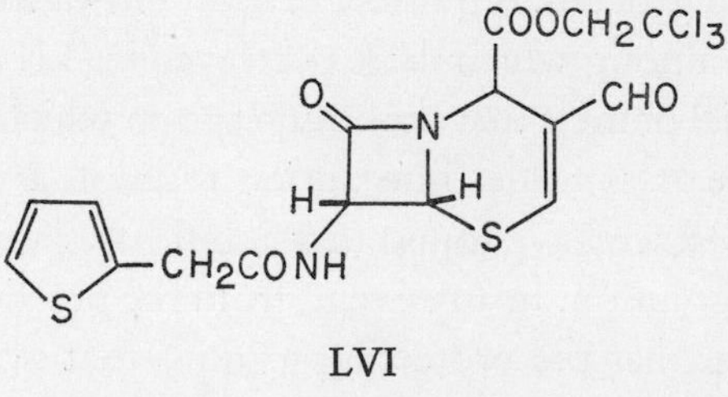

LVI

alcohol (LVII). The latter was acetylated in the normal way with acetic anhydride and pyridine to give isocephalothin β,β,β-trichloroethyl ester

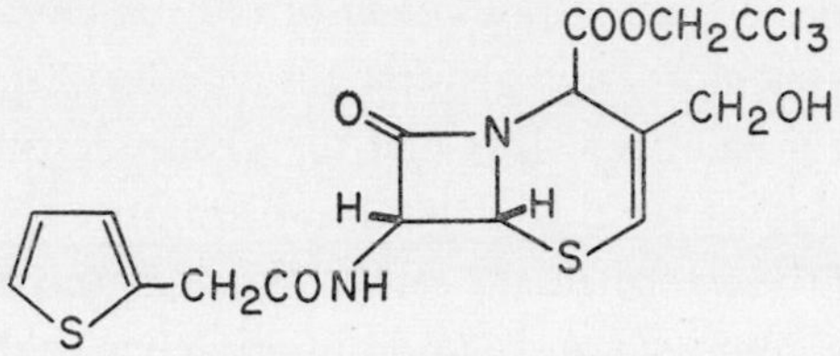

LVII

(LVIII). In its turn this β,γ-unsaturated ester was smoothly equilibrated with the corresponding α,β-unsaturated isomer – cephalothin β,β,β-trichloro

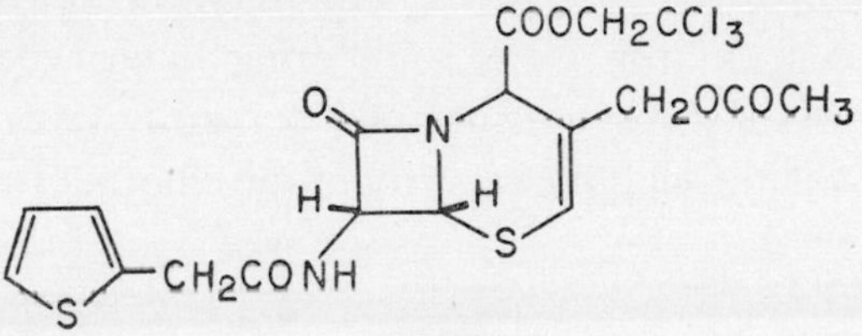

LVIII

ethyl ester (LIX) – when it was allowed to stand in anhydrous pyridine solution at room temperature for three days. Although the β,γ-unsaturated isomer is favored in this equilibrium ($K_{normal/iso} = 1/3$), the two isomers were

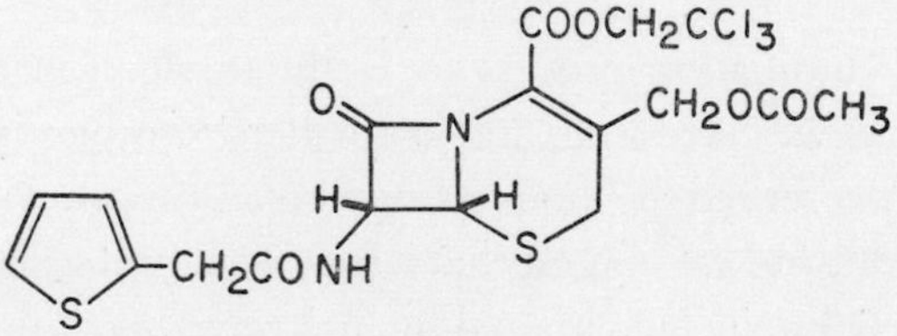

LIX

found to be readily separable by chromatography on silica gel. The conjugated ester was now reduced by zinc dust in 90% aqueous acetic acid at room temperature, and cephalothin (IV ≡ LX) was obtained. The properties of the

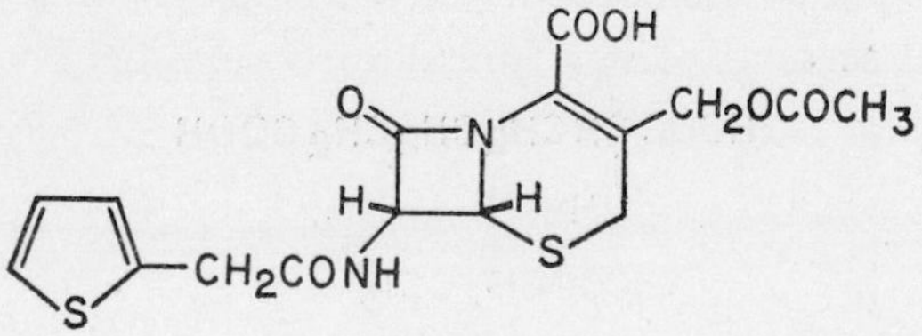

LX

synthetic substance were identical in all respects with those of material prepared from natural cephalosporin C[6].

The final step in our synthesis of cephalothin, namely, the reductive removal of the β,β,β-trichloroethyl grouping is worthy of special comment. In planning our work it had been clear that the group destined to become the free carboxyl function of the final cephalosporin must appear in some protected form during the intermediary stages. Further, the protection must be such that it could be removed without doing violence to the highly sensitive β-lactam ring, which is especially prone to hydrolytic attack. Some years before, in Cambridge, Mr. Robert Kohler, faced with a not dissimilar problem, at my instigation investigated in a preliminary way the action of reductants upon β,β,β-trichloroethyl derivatives, with very encouraging results. The idea had been that an electron source could bring about a concerted elimination process (arrows in LXI) which might be highly favored on statistical grounds, and as well through the capacity of the chlorine atoms not directly

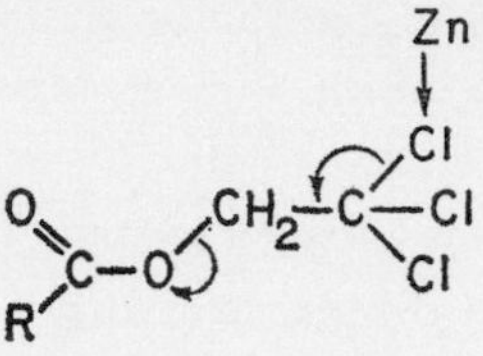

Zn

$R-C(=O)-O-CH_2-CCl_3$

LXI

involved in the elimination process to facilitate electron accession in the transition state. As we have seen, and as further examples in the sequel will show, the grouping served the desired function admirably in our work in the cephalosporin field, and we suggest that it may well find some general utility; indeed, Dr. Fritz Eckstein, encouraged by his knowledge of our early studies, has very recently shown how it can be put to very good effect in work with the nucleotides[8].

We turn now to the completion of the synthesis of cephalosprin C itself. The amino aldehyde (LI) was in this case condensed in tetrahydrofurane solution with *N*-β,β,β-trichloroethyloxycarbonyl-D-(−)-α-aminoadipic acid (LXII) in the presence of dicyclohexylcarbodiimide. The resulting crude

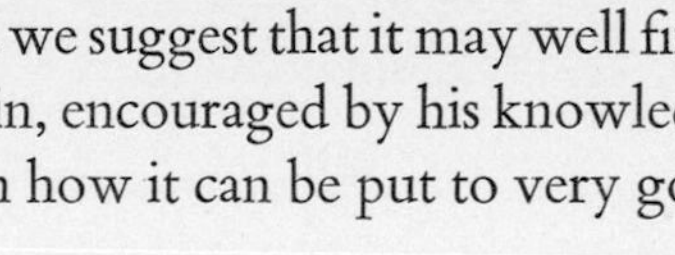

$HOOC\,CH(NH\text{-}COOCH_2CCl_3)CH_2CH_2CH_2COOH$

LXII

reaction mixture was then esterified directly, using β,β,β-trichloroethanol in methylene chloride in the presence of dicyclohexylcarbodiimide and pyridine. This sequence of reactions gave two main products, which were readily separated by chromatography on silica gel using benzene/ethyl acetate (3/1) as eluant. The more polar of the two products was (LXIII), since it was

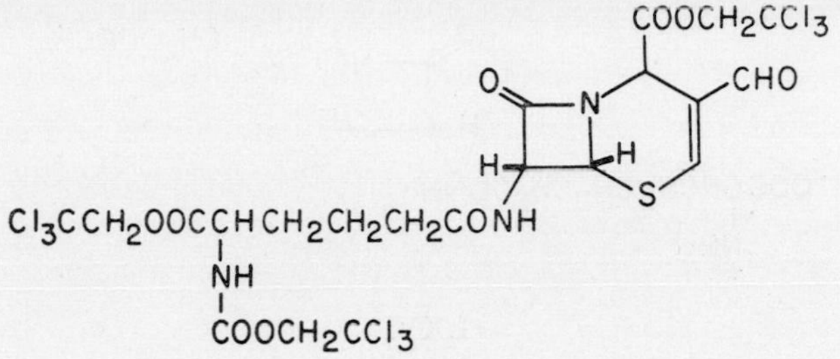

LXIII

converted by reduction in tetrahydrofurane with diborane, followed by acetylation with acetic anhydride/pyridine to the β,γ-unsaturated ester (LXIV).

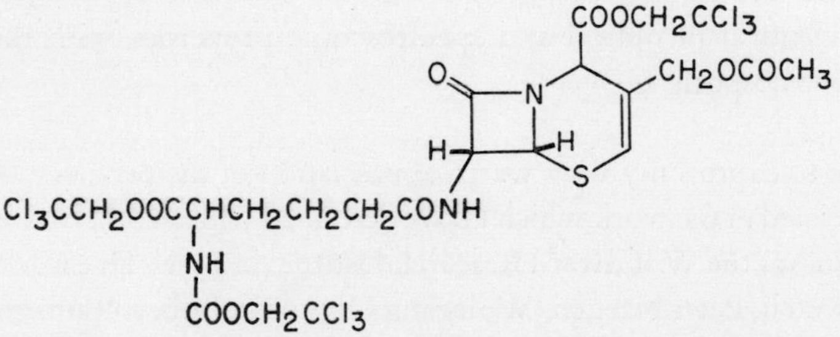

LXIV

As in the cephalothin series, this unconjugated ester was smoothly equilibrated with the conjugated isomer (LXV) when it was allowed to stand in pyridine

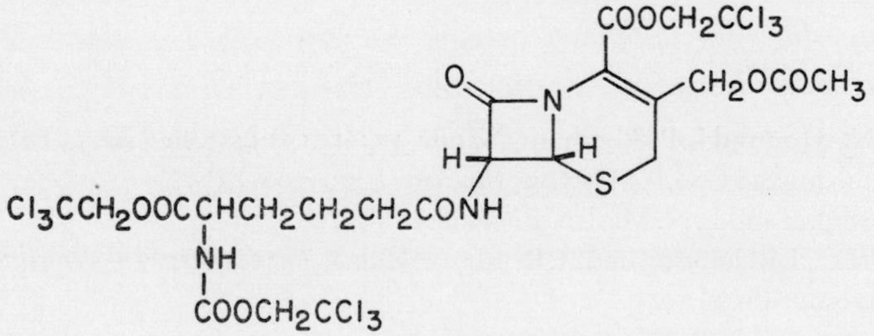

LXV

at room temperature for three days ($K_{normal/iso} = 1/4$). Again, the two isomeric esters were readily separated, and the conjugated isomer was reduced by zinc dust and 90% aqueous acetic acid at 0° during two and one-half hours to synthetic cephalosporin C (LXVI ≡ I). The identity of the synthetic material

LXVI

was in this case established through examination of its paper chromatographic behavior in several systems as well as through observation of its antibacterial activity against *Neisseria catarrhalis*, *Alcaligenes faecalis*, *Staphylococcus aureus* and *Bacillus subtilis*. Further the synthetic crystalline barium salt was identical in optical and spectroscopic properties with the salt of natural cephalosporin C.

It remains to express my very warm appreciation of the privilege of having been associated in the work which I have described with an outstanding group of colleagues at the Woodward Research Institute in Basel. Drs. Karl Heusler, Jacques Gosteli, Peter Naegeli, Wolfgang Oppolzer, Robert Ramage, Subramania Ranganathan, and Helmut Vorbrüggen are those whose high experimental skill and unflagging spirit brought this investigation to its successful conclusion, and I am glad to have this opportunity to express my admiration for their achievement.

1. G.G.F. Newton and E.P. Abraham, *Nature*, 175 (1955) 548; *Biochem. J.*, 62 (1956) 651.
2. E.P. Abraham and G.G.F. Newton, *Biochem. J.*, 79 (1961) 377.
3. D.C. Hodgkin and E.N. Maslen, *Biochem. J.*, 79 (1961) 393.
4. H.T. Clarke, J.R. Johnson and R. Robinson (Eds.), *The Chemistry of Penicillin*, Princeton University Press, 1949.
5. J.C. Sheehan and K.R. Henery-Logan, *J. Am. Chem. Soc.*, 81 (1959) 5838; 84 (1962) 2983.

6. R.R. Chauvette, E.H. Flynn, B.G. Jackson, E.R. Lavagnino, R.B. Morin, R.A. Mueller, R.P. Pioch, R.W. Roeske, C.W. Ryan, J.L. Spencer and E. van Heyningen, *J. Am. Chem. Soc.*, 84 (1962) 3402.
7. G.E. Woodward and E.F. Schroeder, *J. Am. Chem. Soc.*, 59 (1937) 1690.
8. F. Eckstein, *Angew. Chem.*, 77 (1965) 912.

Biography

Robert Burns Woodward was born in Boston on April 10th, 1917, the only child of Margaret Burns, a native of Glasgow, and Arthur Woodward, of English antecedents, who died in October, 1918, at the age of thirty-three.

Woodward was attracted to chemistry at a very early age, and indulged his taste for the science in private activities throughout the period of his primary and secondary education in the public schools of Quincy, a suburb of Boston. In 1933, he entered the Massachusetts Institute of Technology, from which he was excluded for inattention to formal studies at the end of the Fall term, 1934. The Institute authorities generously allowed him to re-enroll in the Fall term of 1935, and he took the degrees of Bachelor of Science in 1936 and Doctor of Philosophy in 1937. Since that time he has been associated with Harvard University, as Postdoctoral Fellow (1937–1938), Member of the Society of Fellows (1938–1940), Instructor in Chemistry (1941–1944), Assistant Professor (1944–1946), Associate Professor (1946–1950), Professor (1950–1953), Morris Loeb Professor of Chemistry (1953–1960), and Donner Professor of Science since 1960. In 1963 he assumed direction of the Woodward Research Institute at Basel. He was a member of the Corporation of the Massachusetts Institute of Technology (1966–1971), and he is a Member of the Board of Governors of the Weizmann Institute of Science.

Woodward has been unusually fortunate in the outstanding personal qualities and scientific capabilities of a large proportion of his more than two hundred and fifty collaborators in Cambridge, and latterly in Basel, of whom more than half have assumed academic positions. He has also on numerous occasions enjoyed exceptionally stimulating and fruitful collaboration with fellow-scientists in laboratories other than his own. His interests in chemistry are wide, but the main arena of his first-hand engagement has been the investigation of natural products – a domain he regards as endlessly fascinating in itself, and one which presents unlimited and unparalleled opportunities for the discovery, testing, development and refinement of general principles.

Prof. Woodward holds more than twenty honorary degrees of which only a few are listed here: D.Sc. Wesleyan University, 1945; D.Sc. Harvard

University, 1957; D.Sc. University of Cambridge (England), 1964; D.Sc. Brandeis University, 1965; D. Sc. Israel Institute of Technology (Haifa), 1966; D.Sc. University of Western Ontario (Canada), 1968; D.Sc. Université de Louvain (Belgium), 1970.

Among the awards presented to him are the following: John Scott Medal (Franklin Institute and City of Philadelphia), 1945; Baekeland Medal (North Jersey Section of the American Chemical Society), 1955; Davy Medal (Royal Society), 1959; Roger Adams Medal (American Chemical Society), 1961; Pius XI Gold Medal (Pontifical Academy of Sciences), 1961; National Medal of Science (United States of America), 1964; Willard Gibbs Medal (Chicago Section of the American Chemical Society), 1967; Lavoisier Medal (Société Chimique de France), 1968; The Order of the Rising Sun, Second Class (His Majesty the Emperor of Japan), 1970; Hanbury Memorial Medal (The Pharmaceutical Society of Great Britain), 1970; Pierre Bruylants Medal (Université de Louvain), 1970.

Woodward is a member of the National Academy of Sciences; Fellow of the American Academy of Arts and Sciences; Honorary Member of the German Chemical Society; Honorary Fellow of The Chemical Society; Foreign Member of the Royal Society; Honorary Member of the Royal Irish Academy; Corresponding Member of the Austrian Academy of Sciences; Member of the American Philosophical Society; Honorary Member of the Belgian Chemical Society; Honorary Fellow of the Indian Academy of Sciences; Honorary Member of the Swiss Chemical Society; Member of the Deutsche Akademie der Naturforscher (Leopoldina); Foreign Member of the Accademia Nazionale dei Lincei; Honorary Fellow of the Weizmann Institute of Science; Honorary Member of the Pharmaceutical Society of Japan.

Woodward married Irja Pullman in 1938, and Eudoxia Muller in 1946. He has three daughters: Siiri Anne (b. 1939), Jean Kirsten (b. 1944), and Crystal Elisabeth (b. 1947), and a son, Eric Richard Arthur (b. 1953).

Chemistry 1966

ROBERT S. MULLIKEN

«for his fundamental work concerning chemical bonds and the electronic structure of molecules by the molecular orbital method»

Chemistry 1966

Presentation Speech by Professor Inga Fischer-Hjalmars, University of Stockholm

Your Majesty, Your Royal Highnesses, Ladies and Gentlemen.

The Greek word for Nature is φ ύ σ ι ζ (fýsis) and for Natural Science φυσιχή (fysiké). Later on, this science became so comprehensive that it was divided into a number of smaller domains, such as Biology, Geography, Chemistry, and Physics in a restricted sense. Subsequently, each of these domains has expanded considerably and developed several special fields. Therefore, it seems as if the different Natural Sciences only continue to diverge like the parts of an expanding Universe. However, the simultaneous deepening of our knowledge has brought about a convergence between the fundamental aspects of the different fields. Especially, Physics and Chemistry have to a great extent drawn closer together. The expression Physical Chemistry and Chemical Physics show that it is no longer possible to draw a sharp borderline between these sciences.

The problem of the nature of the chemical bond evidently belongs to this borderland. By chemical bond we mean the forces that tend to keep together the atoms in a molecule. Already in 1812 Berzelius suggested that these forces originate from positive and negative electrical charges of the atoms. This idea became more firmly founded when in the beginning of the twentieth century Rutherford discovered that each atom consists of a heavy nucleus with positive charge and a swarm of agile electrons totally with an equal amount of negative charge. In 1916 this discovery inspired Lewis to the hypothesis that the chemical bond is caused by two electrons, paired somehow and staying in the domain between the bonded atoms. Although physically questionable, Lewis' theory has exerted a great influence upon the development of Chemistry. In an epoch-making investigation in 1927 Heitler and London also succeeded in casting Lewis' pair theory in a physically more satisfactory form by aid of quantum mechanics. In this shape the theory has highly stimulated chemical thinking, especially under the impact of the further development and the many applications, made by Pauling, who received the 1954 Nobel Prize for Chemistry.

However, there are quite a few chemical questions that are not answered satisfactorily by the electron-pair theory, neither in its original nor in its quantum mechanical shape. Many problems in the Chemistry of unsaturated compounds belong to these questions. To clarify such obscurities in the nature of the chemical bond it was necessary with an entirely new opening. The new move was again inspired by Physics.

To understand how atoms can be bound together to complicated molecules it is first necessary to have a clear idea of the building of an isolated atom. The solution of this fundamental problem of Theoretical Chemistry was given by the Nobel Laureate in Physics Niels Bohr. He showed in 1922 that the electrons in an atom are moving in such a manner that they can be assigned to different shells at various distances from the nucleus. The electrons in the outermost shell are most loosely bonded and mainly responsible for the chemical properties of the element.

Already in 1925 Bohr's principle for atoms was applied to the molecular problem by Robert Mulliken. He assumed a similar building-up principle for molecules as that of atoms, but differing in the respect that the electron shells of a molecule should enclose several atomic nuclei. The electronic motions extended over the whole molecule, was described by Mulliken using a theoretical concept, which he later called a molecular orbital. During the decade after the break-through of the modern quantum mechanics in 1926 these ideas were re-formulated and further developed, mainly by Hund and Mulliken, but with important contributions also from other scientists. The molecular-orbital method means a principally new understanding of the nature of the chemical bond. Previous ideas started from the assumption, most natural from the chemical point of view, that the bonding depends on interaction between complete atoms. The molecular-orbital method, on the other hand, starts from quantum-mechanical interaction between all the atomic nuclei and all the electrons of the molecule. This new view has clarified many molecular properties and reactions. The method has given exceedingly important contributions to our qualitative understanding of the chemical bond and the electronic structure of molecules.

In several connections, however, a qualitative picture is not sufficient but it is necessary to have quantitative, theoretical results for comparison with experiments. Since even small molecules contain many electrons, more extensive quantitative calculations have been possible only during the last decade after the advent of the modern electronic computers. Mulliken realised early the new possibilities offered by these machines. He and his co-workers in Chicago have devoted much energy and tenacity to adapt the molecular-orbital method to computer language. For various reasons it is a difficult numerical problem to make accurate computations of the quantities, representing measurable chemical effects. In spite of this Mulliken's laboratory has very lately succeeded to compute by the molecular-orbital method different molecular properties of small molecules with such an accuracy that the theoretical values only differ by a few per cent from the experimental ones. From these results, highly interesting by themselves, can be derived important complementary information about the nature of the chemical bond. In addition, these results demonstrate that we now have at hand an entirely new possibility to investigate small molecules, inconvenient or inaccessible to experiments. Examples of this are intermediate states of chemical reactions and molecules and molecular fragments of great importance in Space Research.

Significant, theoretical results have also been obtained for large molecules. In such cases it is not yet possible to make purely theoretical, quantitative calculations. But Mulliken has developed the general scheme of an elegant method to combine the theoretical computations with experimental information from small molecules. This kind of calculation is exceedingly illuminating in several connections, for instance for the interpretation of measurements. Just as for small molecules the method has also been used to gain information about molecules, inaccessible to experiments, such as compounds of importance for life processes. In such cases the theoretical results cannot be directly compared with measurements, but they can suggest new kinds of experiments.

It is only after time-consuming, strenuous efforts by many scientists that we now know what an extraordinary instrument for investigations of the properties of matter we have at our command in the molecular-orbital method. The leader of this achievement has been continuously and still is Robert Mulliken.

Professor Mulliken, it is now more than forty years since you started to investigate what the electron is doing in the molecule. Your deep understanding of the physical laws, governing the behaviour of the electron, combined with your intimate knowledge of the problems of Chemistry have greatly advanced our understanding of the properties of molecules. Especially, by the development of the molecular-orbital method you have given us a powerful key to the mechanism of the chemical bond. Your thorough penetration of the inherent possibilities of the method and your unfailing enthusiasm in guiding the development have very lately pushed the progress to the point of accurate, quantitative applications to chemical problems. By these achievements you have opened a road of exciting possibilities for future, theoretical investigations of molecular properties.

On behalf of the Royal Academy of Sciences I extend to you the most hearty congratulations. And now it is my privilege to ask you to receive the Nobel Prize for Chemistry from the hands of His Majesty the King.

ROBERT S. MULLIKEN*

Spectroscopy, molecular orbitals, and chemical bonding

Nobel Lecture, December 12, 1966

I am most deeply appreciative of the 1966 Nobel prize for chemistry awarded for «fundamental work concerning chemical bonds and the electronic structure of molecules by the molecular-orbital method». In the title of my lecture I have added the work spectroscopy, since it was a study of molecular spectroscopy which pointed the way toward molecular orbitals. I think it is appropriate also to remember that in Niels Bohr's classical 1913 papers[1] «On The Constitution of Atoms and Molecules», best known for his theory of the hydrogen atom, and in his 1922 theory of the structure of atoms and the periodic system of the elements, atomic spectroscopy provided essential guide-posts for the path toward the theory.

Let me now ask, what is a molecular orbital? A really adequate answer is unavoidably technical. However, in an effort to make matters as clear as possible, I shall begin this lecture by reviewing a number of things which may be regarded as uninteresting old history, or else as boringly well known, at least by physical scientists. For this approach I beg your indulgence and ask your forgiveness.

Let us first go back to the quantum theory of atomic structure initiated by Bohr but shaped up in further detail by Sommerfeld. In this older quantum theory, Bohr assumed that the electrons move in orbits around the very small but relatively very heavy positive nucleus of the atom, like planets around a sun. Going back historically a step further, it is good to recall that the picture of the atom as containing a small heavy positive nucleus first emerged from Rutherford's work at Manchester, and that Bohr began the development of his theory while he was at Manchester in Rutherford's laboratory[1].

As compared with the motion of planets around a sun, there were of course several important differences in the Bohr–Sommerfeld theory of atoms in

* Distinguished Service Professor of Physics and Chemistry, University of Chicago, Chicago, Ill., and (winters) Distinguished Research Professor of Chemical Physics, Florida State University, Tallahassee, Florida (U.S.A.).

such matters as the sizes of the orbits and the degree to which they are crowded, and the strengths of the forces acting. However, there was one much more radical difference, namely that the possible electron orbits were assumed to be limited to particular sizes and shapes defined by quantum rules. (Bohr in his first papers mentioned elliptical orbits, but in order to get some definite results assumed rings of electrons moving in circular orbits with an angular momentum of $h/2\pi$ for each electron, where h is Planck's constant.)

Given the energies and angular momenta of the electron orbits, the Bohr–Sommerfeld theory continued to use the familiar laws of physics to describe them, in particular the principles of mechanics first set forth by Newton. However, quantum mechanics in 1925–1926 replaced Newtonian mechanics with radically new concepts for small-scale things like atoms and molecules. It showed the necessity in dealing with these of a new way of thinking very different from normal human-scale thinking. Nevertheless, it still allowed us to visualize an atom as a heavy positive nucleus surrounded by negative electrons in something like orbits.

Now to attempt an answer to the question posed earlier, an *orbital* means, roughly, *something like an orbit*; or, more precisely, *something as much like an orbit as is possible in quantum mechanics*. Still more precisely, the term «orbital» is simply an abbreviation for *one-electron orbital wavefunction* or, *preferably*, for *one-electron orbital eigen-function*. The last-mentioned expression refers to any one of the so-called characteristic solutions or *eigen-functions* of Schrödinger's quantum-mechanical wave equation *for a single electron* in an atom or molecule.

According to a picturesque expression once used by Van Vleck, a set of orbitals represents a housing arrangement for electrons. A very strict rule (Pauli's exclusion principle) applies to every orbital, whether atomic or molecular, namely that not more than two electrons can occupy it. In other words, it can be empty, or it can hold one electron, or it can hold two electrons. Every electron has a spin, like the earth on its axis; if there are two electrons in an orbital, their spins are oppositely directed.

An *atomic orbital* (*abbreviated AO*) is best taken to be an eigen-function of a one-electron Schrödinger equation which is based on the attraction of the nucleus for the electron we are considering plus the average repulsion of all the other electrons. Following Hartree, this average field is called a *self-consistent field* because of the way the calculations are made; namely, the orbital for each electron in turn is calculated assuming all the other electrons to be occupying appropriate orbitals.

A *molecular orbital* (*abbreviated MO*) is defined in exactly the same way, except that its one-electron Schrödinger equation is based on the attractions of two or more nuclei plus the averaged repulsions of the other electrons.

An orbital (either atomic or molecular) is, strictly speaking, just a mathematical function in ordinary 3-dimensional space. When an electron is occupying an orbital, the form of the orbital tells us, among other things, what fraction of time the electron in it can spend in different regions of space around the nucleus, or nuclei. Each orbital *favors* some particular regions of space and *disfavors* others, yet all the orbitals in a given atom or molecule extend at least to some small extent throughout all regions of the atom or molecule*. Orbitals differ most strikingly from the orbits of Bohr theory in the fact that they do not specify detailed paths for electrons. They do, however, give some information about average speeds as well as positions of electrons.

I have always felt that a *true* AO or MO (sometimes I have called it a *best* AO or MO) for an electron is one which corresponds to a self-consistent field as above described. Commonly, however, the word orbital is used to refer to mathematical functions which are only *approximations* to these true AO's or MO's. The main reason for this fact is that until recently we have had in most cases, especially for MO's, only a rather roughly approximate knowledge of the true forms.

In the case of MO's, the so-called LCAO approximations are rather generally familiar. However, when people have talked about the electronic structures of atoms or molecules, or of their excited states, in terms of AO's or MO's, they have really been *thinking* in terms of the *true* AO's or MO's which we knew must exist whether or not we knew their exact forms. Thus I would like to maintain that in the concept of an orbital, the proper norm is that of the true accurate self-consistent-field AO or MO**.

Figs. 1–5 show contour maps of the true accurate valence-shell MO's of the oxygen molecule, as obtained by calculations at the Laboratory of Molecular Structure and Spectra at the University of Chicago. In these «portraits» of MO's, each contour is marked with a number which gives the magnitude, not of the MO itself, but of its square. This use of the square is particularly instruc-

* Except for certain infinitely thin «nodal surfaces».

** In view of the fact that a set of orbitals which are only approximate can still correspond to a self-consistent field which is, however, like the orbitals, only *approximate*, many people (including my Chicago colleagues) commonly designate true exact (or almost exact) self-consistent-field orbitals by the name Hartree–Fock orbitals. In the case of atomic or moleculars states with non-zero spin, there are additional complications.

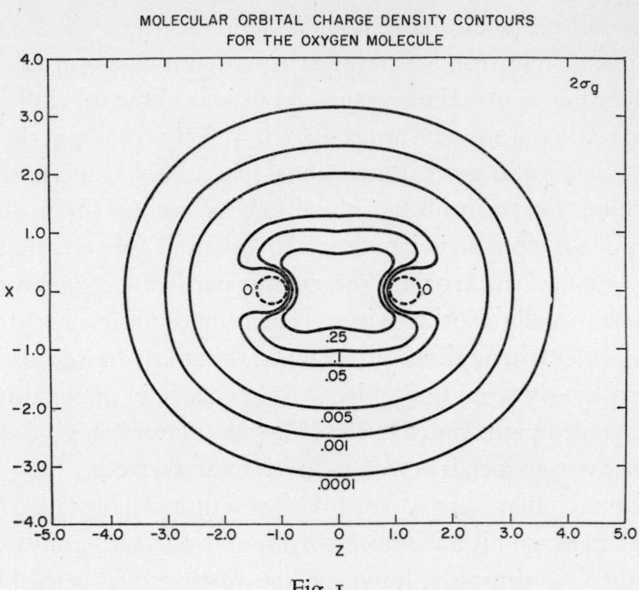
MOLECULAR ORBITAL CHARGE DENSITY CONTOURS
FOR THE OXYGEN MOLECULE
2σg
X
Z
.25
.05
.005
.001
.0001
0
4.0
3.0
2.0
1.0
-1.0
-2.0
-3.0
-4.0
-5.0
5.0

Fig. 1.

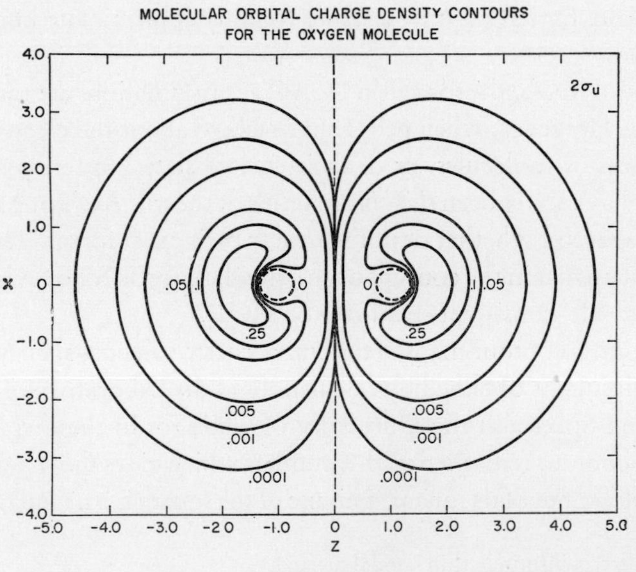
MOLECULAR ORBITAL CHARGE DENSITY CONTOURS
FOR THE OXYGEN MOLECULE
2σu
X
Z
.05
.1
.25
.005
.001
.0001
0
4.0
3.0
2.0
1.0
-1.0
-2.0
-3.0
-4.0
-5.0
5.0

Fig. 2.

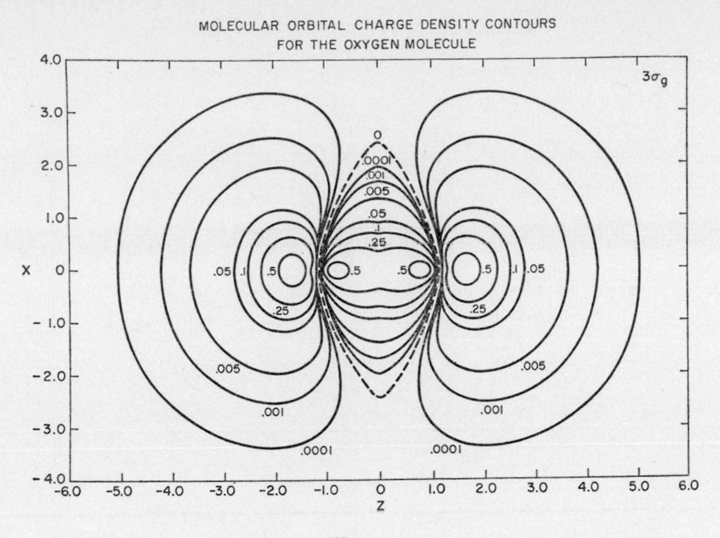
MOLECULAR ORBITAL CHARGE DENSITY CONTOURS
FOR THE OXYGEN MOLECULE
$3\sigma_g$
X
Z
0
.0001
.001
.005
.05
.1
.25
.5

Fig. 3.

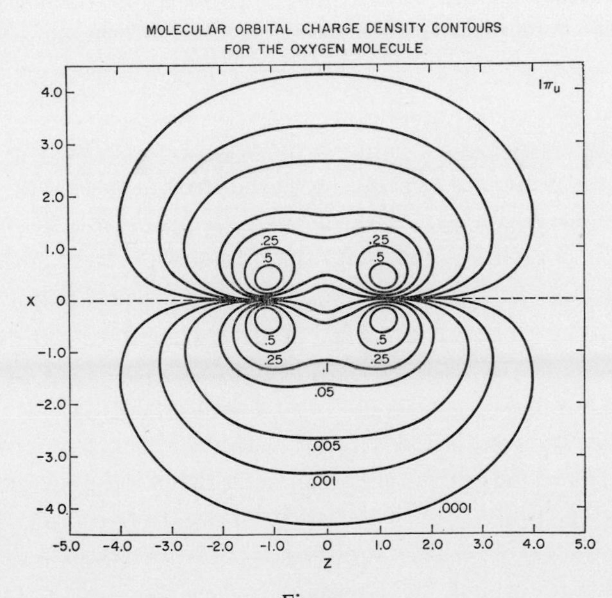
MOLECULAR ORBITAL CHARGE DENSITY CONTOURS
FOR THE OXYGEN MOLECULE
$1\pi_u$
X
Z
.25
.5
.1
.05
.005
.001
.0001

Fig. 4.

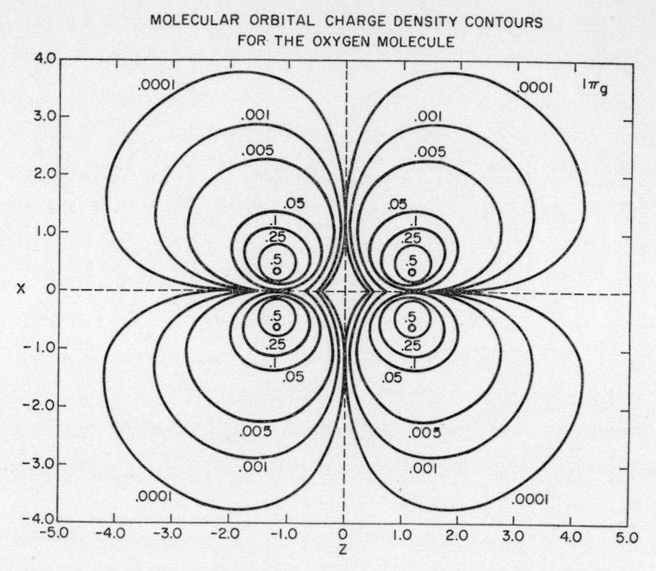

Figs. 1–5. Portraits of valence-shell spectroscopic MO's of the oxygen molecule. Each contour is marked with a number which is the *square* of the value of the MO on that contour. Each dashed line or curve marks a boundary between positive and negative regions of the MO itself. From an as yet unpublished paper by P.E. Cade, G.L. Malli and A.C. Wahl; all rights reserved.

tive because the square of the value of the orbital at any point in space is proportional to the probability of finding the electron at that point when it is in that orbital. Thus what Figs. 1–5 show are really probability density portraits, rather than direct pictures of the MO's. Additional portraits of this kind, and helpful discussion of them, are contained in two recent articles[2]. The necessary calculations which are making MO portraits possible are extremely complex and require the use of large computing machines. They have involved the work of a number of people, of whom I shall speak later.

A definite *energy* is associated with each orbital, either atomic or molecular. The best interpretation of this orbital energy is that it is the energy required to take the electron entirely out of the orbital, out into free space. The lowest-energy orbitals are those which favor regions of space closest to the nucleus, in the case of atomic orbitals, or closest to one or more nuclei in the case of molecular orbitals.

In what I like to call the *normal state*, but most people call the ground state (German «Grundzustand»), of an atom or molecule, the electrons are settled in the lowest-energy orbitals that are available. (Bohr[1] called it the «permanent» state.) Higher in energy than the normal-state orbitals and favoring regions of space farther from the nucleus or nuclei, there are great numbers of *vacant* orbitals, into any one of which, however, any electron can go if given enough extra energy by the right kind of a push or kick. These ordinarily vacant AO's or MO's are called excited orbitals, and when one (or sometimes more) electrons of an atom or molecule have been kicked into excited orbitals, the atom or molecule is said to be *excited*. Excited states of atoms or molecules, with some exceptions, do not last long. Instead, the molecule loses its extra energy, generally either in collisions with other molecules, or by sending it out in the form of electromagnetic radiation: that is, visible or infrared or ultraviolet light or X-rays. A careful study of the spectrum of wave lengths of such radiation gives important information about the forms and energies of the orbitals involved, and about other properties of the atom or molecule both in its normal and in various excited conditions.

A prominent feature of Bohr's 1922 theory of atoms was the Aufbauprinzip (building-up principle) according to which if electrons are fed one by one to an atomic nucleus and the atom is allowed to subside into its normal state, the first electrons fall into the lowest-energy orbits, the next into those next lowest in energy, and so on. In this way Bohr first explained the formation of successive electronic shells and the periodic system of the chemical elements. In the modern quantum mechanics, exactly the same description holds good except that atomic orbitals replace orbits. In the *molecular orbital method* of describing the structure of molecules, an entirely analogous use is made of an Aufbauprinzip in which electrons are fed into molecular orbitals.

However, there is also another way of describing molecules, usually called the valence-bond method. This was initiated by the work of Heitler and London on the hydrogen molecule, and developed further by Slater and Pauling especially. In this method, each molecule is thought of as *composed of atoms*, and the electronic structure is described using atomic orbitals of these atoms. This approach, which I prefer to call the *atomic orbital method**, is a valid alternative to the MO method, which in its most general form regards each molecule as a self-sufficient unit and *not* as a mere composite of atoms.

The AO method at first appealed to chemists because it was much easier to

* To speak of the «valence-bond method» places the emphasis in chemical bonding on a few pairs of electrons holding atoms together in the Heitler–London manner, whereas actually the interactions of many of the other electrons often have very important effects on the stability of molecules.

fit into customary ways of thinking. However, it has become increasingly evident that the MO method is more useful for a detailed understanding of the electronic structures of molecules, especially if extensive theoretical calculations are to be made, as is now increasingly feasible with the help of modern large-scale digital computers. Also the MO method is far better suited for an understanding of the *electronic spectra* of molecules and thus also of their photochemical behavior, a subject which is now receiving increasing (and increasingly understanding) attention.

I have just stated that the AO and MO methods are valid alternatives to each other, although they differ with respect to ease of understanding and of application. But why is not one right and the other wrong? The explanation is, roughly, that both methods correspond only to *approximate* solutions of the *complete* equations which govern the behavior of molecules that contain more than one electron. Starting from either method, further and in general very difficult calculations are needed for an exact understanding.

Why this is true can be seen in comparing an atom with a planetary system. In a planetary system, the sun is vastly larger than any of the planets and the gravitational attractive force it exerts on each planet is exceedingly large compared with the small gravitational forces which the planets exert on one another. Thus the motion of each planet in its orbit can be calculated almost as if it were completely independent of the motions of the other planets, and the small effects of other planets can be calculated to a satisfactory degree of exactness by a method called *perturbation theory*. However, it has not proved mathematically possible to obtain an absolutely exact solution which would be true over very long intervals of time. The same statement holds for every situation in which more than two objects are exerting forces on another. Such a situation is called a many-body problem.

Although with a sun and planets the lack of a solution of the many-body problem is not very serious, matters are very different for an atomic nucleus and its electrons, for two reasons: (*1*) the electrons in an atom, though not really close together, are vastly more crowded than planets in a planetary system; (*2*) the forces between electrons, though not as large as the force exerted by the nucleus on each electron, are nevertheless too large to be treated as small perturbations as could be done for planets and a sun. These difficulties existed in the old quantum theory of electron *orbits*, and similar difficulties still occur for *orbitals* in the modern quantum mechanics. Perturbation theory is valuable in both the Bohr theory and in quantum mechanics, but it does not easily solve the many-body problem. Consequently, it is an exceedingly difficult matter to obtain a reasonably exact understanding of the electronic structure of atoms or, especially, of molecules, except if there is only one electron, as in the hydrogen atom, or (more difficult) in the positively charged hydrogen molecule ion. In these cases, the AO (for H) or the MO (for H_2^+) is an exact solution of Schrödinger's equation. The first fairly exact calculation on the normal state of H_2^+, by Burrau[3], is the earliest example of the nearly exact calculation of the form of a true MO.

Thus we are brought face to face with the fact that when the structure of a typical atom or molecule is described by assigning each electron to an orbital, this description is usually rather far from being exact. It *is* good enough to be extremely useful, and in the case of atoms is sufficient to account for the main features of atomic structure, the periodic system of the elements, and atomic spectra, but it is by no means exact. Analogous comments apply for the description of the structure of a molecule in terms of electrons assigned to MO's.

The description in terms of a single set of orbitals for the electrons is called an *independent-particle model*. It is a kind of model which is very nearly exact in the case of the orbits of planets going around a sun, but is only a rather rough approximation in the much more crowded situation, with much stronger forces, of electrons in an atom or molecule. What is lacking is called *electron correlation*. Because orbitals are based on an allowance only for the *average* forces exerted by other electrons, the simple orbital description needs to be rather strongly corrected for the fact that electrons in their motions are sometimes closer, sometimes less close than their average distance, so that the forces between them vary accordingly, and to an important extent.

Now let us return to the question of how it is possible that both of two seemingly very different methods, the AO and the MO method, can represent useful descriptions of the electronic structures of molecules in their normal states and can help us to understand chemical bonding. The answer lies in the fact that both methods need a considerable amount of correction for electron correlation before their descriptions become accurate. The fact that they differ so strongly from each other is explained by noting that they lie as it were on *opposite sides* of an accurate description, which then lies between them. For a full explanation, however, much more must be said than is possible here. Nevertheless, it seems to be true that the MO method is better suited not only as a basis for rather accurate calculations at the degree of approximation possible in an independent-particle method, but also that it is well suited to going further with the necessary corrections to take electron correlation rather well into account[4].

Now before going further, I would like to return briefly to the historical development of MO ideas in the early 1920's before the time of the modern quantum mechanics. Although Bohr in his early papers[1] proposed molecular models in which pairs of electrons rotating in a circular orbit between two atoms served to form a chemical bond, later calculations based on this model, even for the simplest case of the hydrogen molecule, proved as *unsuccessful* as the Bohr theory of the hydrogen atom has proved successful.

On the other hand, molecular spectroscopists in the early 1920's found that the excited electronic states of diatomic molecules show various features which could be explained by postulating resemblances to those of atoms[5,6]. This experimental evidence suggested that the electrons in molecules, to an extent similar to that of electrons in atoms, are moving in something like orbits and that some sort of Aufbauprinzip is valid for the electronic structures of molecules.

My own work in 1923–25 was at first concentrated on trying to understand the visible and ultraviolet spectra of diatomic molecules, called band spectra, at the Jefferson Physical Laboratory of Harvard University. In learning about this field, which at that time was completely new to me, I had the very kind help of Professor F. A. Saunders in experimental spectroscopy, and Professor E. C. Kemble in quantum theory. I also benefited greatly from correspondence with Professor R. T. Birge of the University of California. It is very interesting at this point to note that in those days basic spectroscopy and the theory of molecular electronic structure were being studied primarily by physicists (my papers until the advent of the *Journal of Chemical Physics* were published in the *Physical Review* or the *Reviews of Modern Physics*). Now, however, these subjects, as well as the newer branches of spectroscopy (nmr, esr, etc.) which were born in physics laboratories, are generally considered to belong primarily to chemistry. These circumstances account for the fact that, although my B. S. and Ph. D. degrees were in chemistry, I have for a long time been a member of physics departments, where I am classified as a molecular physicist. Only rather recently have I become formally associated also with chemistry departments, thereby giving recognition to the migration of molecular spectroscopy and MO's from physics toward chemistry. Nevertheless, the basic facts of these areas of science do still lie in the border region between physics and chemistry.

Now to return to my early efforts to understand diatomic band spectra: the detailed structures of these spectra fell into several distinct types which indicated the existence of several types of molecular electronic states. Moreover, these types appeared to differ, as did the atomic states of Bohr–Sommerfeld theory, in respect to angular momentum properties*. Following a suggestion of Birge, I called them S, P, D states, using the same symbols as for atomic states, although the characteristic described by the symbol was not total orbital angular momentum as in the case of the atomic symbol, but only the axial component of angular momentum.

With the advent of quantum mechanics about 1926, the short-comings of the old quantum theory of atoms and its inability to deal seriously with molecules were quickly removed. Among other changes, atomic electron orbits were replaced by atomic orbitals, although the *name* orbital was given only later, in 1932. My friend Friedrich Hund, whom I first met in Göttingen in 1925, and with whom I had many discussions then and in 1927 and later, applied quantum mechanics to a detailed understanding of atoms and their spectra, and then to the spectra and structure of molecules[5]. Using quantum mechanics, he quickly clarified our understanding of diatomic molecular spectra, as well as important aspects of the relations between atoms and molecules, and of chemical bonding. It was Hund who in 1928 proposed the now familiar Greek symbols Σ, Π, Δ, for the diatomic molecular electronic states which I had been calling S, P, and D. Molecular orbitals also began to appear in a fairly clear light as suitable homes for electrons in molecules in the same way as atomic orbitals for electrons in atoms. MO theory has long been known as the Hund–Mulliken theory in recognition of the major contribution of Professor Hund in its early development.

I have emphasized already that a *true* AO or MO is properly considered as one which is appropriate for an electron assumed under the influence of the *average* electric field of the other electrons, all in accurate self-consistent-field orbitals. However, for MO's there are also several very useful approximations to the exact method. These approximations can be briefly characterized as corresponding to varying degrees of localization or *delocalization*. The purest and most accurate MO method, yielding true MO's, involves the maximum amount of delocalization, with every MO spread to some extent** over the whole molecule. These pure MO's I like to call *spectroscopic MO's*, since it is

* The major structural features of diatomic spectra are dominated by the existence of molecular vibrations and rotations, but the detailed structures depend on the interaction of molecular rotation with electronic orbital and spin angular momenta.

** To be sure, often some (or even most) of them turn out to be mainly (or, in some cases, almost wholly) concentrated near particular atoms or groups of atoms.

they which are particularly important for understanding electronic spectra in molecules. They are also of especial importance for understanding ionization processes; each of the simplest ionization processes corresponds to removal of an electron from a particular spectroscopic MO.

Several types of more or less localized MO methods, although they represent somewhat less accurate descriptions, are also useful, especially in understanding and describing chemical bonding. On starting with a localized-MO description, and afterward proceeding to one or more successive steps of delocalization, and then finally introducing electron correlation, we can often gain much added insight by this step-wise approach into the chemical consequences of what the electrons are doing. The most fully localized sets of MO's include *bond* MO's localized between two or sometimes three or four atoms, taken together with some MO's so strongly localized that they are just AO's (but often *hybrid* AO's) on single atoms; we note here that, after all, an AO can be considered as a special type, the simplest possible type, of MO. These localized MO's I like to call *chemical MO's* (or just *chemical orbitals* because of the fact that some of the orbitals used are now really AO's). In simple molecules, electrons in chemical MO's usually represent the closest possible quantum-mechanical counterpart to G. N. Lewis' beautiful pre-quantum valence theory with its bonding electron pairs, lone pairs, and inner shells. It is the inner-shell and the lone-pair electrons which are in AO's when chemical MO's are used.

It was Hund in a paper on chemical bonding[7] who first referred to σ and π bonds: a single bond is a σ bond, a double bond is a σ plus one π bond**, a triple bond is a σ plus two π bonds, and each bond corresponds to a pair of electrons in a bond MO localized around the two atoms of the bond. While, as I have already said, it is necessary for a thorough understanding to take the effects of all the electrons into account, a consideration just of electrons assigned to localized bond MO's does give a useful approximate understanding of important aspects of chemical bonding, – for example, the existence of nearly free rotation around single bonds but restriction of the bonded and neighboring atoms to a plane in the case of double bonds.

* According to their original definition, valid for diatomic (or linear) molecules, π orbitals are two-fold degenerate. That is, there are two varieties of π orbitals which differ only by a 90° rotation around an axis of cylindrical symmetry. But in the case of double bonds, only one of these is used and the other no longer exists as such but is mixed with σ orbitals. The «π» orbitals of double bonds really ought to have a different name.

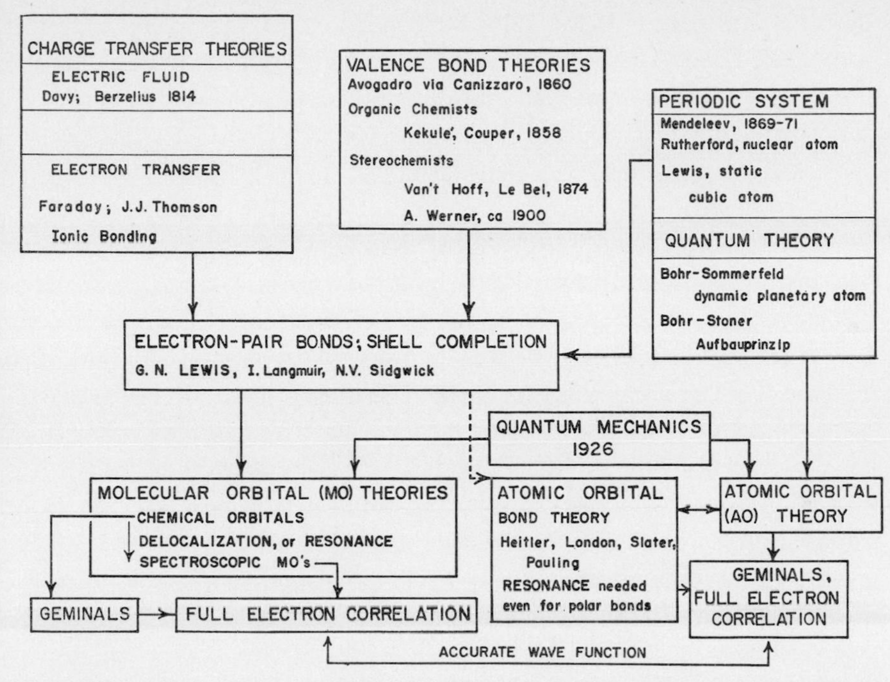

Fig. 6. Historical flow diagram of ideas leading to MO and AO theories of molecular electronic structure, and on to accurate wave functions. From the author's 1965 Silliman Lectures at Yale University.

Before going further, I should like to show four slides* to illustrate the relation of G.N. Lewis' theory to MO theory using chemical orbitals, and also to summarize some other historical relationships. Fig. 6 is more or less selfexplanatory. It is designed to show, first of all, how Lewis resolved the long-standing conflict between, on the one hand, ionic and charge-transfer theories of chemical bonding and, on the other hand, the kind of bonding which is in evidence in bonds between equal atoms, for example, in H_2, or C–C in C_2H_6: Lewis represented each bond by a pair of electrons placed between the two bonded atoms, with the electron pair located closer to one atom than to the other in the case of polar bonds. Figs. 7 and 8 show some examples of Lewis structures, including examples of coordination compounds like $H_3N \cdot BH_3$ or (as pointed out later by Sidgwick) $CO(NH_3)_6{}^{3+}$. In coordination compounds, Lewis lone pairs belonging to electron-donor molecules, for example NH_3, are shared to some extent with electron-deficient molecules or ions like BH_3 or CO^{3+}, forming «dative bonds» or partial dative bonds.

* These slides have been borrowed from my 1965 Silliman lectures at Yale.

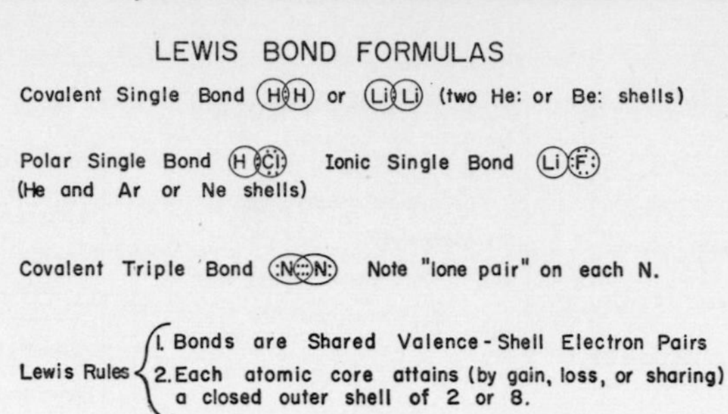

Fig. 7.

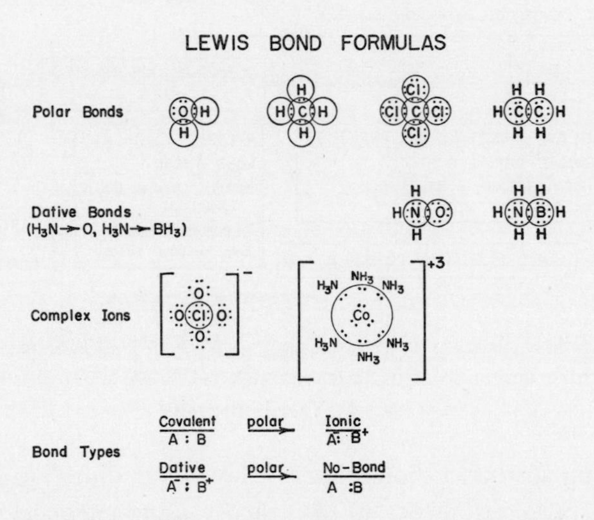

Fig. 8.

Figs. 7 and 8. Examples of G. N. Lewis bond formulas, showing also pair or octet completion by sharing. From the author's 1965 Silliman Lectures at Yale University.

Lewis made use of an Aufbauprinzip in terms of electron shells (pairs and octets mainly) which could in part be obtained by *sharing*, so that the same pair of electrons could be counted in the shells of both of two atoms, as suggested by circles in Figs. 7 and 8 (not shown in every case). For individual atoms, Lewis' electron shells were three-dimensional, in contrast to Bohr's planar electron orbits, in this respect being closer to the present quantum mechanics than the Bohr theory. However, of course Lewis' theory was empirical, schematic, and purely qualitative, and gave no explanation of how the electrons might be moving or why and how they should station themselves between atoms to form bonds, or in pairs or octets in shells. Bohr's early papers[1] included some pictures of pairs of electrons (the electrons of each pair moving on opposite sides of a single circular orbit) forming chemical bonds between atoms, for example in H_2 and in CH_4 in three-dimensional arrangements.

The Heitler–London AO theory of the chemical bond is rather generally regarded as the quantum-mechanical counterpart of Lewis' electron-pair bond. However, a pair of electrons in a bond MO represent an approximately equally good counterpart in the case of a symmetrical (homopolar) bond, while for a polar bond (as in HCl, or in H_2O) they represent a much better counterpart.

The justification for this last statement can be seen most easily by writing the *bond MO* in the LCAO approximate from $\alpha\chi_a+\beta\chi_b$, where χ_a and χ_b are AO's of atoms *a* and *b*. For homopolar bonds, $\alpha=\beta$, but for polar bonds α and β are unequal to an extent which matches the polarity of the bond, for example, $\alpha < \beta$ for an HCl or NaCl molecule, if atom *a* is H or Na, with a much greater inequality for NaCl than for HCl. (Pure ionic NaCl would be represented by $\alpha=0$, $\beta=1$, but actually the NaCl molecule is not quite pure ionic, or if one wants to call it pure ionic, it is necessary to say also that the Cl^- ion is strongly polarized.) Thus the chemical-MO theory has the same flexibility as the Lewis theory in representing polar bonds, while the AO theory has to assume mixtures of Heitler–London and ionic bonding to represent polar bonds. The chemical-MO theory also furnishes the counterpart of Lewis' lone pairs (NH_3 is a good example), and also shows they can be modified into pairs occupying polar bond MO's in coordination compounds.

Fig. 6, after illustrating Lewis' synthesis of earlier ideas, goes on to show how the intervention of quantum mechanics in 1926 permitted further progress by MO theories, and then indicates the necessity of the final step of electron correlation for accurate descriptions. It also shows the alternative route *via* AO theories of atoms and AO bond theory, followed by electron correlation again as a final step to give accurate wave functions identical in content, if not necessarily in form, with those obtained *via* MO theory.

Fig. 9 illustrates for the CH_4 molecule how chemical MO's and spectroscopic MO's are related, and shows also how chemical MO's could be used instead of spectroscopic MO's for an atom like neon. The figure depicts for the neon atom just one of four localized or chemical AO's which are identical except for their orientation; each one is symmetrical about a line directed toward one of the four corners of a tetrahedron, marked 1, 2, 3, and 4. The four BMO's (bond MO's) of CH_4 if they were depicted (instead a pair of dots as in a Lewis formula is shown for each) would be similar in appearance to the chemical AO's of neon except that each one would spread out somewhat

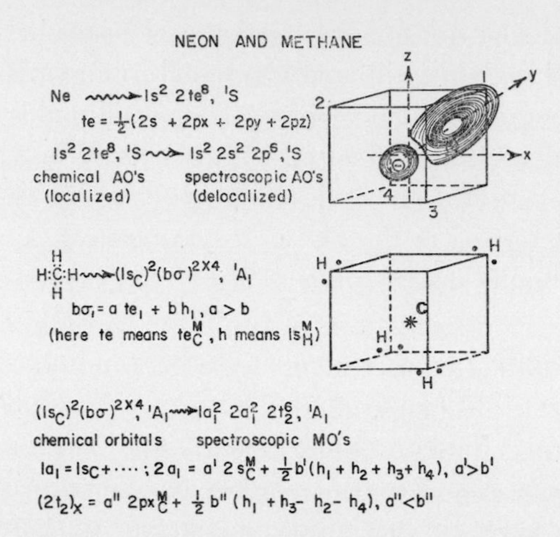

Fig. 9. Chemical and spectroscopic MO's, illustrated by the neon atom and the methane molecule. From the author's 1965 Silliman Lectures at Yale University.

more around its appropriate H nucleus. LCMAO expressions (the extra M means *modified*, – see discussion near the end of this paper) are given for these BMO's, and also for the spectroscopic MO's of CH_4.

Next I must mention the simple LCAO (linear combination of AO's) procedure which for many years represented the usual way of trying to approximate the forms of MO's, whether of localized or delocalized type[6]. For metals, Bloch in 1928 used fully delocalized MO's extending throughout the metal, constructed approximately as linear combinations of valence-electron AO's of all the atoms. Lennard-Jones in 1929 pointed out the general usefulness of simple LCAO expressions in approximating valence-shell diatomic MO's; for inner shells he used AO's*. Herzberg then emphasized that the number of bonds in a diatomic molecule can be set equal to half the difference between the number of electrons in bonding MO's (which have *additive* LCAO forms, that is, α and β of the same sign in $\alpha\chi_a + \beta\chi_b$) and the number in *antibonding* MO's (which have *subtractive* LCAO forms, that is, α and β of opposite signs). Hückel developed his very simple LCAO treatment for the π electrons in unsaturated and organic molecules, a procedure which while rough and subject to some serious limitations, has been very useful to the organic chemists for many years. Before they became much interested, how-

* Perhaps the first example of the LCAO type of description was its use by Pauling for H_2^+, – which, however, can be considered as an example of the AO equally as well as of the MO method.

ever, the subject had been developed further by Coulson, Longuet-Higgins, Dewar, and others, and presented rather forcefully to them by Dewar.

I must not take too long for matters of historical interest, since I want to say something about current developments and about the future, so I will now give only a condensed account of some of the further developments of MO theory.

I have already mentioned the use of the LCAO method for the rather roughly approximate visualization of MO's of diatomic molecules and of metals, and also (although until recently for π electrons only) for many of the important molecules of organic chemistry. In the years 1932–1935 I turned my attention to the exciting possibilities of understanding the electronic structures and spectra of small polyatomic molecules, many of them as prototypes of larger organic and inorganic molecules. In so doing, I used molecular symmetry properties and the LCAO method. At that time J.H. van Vleck called my attention to the applicability of Bethe's group-theoretical determination of the irreducible representations for the orbitals of an atom in a crystal to the classification of MO's. For the different species of MO's I then adopted a system of symbols nearly like that in a paper on Raman spectra by Placzek of which I secured a proof copy[8]. A particularly appealing type of molecules for understanding by the MO method was that of the complex ions of high symmetry, but after a brief mention, I postponed going into these, saving them as a choice tidbit for some future occasion which, however, got indefinitely postponed. In recent years others have done full justice to this subject of ligand-field MO theory.

Among later aspects of my own work was an interest in the absolute intensities of molecular spectra, in particular intramolecular charge-transfer spectra. Growing out of that were some ideas on conjugation and hyperconjugation in organic-chemical molecules. Later on, in trying to explain some new spectra of iodine in solutions in benzene or other related compounds, I became interested in the interaction of molecules with one another involving the partial transfer of an electron from a donor to an acceptor molecule to form a molecular complex. In this connection I got into a study of intermolecular charge-transfer spectra, of which the benzene–iodine spectrum of Benesi and Hildebrand is the classic example.

I mention all these things together because they are all concerned with what happens to our understanding of molecules when less and less localized approximate MO's are used. I have said earlier that the best or truest MO's are those which are *fully* delocalized. However, it is very instructive to start with localized MO's, and see what we can learn at each step as the theoretical description is made more accurate in successive steps of delocalization. Consider for example the 1,3-butadiene molecule. The simplest set of chemical orbitals would here consist of a K shell (1s) MO on each of the four carbon atoms, six σ-bond MO's for carbon–hydrogen bonds, three carbon–carbon σ-bond MO's and two carbon–carbon π-bond MO's corresponding to the chemical formula

```
                 H       H
                  \ 3   4/
         H         C=C
          \ 1   2/     \
           C=C          H
          /     \
         H       H
```

The π-electron part of the electron configuration would then consist in two pairs of electrons occupying two bond MO's $\varnothing_{12}$ and $\varnothing_{34}$ describable in LCAO approximation as follows:

$$\varnothing_{12}=a(\chi_1+\chi_2) \qquad \varnothing_{34}=a(\chi_3+\chi_4)$$

where the four χ's are π AO's on atoms 1, 2, 3, and 4 of the chemical formula. The pair in the MO $\varnothing_{12}$ forms a π bond between atoms 1 and 2, that in $\varnothing_{34}$ forms an exactly similar π bond between atoms 3 and 4.

However, a distinctly improved approximate wave function[9] is obtained if the two localized π-bond MO's $\varnothing_{12}$ and $\varnothing_{34}$ are replaced by two new fully delocalized MO's (spectroscopic π MO's) as follows, each extending over all four atoms:

$$\varnothing_{\mathrm{I}} = b\chi_1+c\chi_2+c\chi_3+b\chi_4$$
$$\varnothing_{\mathrm{II}} = b'\chi_1+c'\chi_2-c'\chi_3-b'\chi_4$$

with c somewhat larger than b and c' somewhat smaller than b' (b, c, b', c' all positive). Both $\varnothing_{\mathrm{I}}$ and $\varnothing_{\mathrm{II}}$ give π bonding between atoms 1 and 2, and between 3 and 4, so that the total π bonding in the original bonds is not much changed, but now the electrons in $\varnothing_{\mathrm{I}}$ give π bonding between atoms 2 and 3, while those in $\varnothing_{\mathrm{II}}$ give antibonding between 2 and 3, but because $c>b$ and $b'<c'$, the net bonding effect of the pair of electrons in $\varnothing_{\mathrm{I}}$ outweighs the antibonding effect of those in $\varnothing_{\mathrm{II}}$, giving some net π bonding between atoms 2 and 3. Without delocalization, the π electron pairs in $\varnothing_{12}$ and $\varnothing_{34}$ would have created a net antibonding effect[9] (repulsion) between atoms 2 and 3.

Delocalization here results in a small decrease in the calculated energy but, more important, it accounts for the stability of the arrangement of the atoms in one plane, and predicts certain differences in chemical properties as compared with those which would be expected if there were no delocalization. When two double bonds are separated by one single bond, as in butadiene, the double bonds are called conjugated. Conjugated π-electron molecules are characterized by special properties which are understandable by MO theory in terms of π-electron delocalization, as just described. Of course the *actual* molecule shows those properties which correspond to conjugation or π delocalization; the localized description is an approximation which much less accurately describes the character of the actual molecule.

Finally, in a completely delocalized description the various localized σ MO's are replaced by spectroscopic MO's, which are fully delocalized σ MO's of various symmetry types extending over the whole molecule. This final stage of delocalization can be categorized as a variety of hyperconjugation, although not one of the most typical or important kinds.

In the 30's I tried to deduce all I could about MO's from qualitative considerations of energy and symmetry taken together with empirical evidence from molecular spectra and other properties. During this period and up to the time of the war in the early 40's, molecular spectroscopy was a major activity in our laboratory, under the able guidance in particular of Dr. Hans Beutler and then of Dr. Stanislaus Mrozowski. Toward the end of this period the enthusiastic assistance of Mrs. C. A. Rieke made possible many desk-machine calculations on hyperconjugation and on π-electron systems using the Hückel LCAO method. The contents of several notebooks from this work were never published because of the break caused by war-time activities.

Another subject of our interest beginning at this time was the importance of overlap integrals, which in the Hückel method until then everyone was neglecting because they made the calculations more complicated. After the war I took up this matter again, and wrote about various relations of overlap integrals to chemical bonding. I began also to be very dissatisfied with other inadequacies of Hückel-method calculations[6].

C. C. J. Roothaan had come to me as a graduate student in physics in 1947, already so well prepared in his studies with R. L. Kronig at Delft that I could only make some suggestions to him about problems on which calculations would be interesting. One study that he made by the Hückel method dealt with the structure of the ethylene molecule and its excited states and their behavior on twisting the molecule. The theoretical calculation confirmed the qualitative conclusion that twisted ethylene is strongly stabilized by hyperconjugation.

I tried to induce Roothaan to do his Ph.D. thesis on Hückel-type calculations on substituted benzenes. But after carrying out some very good calculations on these he revolted against the Hückel method, threw his excellent calculations out the window, and for his thesis developed entirely independently his now wellknown all-electron LCAO SCF self-consistent-field method for the calculation of atomic and molecular wave functions, now appropriately referred to, I believe, as the Hartree–Fock–Roothaan method. After a short period at Catholic University, Roothaan returned to our laboratory, where he expressed an unquenchable ambition to conquer the calculation of some of what then seemed almost incredibly difficult electron-repulsion integrals, and which had been one of the main obstacles to converting the molecular-orbital theory from a descriptive and semi-empirical to a more nearly quantitative theory. Another very important contributor to this endeavor at that time was Klaus Ruedenberg, who since then has added very much to our insights into the nature of chemical bonding.

I shall return shortly to the theme of the purely theoretical calculation of molecular structures and properties, but first wish to mention another development which has been very fruitful. Robert Parr spent the summer of 1949 at Chicago, and together we worked out some interesting applications of the semi-empirical π-electron-only LCAO SCF method (pioneered by Goeppert-Mayer and Sklar in 1939) for π-electron organic molecules. Somewhat later Parr and his student Pariser developed the Pariser–Parr method to deal with π-electron molecules in a way which (along with the rather similar Pople method) represented a great improvement on the Hückel method, and which has proved extremely fruitful for an improved understanding of molecules of importance in organic chemistry and biology.

In the late 40's it was not yet clear that really accurate theoretical calculations on molecules would be feasible, and we were happy to make progress by semi-empirical methods[10]. We did not realize that the big modern digital computers would become available and be rapidly improved in size and flexibility, and would transform theoretical computations into a tool which has already begun to compete with or in some cases even to go beyond experimental work in the laboratory. The rest of my speech will be devoted to some of the progress which has already been made in that direction.

What I shall now present to you will not be my own work, but that of those who have been my associates in our group at Chicago. Let me also say that it is only because of lack of time that I shall say very little about the many others at other institutions in various countries who have also made major contributions; I hope they will forgive me for this omission.

Here let me quote briefly, with minor changes, from a 1958 paper[11] which is already out of date because of the rapid development of bigger and better computers.

«Dirac once stated that, in principle, the whole of chemistry is implicit in the laws of quantum mechanics, but that in practise, prohibitive mathematical difficulties stood in the way...»

«In the early days of quantum mechanics, many of the world's best theoretical physicists engaged in calculations on molecules using the then new tool of quantum mechanics, in the hope of understanding and explaining molecular properties. But except in the simplest cases, those of the helium atom and the hydrogen molecule, the computations proved to be complicated and laborius without yielding more than roughly approximate results. Frustrated and repelled, many of the theorists turned to other problems.»

«Perhaps the most forbidding difficulty was that of the evaluation and numerical computation of certain integrals representing the energies of repulsion between electrons in different orbitals. After the early years of quantum mechanics, the work of a number of Japanese, English, American, and other investigators was directed toward breaking the bottleneck of the difficult integrals but it was only in the 50's that really substantial progress was made. Among the most active workers were Kotani and his group in Japan, Boys in Cambridge, Coulson and his group at Oxford, Löwdin and his group in Uppsala, Slater's group at M.I.T. and our own group at Chicago.»

«A major and indeed crucial step beyond the development of formulas for molecular integrals was the programming for large electronic digital computers of the otherwise still excessively time-consuming numerical computation of these integrals, and of their combination to obtain the desired molecular wave functions and related molecular properties. The pioneering work in this field was that of C. F. Boys at Cambridge, England.»

Now let me turn to the work at Chicago in the area of large-scale machine computations, for which my colleague Roothaan has been primarily responsible together with his students and co-workers. This work has gone through successive stages of development with increasingly powerful machines. The calculations to which I shall refer are so-called non-empirical, in other words, purely theoretical, calculations in which *all* the electrons in the atom or molecule are included, as contrasted with the still extremely valuable semi-empirical methods already mentioned which took specific account only of the π-valence electrons. A major improvement in depth of understanding is added when all the electrons, inner shell as well as outer shell, σ as well as π, are included in the calculation.

The first all-electron calculation at Chicago was done by C. W. Scherr for his Ph. D. thesis published in 1955; it was a Roothaan-type LCAO SCF calculation on the nitrogen molecule using, however, only what might be called a skeleton crew of AO's in his LCAO expressions, a so-called *minimal basis set*. This calculation done by Scherr on desk computers with the help of two assistants took him two years. The same computation could now be repeated in about two minutes with the largest computers now available, - provided of course, that the preliminary work of writing the machine program had been done.

Writing a good machine program for molecular electronic structure calculations is, however, a very difficult and time-consuming operation. Two generations of machine programs have been developed at Chicago under Roothaan's direction, and a third is now being prepared.

Extensive computations have been made using the first two of these programs for diatomic molecules, especially by Dr. B. J. Ransil and associates with the first program, and by Dr. Paul E. Cade, Dr. Winifred M. Huo, and Dr. A. C. Wahl and associates with the second program, - for whose preparation Wahl, Huo and others were largely responsible. There were also machine programs and very extensive computations for *atoms*.

In the second machine program, provision was made for building up the MO's from a *large number* of Slater-type orbitals, or better stated (as proposed by Roothaan and others), Slater-type *functions* (STF's). This LC·STF approach represents the use of the Roothaan method in its general form to build up MO's.

LCAO calculations until recently have for the most part been minimal-basis-set calculations, in which the number of AO's used in constructing MO's has been equal to the number of occupied AO's of the atoms from which the molecule can be formed, or at most includes one more valence-shell AO per atom. For example, the minimal basis set for LCAO MO's of Li_2 of 1s and 2s AO's, just as in the Li atoms, *plus* a 2pσ AO, which is not used in the free atom in its normal state. Inclusion of the 2pσ AO permits 2s–2pσ hybridization, which is important if reasonably good LCAO MO's are to be obtained. For N_2, the minimal basis set consists of 1s, 2s, 2pσ, and 2pπ AO's for each atom, all of which are occupied in the normal state of the free atom. In the earlier calculations, each AO in an LCAO MO expression was approximated by a single Slater AO (which is an STF of a size governed by certain very useful simple empirical rules which Slater set down in 1930)[12]. Much more accurate MO's can be obtained if SCF AO's are used instead of Slater AO's in the LCAO expressions; these might be called LC·SCFAO MO's. Such LCAO expressions are, however, not yet adequate to describe really accurate true SCF MO's. But if in the usual LCAO expressions, suitably *modified* SCF AO's, which can be called MAO's are used, it is possible to reproduce[13] the true SCF MO's. The required modifications consist of scaling,–shrinking or expanding the size,–and polarization or hybridization. One can then think of the true SCF MO's as being described by simple LCMAO, instead of by simple LCAO expressions. For computational purposes, however, extended linear combinations of a rather large number of STF's, or of GF's (Gaussian functions), are used. Nevertheless, for *conceptual* purposes, *simple LCMAO expressions* are especially illuminating.

In the actual computations, each MAO is, in effect, expanded into a linear combination of, in general, a number of STF's or GF's. Roothaan's method is thus really a LC·STF method using extensive linear combinations of STF's. In this way it became possible to obtain almost perfectly the forms of the true or spectroscopic MO's of which I have spoken earlier. And from the corresponding SCF–MO wave functions, the values of several molecular properties, some of them hitherto not generally known from experimental work, have been computed with a considerable degree of accuracy. I have already shown pictures (Figs. 1–5) of the valence-shell MO's of the oxygen molecule, as determined by the calculations of Cade, G.L. Malli, and Wahl, and wish now to show three figures (Figs. 10–12) to illustrate some of the results of the computation of molecular properties from SCF–MO wave functions. I am indebted especially to Dr. Cade for permission to reproduce these figures.

Fig. 10 shows dipole moments for all the first-row and second-row diatomic hydride molecules, as computed by Cade and Huo from their accurate SCF–MO wave functions[14]. One sees that the agreement of computed with experimental values in the five cases where the latter are known (LiH, HF, HCl, OH and CH) is very good, giving considerable confidence that the computed values in the other cases are also rather accurate. Most of these other cases are *radicals*, for which measurement is difficult. Here we have a good example of a situation that will become increasingly frequent, namely that molecular properties, especially for radicals, may be more easily obtainable from theoretical calculations than from experiments.

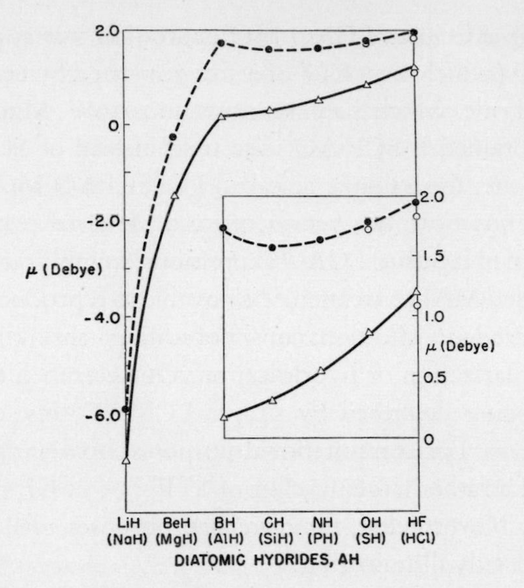

Fig. 10. The dipole moment for the ground state of the first- and second-row hydrides; first-row calculated values (●), second-row calculated values (Δ), and experimental values (○). Right-hand scale for small inset figure and left-hand scale for large figure. Reproduced by permission of the authors[14] and of the American Institute of Physics.

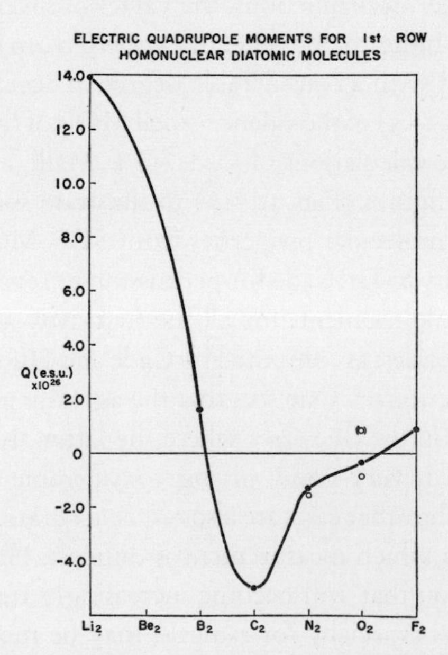

Fig. 11. Reproduced by permission of Dr. Paul E. Cade, from unpublished work.

Fig. 11 shows quadrupole moments of first-row homopolar diatomic molecules and radicals, as computed (Cade, unpublished) from accurate SCF–MO wave functions obtained by several investigators (Cade, Wahl, Malli, K.D. Sales and J.B. Greenshields) at Chicago by the use of the second machine program. Here the sign of the quadrupole moment is sometimes positive, sometimes negative, and until recently was not known experimentally in any case. In Fig. 11, the black circles are the computed values and the white circles are recent experimental values: for N_2 by A.D. Buckingham, for O_2 by microwave work, with sign uncertain; however, recently Buckingham has obtained an experimental value for O_2 which on the scale of the curve is coincident with the computed value.

Fig. 12 shows computed electric field gradients (nuclear quadrupole coupling constants) at the A and H (or D) nuclei in the first-row hydride molecules and radicals (from a forthcoming paper by Cade and Huo). Here experimental data are available for q_D in LiD and HD, but only indirectly from measured eqQ values and the nuclear quadrupole moment of the Deuteron, from accurate measurements and calculations on HD or D_2.

Having accurate SCF-wave functions with spectroscopic MO's, one can ask, how well do these answer the questions with which chemists are con-

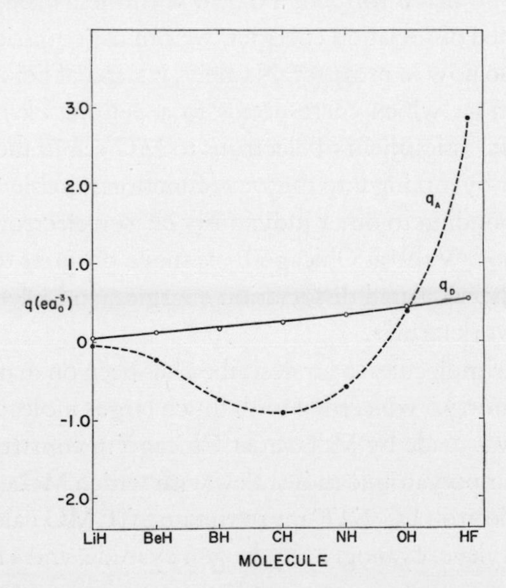

Fig. 12. Electric field gradients, q, at A and H (or D) nuclei for first-row hydrides. Reproduced from a forthcoming paper by P.E. Cade and W. Huo by permission of the authors;

cerned? I have already referred to dipole moments and quadrupole moments, where agreement with true values is within 5 or 10% in the examples where comparison was possible*. A similar degree of agreement is found for ionization potentials. Moreover, *all* the ionization potentials which correspond to removal of a single electron from any outer or inner shell can be computed. In all these cases we are dealing with properties that depend on one electron at a time. For such cases there is a theorem which states that values computed from accurate SCF wave functions should be correct to first order.

But how about binding energies (dissociation energies) of molecules? These are of very special interest to chemists. Here we can subtract the SCF energy (the calculated energy of the SCF molecular wave function) from the sum of the SCF energies of the component atoms, and one might think that the difference should be the dissociation energy. However, the agreements are generally poor; the calculated dissociation energies are often only half as large as they should be, or occasionally even come out less than zero.

There is a good reason for these disagreements, namely the fact that the electron-correlation energy of which I have spoken earlier is generally larger in a molecule than in the corresponding atoms. In fact, as Clementi has pointed out, there is a more or less standard *extra correlation energy* in a molecule for each chemical bond that is formed. To deal with these needed corrections to the SCF-computed dissociation energies, we can use empirical estimates, but a better way is also now in prospect. Namely, instead of being satisfied with a SCF-wave function, which corresponds to a definite *electron configuration*. that is, one definite assignment of electrons to MO's as in the Aufbauprinzip, we can go further by mixing into the wave function suitable amounts of wave functions corresponding to other judiciously chosen electron configurations. In this way Das and Wahl[4] at Chicago have made progress toward obtaining good theoretically calculated dissociation energies, and Clementi and others are pushing this work farther.

All the work on molecules so far described has been on diatomic molecules. But most of chemistry is concerned with much larger molecules. Some interesting progress was made by McLean at Chicago in constructing a machine program for linear polyatomic molecules, with which McLean and Clementi made some all-electron LC·STF approximate SCF MO calculations on carbon dioxide, acetylene, cyanogen, hydrogen cyanide, and a number of other molecules.

* Of course an agreement in terms of *percentage* is no longer relevant in cases where the value of the quantity is near zero.

Recently several groups have been using linear combinations of a different type of basis functions, namely Gaussian functions, instead of STF's, to build up approximate SCF–MO's. For comparable accuracy, at least twice as many Gaussians as STF's must be used. This procedure was first proposed by S.F. Boys of Cambridge, England. Although Gaussians are intrinsically much poorer building blocks than STF's for constructing true MO's, calculations with them are easier, and it has proved possible to use them successfully to get rather good approximations to true SCF–MO wave functions. Among those who have been using Gaussians recently are Moskowitz and Harrison using a machine program which was constructed by Harrison while working with Slater at M.I.T.; Allen and associates at Princeton; and recently Clementi, who spent most of the year 1966 at Chicago on leave from IBM's San José Research Laboratory.

During 1966 Clementi, with some cooperation of others, and with the use of copious amounts of machine time mostly at IBM's Yorktown laboratory, has carried through all-electron SCF–MO calculations of considerable accuracy on a notable array of molecules: ammonia, ethane, pyrrole, benzene, pyridine, and pyrazine. Further, he has examined in detail what happens to MO's, to energies, and to charge distributions when a hydrogen chloride molecule approaches an ammonia molecule.

This last is of particular interest, but let me first mention the topic of population analysis[15]. That technique makes it possible in a fairly meaningful way to calculate how the total population of electrons is distributed among the atoms in a molecule. Among other things the procedure gives for each atom a number which can be identified as the electrical charge on that atom. It also yields so-called *overlap* populations, which are found to be well correlated with the strengths of chemical bonds. I am sorry there is no time to explain the method here. In a way it seems to contradict my basic theme that a molecule can better be thought of as an individual rather than a collection of atoms. However, the molecule does contain atomic *nuclei*, and the so-called charge on each atom in a molecule can be considered as an old-fashioned and familiar terminology for describing how electrical charge is distributed in the neighborhood of each nucleus.

The SCF–MO wave functions obtained at Chicago and elsewhere have yielded many interesting results when subjected to population analysis, but I will refer here only to one particularly interesting example, based on Clementi's calculation of what happens when an HCl approaches an NH_3 molecule. This case can be considered as an example of the formation of a molecular complex of a type which is particularly interesting, and is very important in biological systems, namely a hydrogen-bonded complex. It has long been a moot question as to how the distribution of electrical charges changes during the approach of the two partners in a hydrogen-bonded complex. Clementi's wave functions, when subjected to a population analysis, give an answer to this question, and his calculations also show how the energy changes during the approach.

What Clementi calculated were spectroscopic or true MO's for the *combined system* $NH_3 + HCl$. This procedure, the whole-complex MO method, which could also be used with equal validity in understanding the electronic structure of any electron-donor electron-acceptor molecular complex*, represents another example of the improved understanding and accuracy which can result in going over from localized to delocalized MO's, – in this case from MO's of the two molecules to MO's of the complex as a whole.

To justify the use of whole-complex MO's here, we note that each of the separate molecules NH_3 and HCl has, in terms of MO's, a closed-shell electron configuration. Now when two *atoms* in closed-shell AO configurations approach, for example, two helium atoms, the SCF–MO approximation remains a good approximation at all distances of approach. In other words, there is then no large increase in electron correlation energy such as occurs when two atoms with unpaired valence electrons, for example two H atoms, or two N atoms, approach to form a molecule; it will be recalled that this increase in correlation energy on typical molecule formation had as a result that SCF–MO energies compared with SCF–AO energies do not give good values for dissociation energies. By analogy with the case of two closed-shell atoms, it appears, however, that the SCF–MO approximation should be valid without any strongly varying electron correlation corrections when molecules with MO closed shells come together; the fact that two He atoms do not form a stable molecule does not matter for the present argument. Thus Clementi's SCF–MO calculations on $NH_3 + HCl$ should throw important light on the changes which occur in hydrogen bonding. Actually, Clementi's calculations

* Up to now I have used mainly a different procedure[16] with a wave function corresponding partly or largely to an electron configuration of MO's of the free donor and acceptor, but with some mixing in of a second configuration in which an electron has been transferred from the donor to the acceptor. For loose complexes the two procedures are roughly equivalent, but the whole-complex method is becoming advantageous now that all-electron SCF computations are becoming feasible for relatively large molecular systems.

show a gradual transfer of charge from the NH_3 to the Cl atom, accompanied by some stretching of the H–Cl distance, until at equilibrium a structure approaching that of an $NH_4^+Cl^-$ ion-pair, but with considerable polarization of the Cl^- (H-bonding of NH_4^+ to Cl^-) is attained. The NH_3 + HCl system is thus apparently an example of ion-pair formation rather than ordinary loose hydrogen bonding; however, the changes in charge distribution during the early stages of approach of the HCl and NH_3 should probably be similar to those in ordinary H-bonding, and thus instructive for the latter.

In conclusion, I would like to emphasize strongly my belief that the era of computing chemists, when hundreds if not thousands of chemists will go to the computing machine instead of the laboratory for increasingly many facets of chemical information[11], is already at hand. There is only one obstacle, namely that someone must pay for the computing time. However, it seems clear that the provision of adequate funds by government and other organizations for computing molecular structures has at least as high an order of justification as the provision of adequate funds for the cyclotrons, betatrons, and linear accelerators used in studying nuclear structure and high-energy particles, or for rockets to explore the moon, planets, and interplanetary space. Chemistry, together with the physics of solid matter on the earth, deal with the foundations of the material world on which all our life is built. Yet at the present time the rapid progress which could be made even with existing machine programs is *not* being made, simply because available funds to pay for machine time are far too limited. Computing time is rather expensive, yet the amounts of time needed to make adequate use of existing and future machine programs would be trivially small compared with the amounts now being spent on nuclear and high-energy problems and on outer space.

1. The reader is referred to the reprinted 1913 papers with a valuable historical introduction and discussion by L.Rosenfeld, published by Munksgaard, Copenhagen and Benjamin, New York, in 1963.
2. A.C.Wahl, *Science*, 151 (1966) 961; W.M.Huo, *J.Chem.Phys.*, 43 (1965) 624.
3. Ø.Burrau, *Kgl.Danske Videnskab.Selskab*, 7 (1927) 1.
4. For some simple diatomic examples, see G.Das and A.C.Wahl, *J.Chem.Phys.*, 44 (1966) 87.
5. Details and references are given in an article entitled «Molecular Scientists and Molecular Science: Some Reminiscences», *J.Chem.Phys.*, 43 (1965) S2–11.

6. See the introduction to «Report on Molecular Orbital Theory», *J.Chim.Phys.*, 46 (1949) 497, and references given there.
7. F.Hund, *Z.Physik.*, 73 (1931) 1.
8. On the classification of molecular states, see *J.Chem.Phys.*, 23 (1955) 1997–2011.
9. R.S.Mulliken, *J.Phys.Chem.*, 66 (1962) 2306.
10. R.S.Mulliken, *Chem.Rev.*, 41 (1947) 201.
11. R.S.Mulliken and C.C.J.Roothaan, «Broken Bottlenecks and the Future of Molecular Quantum Mechanics», *Proc.Natl.Acad.Sci.* (*U.S.*), 45 (1959) 394–398.
12. For a review of Slater's work, see R.S.Mulliken, *Quantum Theory of Atoms, Molecules, and Solid State*, Academic Press, New York, 1966, pp. 5–13.
13. See R.S.Mulliken, *J.Chem.Phys.*, 36 (1962) 3428, Sec.II, 1–3, and 43 (1966) S39.
14. P.E.Cade and W.M.Huo, *J.Chem.Phys.*, 45 (1966) 1063; detailed papers to be submitted for publication shortly.
15. R.S.Mulliken, *J.Chem.Phys.*, 23 (1955) 1833–1840, 1841–1846, 2338–2342, 2343–2346.
16. R.S.Mulliken, *J.Am.Chem.Soc.*, 74 (1952) 811; *J.Chem.Phys.*, 60 (1964) 20.

Biography

Robert Sanderson Mulliken was born in Newburyport, Massachusetts, on June 7, 1896, the son of Samuel Parsons Mulliken, Professor of Organic Chemistry, and Katherine W. Mulliken. He married Mary Helen von Noè, December 24, 1929. Their children are Lucia Maria (Mrs. John P. Heard) and Valerie Noè.

Mulliken took a B. Sc. Degree in 1917 at the Massachusetts Institute of Technology, Cambridge, Mass., and a Ph. D. Degree at the University of Chicago, Ill., in 1921.

Mulliken has been deeply interested in valence theory and molecular structure. His earlier work on isotopes and on diatomic band spectra was followed by theoretical systematization of the electronic states of molecules, mainly in terms of the idea of molecular orbitals. This included work on electronegativities, dipole moments, and valence-state energies. A subsequent series of papers on the theoretical interpretation of absolute intensities of electronic spectra led him to computations on conjugated organic molecules, and to the quantum-mechanical statement of the concept of hyperconjugation. There followed work on quantum-mechanical questions underlying molecular-orbital theory and on the use of interatomic overlap integrals as measures of bond energies. His more recent work has dealt extensively with the structure and spectra of molecular complexes, on the one hand, and on the other hand (extending and developing earlier work) with the structure and spectra of hydrogen, helium, nitrogen and other small molecules.

Mulliken was National Research Council Fellow, University of Chicago, and Harvard University, 1921–1925; Guggenheim Fellow, Germany and Europe, 1930 and 1932–1933; Fulbright Scholar, Oxford University, 1952–1954; Visiting Fellow, St. John's College, Oxford, 1952–1953; Junior Chemical Engineer, Bureau of Mines, U. S. Department of Interior, Washington, D. C. 1917–1918; Assistant in Rubber Research, New Jersey Zinc Company, Pennsylvania, 1919.

His academic career includes the following positions: Assistant Professor of Physics, Washington Square College, New York University, 1926–1928;

Associate Professor of Physics, University of Chicago, 1928–1931; Professor of Physics, University of Chicago, 1931–1961, and Chemistry, 1961; Ernest de Witt Burton Distinguished Service Professor, University of Chicago, 1956–1961; Distinguished Service Professor of Physics and Chemistry, University of Chicago, since 1961; Distinguished Research Professor of Chemical Physics, Florida State University (Jan.–March), since 1964. Other professional positions held: Director, Editorial Work and Information, Plutonium Project, University of Chicago, 1942–1945; Scientific Attaché, U.S. Ambassy, London, 1955; Baker Lecturer, Cornell University, 1060; Silliman Lecturer, Yale University, Spring, 1965.

Mulliken received honorary degrees at Columbia University, 1939 (Sc.D.); the University of Stockholm, 1960 (Ph.D.); Marquette University, 1967 (Sc.D.); Cambridge University, 1967 (Sc.D.); and he holds several professional awards and honours of which a few are listed here: Bronze Medal Award, University of Liége, 1948; Peter Debye Award, California Section of the American Chemical Society, 1963; Willard Gibbs Medal, Chicago Section of the American Chemical Society, 1965; Gold Medal Award for Scientific Achievement, City College Chemistry Alumni Association, and 15th Bicentennial Lecturer, City College of New York, 1965.

He is a Member of the American Academy of Arts and Sciences, American Chemical Society, American Philosophical Society, National Academy of Sciences, Cosmos Club (Washington, D.C.), Quadrangle Club (Chicago, Ill.); a Fellow of the American Physical Society and the American Academy for the Advancement of Science; an Honorary Fellow of the Chemical Society of Great Britain (London) and the Indian National Academy of Science; a Foreign Member of the Royal Society of Great Britain; an Honorary Member of the Société de Chimie Physique; and a Corresponding Member of the Société Royale des Sciences de Liége.

His recreational interests include: driving a car, oriental rugs, and art.

Chemistry 1967

MANFRED EIGEN

R.G.W. NORRISH

GEORGE PORTER

«for their studies of extremely fast chemical reactions, effected by disturbing the equilibrium by means of very short pulses of energy»

Chemistry 1967

Presentation Speech by Professor H. A. Ölander, member of the Nobel Committee for Chemistry of the Royal Swedish Academy of Sciences

Your Majesty, Your Royal Highnesses, Ladies and Gentlemen.

The chemists of older times were chiefly interested in how to produce substances from natural products which might prove useful; for example, metals from ores and the like. As a matter of course, they were bound to notice that some chemical reactions took place rapidly, while others proceeded much more slowly. However, systematic studies of reaction velocities were hardly undertaken before the mid-19th century. Somewhat later, in 1884, the Dutch chemist, Van 't Hoff, summarized the mathematic laws which chemical reactions often follow. This work, together with other achievements, earned for Van 't Hoff the first Nobel prize for Chemistry in 1901.

Almost all chemical reactions will proceed more rapidly if the mixture is heated. Both Van 't Hoff and Svante Arrhenius, who for other discoveries was awarded the third Nobel prize for Chemistry in 1903, set up a mathematical formula which describes how the velocity of a reaction increases with temperature. This formula could be interpreted by the assumption that when two molecules collide, they usually part again and nothing happens; but if the collision is sufficiently violent, the molecules disintegrate and their atoms recombine into new molecules. One could also envisage the possibility that the molecules moved towards each other at moderate velocity, but that the atoms in one molecule oscillated violently so that no severe impact would be required for that molecule to disintegrate. It was already then realized that higher temperature implied two things: the molecules moved faster, and the atoms oscillated more violently. It was also realized that when a reaction velocity could be measured, only the merest fraction of the collisions involved really resulted in a reaction.

How fast were the reactions that could be measured in the old days? Considering that the substances first had to be mixed, after which samples had to be removed at specified times and then analyzed, the speed of the procedure was necessarily limited. The best case was if one could observe the change in some physical property such as colour; then it was not necessary to remove samples. The chemists had to read off his clock and measuring instrument, and then to make entries in his laboratory journal. If he was quick, he could keep up with a reaction which had run half its course in a few seconds.

How slow were the reactions one could measure? Eigen has said that this is determined by how long a time a young man wants to devote to his doctoral dissertation. If as a practical maximum we say that half the reaction is completed after three years, that comes to around 100 million seconds. Naturally, there are even slower reactions.

Many reactions were of course known to proceed at velocities so great as to defy measurements. For example, no one had succeeded in measuring the velocity of the reaction between an acid and an alkali. In such cases it was understood that the molecules reacted without the collision being very violent. In the study of reactions where a large number of molecules take part, it turned out that the velocity often depended on the quantities of substances used in such a manner that a step-by-step sequence had to be assumed for the reaction: one of these steps was slow and hence determined the overall course of the reaction, while the other steps were immeasurably fast. The German chemist, Max Bodenstein, studied many such reactions at the beginning of this century.

A major advance was achieved in 1923 by the Englishmen, Hartridge and Roughton, who let two solutions arriving through separate tubes meet and be mixed, and then caused the mixture to flow swiftly through an outlet tube, in which the reaction could be observed as it proceeded. This method permitted measurement of reaction times down to thousandths of a second. But there are still many reactions that proceed still more rapidly. They could not be studied by this method for the simple reason that the substances cannot be mixed fast enough.

When nitric acid gets to react with a number of substances, a brown gas, nitrogen dioxide, is formed. This gas has certain properties which were interpreted by assuming that the brown molecules could form pairs, thus doubling their size. This was a typical example of a high-velocity reaction that no one has succeeded in measuring.

In 1910 a student studying for the doctorate with Walter Nernst investigated the velocity of sound in several gases, among them nitrogen dioxide. He found that the equilibrium between the single and double molecules was accomplished much more rapidly than the sound oscillations. But he perceived that the speed of sound ought to be modified if one used sufficiently high-pitched tones – far beyond the capacity of the human ear to hear. No less a person than Albert Einstein carried out a theoretical study of this phenomenon in 1920. However, many years were to elapse before instruments could be devised to measure it. A complication was found to be involved here in that the sound is absorbed by the gas. None the less, the principle is important; the essential point here is that one is not going to mix two things, but rather to start off from a chemical system in equilibrium and to disturb this equilibrium, in this case by exposing the gas to the condensations and attenuations which constitute sound.

The fact that light produces chemical reactions has been known since time immemorial. Thus it bleaches colours and alters silver salts, which action is the very basis of photography. The ability of light to produce a chemical reaction depends on its absorption by a molecule, which then becomes so excited that it can react. Investigations of the energy states thus acquired by molecules were begun some fifty years ago. One of the findings was that the atoms of a molecule oscillated at rates of the order of billionths of a second. Chemical reactions inevitably take longer, for time must be allowed for the atoms to dissociate and re-combine into new molecules. For these purposes the times required come to, say, one tenthousandth part of a millionth of a second. In other words, such are the times for the fastest chemical reactions. They amount to one-tenth of one-millionth of the times Hartridge and Roughton were able to measure with their method. To convey an idea of what one ten-thousandth part of a millionth of a second means, it can be said to form the same part of one second as one second is of three hundred years.

The 1967 Nobel laureates in Chemistry have opened up the whole of this vast field of reaction kinetics for research. They did so by applying the principle I have just mentioned: to start from a system in equilibrium and to disturb this equilibrium suddenly by one means or another.

If a molecule has absorbed light so that it can react, it usually does this so fast that too few of these activated molecules are present at any one time to reveal their existence by any known method of analysis.

Ever since the 1920's, Professor Norrish has been studying reaction kinetics and he was one of the leading scientists in this field. A younger associate joined him in the late 1940's in the person of George Porter. They decided to make use of a flash lamp, the kind you have seen photographers use. The only difference was that they made their lamp thousands of times more powerful. Indeed, subsequent refinements have led to the construction of such lamps with an effect greater than the total effect which the whole city of Stockholm consumes on a winter afternoon with the lights turned on and the factories still humming before closing time – and that is 600000 kilowatts. There is just one catch, however; this enormous effect in the lamp lasts no more than one-millionth of a second or so. Still, in this way much if not most of a substance in a tube next to the flash lamp can be converted into an activated form, or the molecules broken up so as to yield atom groups with a high reactivity. It then becomes possible to study these newly formed molecules spectroscopically, but since they react so readily, this must be made extremely fast. Thanks to modern electronic equipment, however, these rapid processes can be recorded.

The new method developed by Norrish and Porter enabled them to study at first hand many fast reactions which one had previously only guessed that they took place. I cannot begin to enumerate even a sample of the reactions which Norrish and Porter, not to mention a great many other scientists, have investigated with this method. Suffice it to say that, in an earlier day, the study of these short-lived high-energy molecules and their chemical characteristics could hardly even have been contemplated as a wild dream.

The flash photolysis method of Norrish and Porter inflicts a drastic change of behaviour on the molecules. By contrast, Eigen treats his molecules more leniently. In 1953 he and two associates published a study on the absorption of sound in a number of salt solutions. The theoretical part of their report demonstrated how this absorption could be used to estimate the velocity of fast reactions which take place in the solution. Thus a solution of magnesium sulphate contains ions of magnesium and sulphate, as well as undissociated salt molecules. Equilibrium sets in after about 1/100000 of a second. This causes that sound which oscillates 100000 times a second is absorbed by the solution.

Eigen has invented several methods, however. If, say, a solution of acetic acid is subjected to a high-tension electric pulse, more molecules of this substance are dissociated than else would be the case in an aqueous solution. That takes a certain length of time. When the electric pulse is turned off, the solution goes back to its former equilibrium; this also takes some time, and that relaxation can be recorded.

The shock current caused by the application of the high-tension pulse will heat the solution a few degrees. Most chemical equilibria are slightly displaced when the temperature is changed, and the rapid establishment of the new equilibrium after heating can be recorded.

Eigen has also specified other methods for starting fast reactions in a solution formerly in equilibrium.

Whereas the study of electrolytic dissociation equilibria was already commenced in the 1880's by Svante Arrhenius, it is now possible to measure the reaction velocities at which these equilibria are established. A large number of extremely fast reactions can now be studied, involving all kinds of molecules ranging from the very simplest ones to the most complex that the biochemists work with.

Although Eigen starts his reactions in another way than that employed by Norrish and Porter, the instruments that record the fast reactions are largely identical for both research groups.

The chief importance to chemists of the methods worked out by Eigen, Norrish and Porter is their usefulness for the most widely diverse problems. A great many laboratories round the world are now obtaining hitherto undreamt-of results with these methods, which thereby fill what used to be a severely-felt gap in the means of advance available to Chemistry.

Professor Dr.Manfred Eigen. Although chemists had long been talking of instantaneous reactions, they had no way of determining the actual reaction rates. There were many very important reactions of this type, such as the neutralization of acids with alkalis. Thanks to you, chemists now have a whole range of methods that can be used to follow these rapid processes, so that this large gap in our chemical knowledge has now been filled.

May I convey to you the warmest congratulations of the Royal Swedish Academy of Sciences.

Professor Ronald George Wreyford Norrish, Professor George Porter. Photo-reactions have been studied by chemists for more than two hundred years, but the detailed knowledge of the behaviour of the activated molecules was meagre and most unsatisfactory. By your flash photolysis method you have provided us with a powerful tool for the study of the various states of molecules and the transfer of energy between them.

May I convey to you the warmest congratulations of the Royal Swedish Academy of Sciences.

Professor Eigen. May I ask you to come forward to receive the Nobel Prize for Chemistry from His Majesty the King.

Professor Norrish, Professor Porter. May I request you to receive the Nobel Prize for Chemistry from the hands of His Majesty the King.

MANFRED EIGEN

Immeasurably fast reactions*

Nobel Lecture, December 11, 1967

1. «*Prejudice and Pride*»

«The rate of true neutralization reactions has proved to be immeasurably fast». I found this quotation in Eucken's *Lehrbuch der Chemischen Physik*[1] while I was preparing for my doctor's examination. Although as a student of Eucken, this book was for me the «bible of physical chemistry», I was then at the age when one accepts practically nothing unquestioned, and so I started to reflect on just how fast an «immeasurably fast» reaction might be.

Clearly, the two reagents – the H^+ and OH^- ions in the case of neutralization – must come together before they can combine to form the reaction product: the H_2O molecule. The maximum rate will be determined by the frequency of such encounters or collisions, and this in turn will depend only on the diffusion rate and the mean separation of the reaction partners. This problem had already been analysed by Langevin[2], von Smoluchowski[3], Einstein[4], Onsager[5], and Debye[6], all of whom had derived expressions for the frequency of ionic collisions. It was fortunate that I was unable at that time to find in Göttingen Debye's paper in which this problem is solved in general form, for this meant that I had to sit down and derive again for myself the expression for the rate constant of diffusion-controlled reactions. The result was a theory which also applies to the mechanism of reactions in which not every encounter is successful. In such reactions the «lifetime of the collision complex», *i.e.* the time after which the partners diffuse apart once more if the encounter has not resulted in reaction, plays an important part. This time worked out as

$$\tau = \frac{R^2}{3(D_+ + D_-)} \frac{1 - e^{-\varphi}}{\varphi} \qquad (1)$$

* The investigation of fast reactions particularly the more recent applications in biological chemistry, has been dealt with in more detail in a number of recently published reviews[55,56,66,72,75]. The present lecture is not intended to duplicate these papers, but represents an attempt to record the development of a scientific idea from a personal point of view.

where R is the separation at closest approach, D_+ and D_- are the diffusion coefficients of the reagents, and φ is the energy of electrostatic interaction divided by kT (ref. 15).

If in this expression we expand the exponential term for the electrostatic interaction of the reaction partners, which is generally relatively small, and use Einstein's correlation between the diffusion coefficient and the «mean square displacement», we find that the encounter time is simply the time required by the reaction partners to diffuse a distance corresponding to their separation at closest approach. Using the known values for D (10^{-4}–10^{-5} cm^2/sec), this time works out at

$$\tau = 10^{-10}–10^{-11} \text{ sec}$$

If the actual «chemical» change takes place more quickly than this, the reaction rate is determined entirely by the collision frequency, and if it takes place more slowly, the reaction rate is independent of the rate of diffusion of the reaction partners (and hence of the viscosity of the solvent). It is then simply given by the rate of the chemical change (determined by a frequency factor and an exponential term containing the activation energy) multiplied by the statistical probability of the formation of the reaction complex.

At that time a reaction time of the order of magnitude of 10^{-10} sec did in fact seem «immeasurable».

We generally describe as «fast» anything that takes place quickly compared to the rate of resolution of our sense perceptions. However, since our perceptions are in turn based on chemical processes–including the charge neutralization reactions that have just been mentioned–these processes must necessarily be «fast»– indeed «extremely fast».

But what does «immeasurable» mean? If we wish to measure the rate of a chemical reaction, we must do two things: (*1*) induce the reaction; and (*2*) follow the course of the reaction in time.

Chemists generally induce a reaction by mixing the reagents with each other. This must be done very quickly, and here the first difficulty appeared: in many cases the mixing takes longer than the entire reaction under investigation. Attempts were therefore made to reduce the mixing time, and Hartridge and Roughton[7] succeeded by using a flow method in which the partners flow together at high velocity and are mixed together within about one-thousandth of a second. At the same time, Hartridge and Roughton also solved the second problem, that of the rapid observation of the course of the reaction. In the observation tube the time sequence of the reaction appears as a sequence in space, since the mixture flows at a constant velocity. The distance from the mixing chamber (in which the reaction is induced) is a measure of the age of the reaction mixture. This method was undoubtedly a great step forward, and it was subsequently refined and developed into an extremely effective instrument, especially for the enzyme chemist, by Britton Chance and his school in particular, but the reaction times were restricted to the range of milliseconds. Neutralization reactions thus still proved to be «immeasurably fast». If, as has just been shown, we had to assume that the shortest reaction time that might have to be observed was of the order of 10^{-10}–10^{-11} sec, there was a gap here of some 7 or 8 orders of magnitude. On a logarithmic scale, this is the same span as between a millisecond and a day or between a second and the duration of the studies for a doctor's degree, or again the span between the duration of the shortest «measurable» reaction at that time and the time spent in the laboratory to carry out the experiment. This was the situation around 1950, and in this sense neutralization reactions would still have to be regarded as «immeasurably fast» today.

However, there are always two possible courses when one is faced by an obstacle. One can either attempt to overcome the obstacle, as was done by Hartridge and Roughton, or to get around it.

In this case, the second course achieved our objective more quickly: we tried to get around the obstacle. This can be done in the following way: We know that chemical reactions are never entirely completed, and that an equilibrium between the various reagents is always established. Such chemical equilibrium is not static – it does not imply that the individual changes comes to a halt. The situation is, rather, that an opposing reaction sets in and ensures that eventually the number of products formed in unit time becomes equal to the number breaking down in that time. The number of units of the various partners will then on average remain constant. In the case of neutralization, «equilibrium» this means that H_2O molecules are continually breaking down into H^+ and OH^- ions, but that these ions very rapidly combine again to form water molecules, so that on average there are very few H^+ and OH^- ions «in equilibrium» with very many H_2O molecules. In pure water the ratio of the number of H_2O molecules to be number of H^+ ions is almost 10^9. Unfortunately, it is not generally possible to see the rapid «to and fro» occurrence of these reactions in equilibrium, even if one attempts to «observe» the processes with «rapid resolution» methods. The individual reaction signals average out precisely*. It would be necessary to seek to «align» or «synchronize» these mutually compensating individual signals. This can be done by rapidly disturbing the equilibrium, thus producing an «excess reaction» in one particular direction for a brief period (this reaction naturally vanishes when equilibrium is re-established). The only further requirement is that we must look quickly enough to see the decay of the excess reaction. We are therefore once again faced with the problem of inducing the reaction and observing it, but it is no longer necessary for us to mix the reaction partners. We need only transmit the inducing signal into the reaction system and then receive the reaction signal from the system. Since the signal must act through a large number of molecules, it must always travel a certain distance, and since there is a limit to the velocity of the signal it is not possible to measure «infinitely» short times. Where then does the limit lie?

The inducing signal can be either mechanical or electrical. A mechanical signal, for instance a pressure wave such as is observed in an explosion, propagates at about the velocity of sound, *i.e.* at about 10^5 cm/sec in condensed phases. An electric signal, for example a travelling electric wave, is about one hundred thousand times as fast, and its rate of propagation (in condensed phases) is only slightly less than the velocity of light in vacuum. If such signals are made to travel distances of 1–10 mm (the diameter of a measurement cell), we can attain «resolution times» of 1 μsec in the case of mechanical signals and of less then 10^{-10} sec with electrical signals. It is naturally also necessary to use inertia-free, *i.e.* optical or electrical, signals to observe the reaction. We shall see that the main difficulty will lie in recording such signals. In the case of chemical equilibrium reactions, the signals emitted are so weak that they are only slightly above the «thermal noise». However, we can see that in principle we have already solved our problem, namely that of following chemical reactions down to within the time range of 10^{-10} sec. «Immeasurably» fast reactions should therefore be measurable after all.

However, I must now make two reservations: (*1*) When a problem has been solved in principle it is still far from having been solved specifically. (*2*) New discoveries are not generally made in this deductive manner, even if it appears so in retrospect.

First, chance had to come to our aid.

* We shall for the moment omit from consideration a fairly new class of methods in which the line width of resonance signals emitted by the reaction partners provides a certain amount of information about these individual processes.

2. *A Trip to the Seaside*

At the Third Physics Institute of the University of Göttingen, my colleagues Konrad Tamm and Günther Kurtze had investigated the absorption of sound by aqueous electrolyte solutions. Behind these investigations there was a technical problem, that of measuring distances in sea water by means of acoustic probes. When an acoustic signal is emitted in the sea, it does not travel far; it is absorbed. Now, it was found that in certain frequency ranges sea water absorbs sound even more strongly than distilled water.

What was the reason for this phenomenon?

Tamm and his colleagues – and also some groups of workers in the United States[9,10] – had found that it is the magnesium sulphate present in the sea water which is essentially responsible for the absorption. However, it was not clear to what interaction the energy loss of the sound waves was due. Was it the hydrodynamic properties of the ions, was it their interaction with the water, *i.e.* hydration of the ions, or was it some interaction between the ions themselves? Since I had considered such interactions in some detail in my dissertation, Tamm and Kurtze turned to me, and we started a joint programme of measurements. It was very quickly found that the absorption could not be caused solely by the interaction between the Mg^{2+} and SO_4^{2-} ions and the water, for neither magnesium chloride nor sodium sulphate dissolved on their own produced comparable effects. On the other hand, neither could a simple inter-ionic interaction be the explanation, either in terms of the Debye–Hückel ion clouds, for which we would expect a broad continuum of absorption at high frequencies[11], or in terms of ionic association as described by Nernst[12a] or Bjerrum[12b], which should give a single absorption maximum. Two separate maxima were in fact found in the ultrasonic region, one at about 10^5 Hz and the other at about 10^8 Hz, which had to be in some way related to each other because of the way in which they were found to depend on concentration[13]. In short, it appeared that there was an interaction between magnesium ions, sulphate ions, and water molecules in the form of a sequence of linked reactions. Here I should anticipate what will be said in a later section (Section 8a, p. 193) by stating that this was in fact the correct explanation: the interaction involves a stepwise substitution of the water molecules bound in the coordination shells of the magnesium–aquo complex by the sulphate ion, with the absorption continuum of the ion-cloud interactions appearing only at higher frequencies (which could not be reached at that time).

3. *Back to Physics*

At this point I feel I should interrupt my account of the historical course of events, and show how chemical interactions can cause the absorption of sound in the first place.

A sound wave represents an adiabatic pressure change progressing periodically in space and time. In water, as a result of the density maximum at 4°C, the simultaneously appearing temperature wave is very small in magnitude compared to the pressure wave. Thus, in the present case essentially we have to consider only the effect of the pressure change on the chemical equilibrium. A chemical equilibrium is always pressure-dependent whenever the reaction partners (in equilibrium with each other) differ in volume. When this is the case, a pressure change will induce a chemical excess reaction which takes place at a finite rate and leads to adaptation to the particular equilibrium state concerned. If the periodic pressure change takes place very rapidly in relation to the chemical reaction, the system will practically not «notice» these changes: the rapid positive and negative disturbances average out before the onset of any appreciable reaction (*cf.* Fig. 1a).

On the other hand, if the pressure change takes place very slowly compared to the chemical reaction, the system follows these changes with practically no lag. The sound then merely propagates at a slightly lower velocity, for the compressibility of the medium contains a contribution from the state of the chemical equilibrium (*cf.* Fig. 1b). Now, the interesting case is that in which the rate of re-establishment of equilibrium is comparable to the rate of the pressure change (*i.e.* when the time constant for the establishment of chemical equilibrium is of the same order of magnitude as the period of the acoustic wave). In this case the system tries to adapt continuously to the pressure change but does not quite succeed, so that it lags behind the pressure change by a finite phase difference. The chemical state is characterized by the concentrations of the reaction partners or the reaction variable. Because of the finite volume difference between the reaction partners in equilibrium, a volume increment characteristic of the chemical change follows the pressure change with a certain phase lag. In all fields of physics where there is this kind of phase difference between «conjugate» variables there is a transfer of energy (in this case a reduction in the amplitude of the sound waves). For a finite phase difference, the integral $\int P\mathrm{d}V$ is different for the compression and dilatation periods.

The physical situation just described can also be given mathematical expression. The essential steps in the calculation are shown in Tables 1A and 1B.

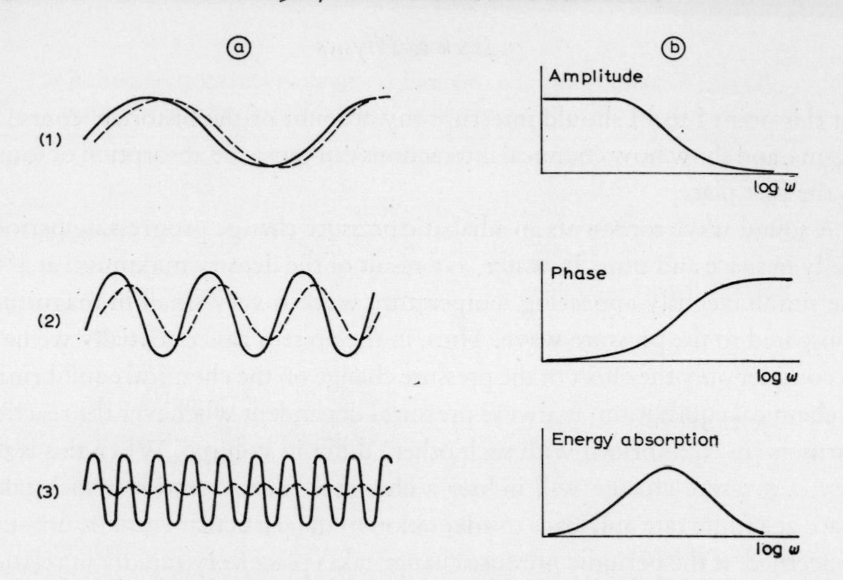

Fig. 1a. Periodic disturbances of a chemical equilibrium. Continuous line, disturbing parameter (*e.g.* sound pressure, affinity). Broken line, instantaneous value of an «internal» system parameter (*e.g.* concentration, reaction variable, volume increment of the chemical change). (1) The disturbance occurs slowly in relation to the chemical change. (2) The period of the disturbance and the reaction time are of the same order of magnitude. (3) The disturbance is rapid in relation to the reaction.

Fig. 1b. Amplitude and phase of the equilibrium disturbance and the energy absorption during one period, all as functions of the angular frequency ($\omega = 2\pi\nu$) in the relaxation range.

The starting point is the reaction rate of a particular system. For small disturbances the relevant differential equation can always be reduced to linear form. The concentration variable is then δc, *i.e.* the deviation of the concentration at any instant from a given reference value (*e.g.* the mean in the case of periodic disturbances); $\delta\bar{c}$ is then the deviation of the hypothetical equilibrium value (relative to the instantaneous value of the pressure) from the same reference value. The deviation of the instantaneous value of the concentration from its equilibrium value is then $(\delta c - \delta\bar{c})$. This difference is proportional to the rate of the chemical change. In the case of periodic disturbances, the differential equation has a complex solution. The imaginary part ultimately re-appears in the absorption coefficient (*cf.* also Table 1B and Fig. 2). The decreases in amplitude, phase difference, and energy absorption per wavelength ($2\,\alpha\lambda$) are shown as functions of frequency in Fig. 1b. The chemical contribution to

Table 1

Dispersion and absorption of sound as a result of chemical relaxation
(A comprehensive theoretical treatment with bibliography is given in ref. 32)

(A)

Relaxation equation $\frac{d(\delta c)}{dt} = -\frac{\delta c - \delta \bar{c}}{\tau}$

Periodic disturbance $\delta c = A\, e^{i\omega t}$

$$\delta c = \frac{\delta \bar{c}}{1 + i\omega\tau}$$

Compressibility $-\frac{1}{V}\frac{\delta V}{\delta P} = -\frac{1}{V}\left(\frac{\delta V}{\delta P}\right)_{\delta c = 0} - \underbrace{\frac{1}{V}\left(\frac{\delta V}{\delta c}\right)_P \frac{\delta c}{\delta P}}_{\text{complex}}$

Wave equation $\frac{\delta^2 P}{\delta t^2} = U^2 \frac{\delta^2 P}{\delta x^2}; \quad \frac{1}{U^2} = -\frac{S}{V}\left(\frac{\delta V}{\delta P}\right)_S$

Reciprocal of the velocity of sound $\frac{1}{U} = \left(\frac{1}{U}\right)_{\text{real}} + \left(\frac{1}{U}\right)_{\text{imag}}$

Reciprocal of the phase velocity $-\frac{i\alpha}{\omega}$

(B)

$$\frac{\delta c}{\delta P} = \frac{\delta \bar{c}/\delta P}{1 + i\omega\tau} = \frac{\delta \bar{c}}{\delta P}\left[\frac{1}{1 + \omega^2\tau^2} - \frac{i\omega\tau}{1 + \omega^2\tau^2}\right]$$

$\alpha\lambda \sim \frac{\omega\tau}{1 + \omega^2\tau^2}$ Absorption within one wavelength

compressibility has been assumed to be small compared to the contribution of the solvent.

The calculations shown in Table 1 formed the starting point for our investigations on fast chemical reactions[14,15]. It can be seen that the relaxation time of the chemical equilibrium state can be read off directly from the variation of the absorption with frequency *e.g.* as the reciprocal of the frequency at the maximum absorption per wavelength. This relaxation time is simply related to the kinetic constants of the reaction system, as shown by the example in Fig. 3. This gives us a direct approach to the kinetics of fast reactions – even reactions that were hitherto «immeasurably» fast.

Of course this simple theory of single-stage relaxation processes cannot be applied directly to magnesium sulphate, but a sufficiently simple system was

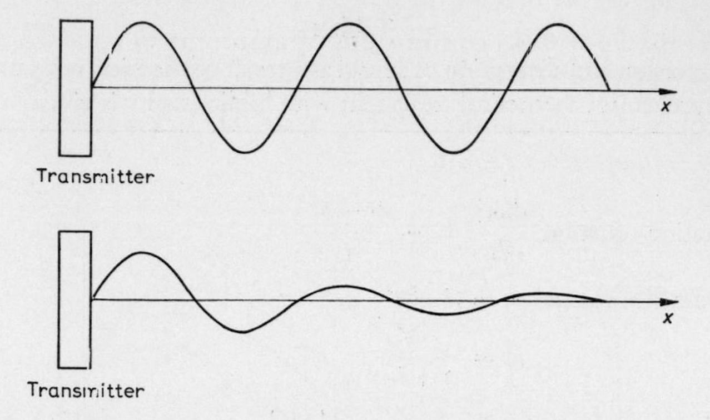

Fig. 2. Schematic representation of a plane propagating sound wave. Top, undamped (U real); below, damped (U complex).

$$P = P_0 \, e^{i\omega(t - x/U)}$$

$$\frac{1}{U} = \frac{1}{U_{Ph.}} - \frac{i\alpha}{\omega}$$

$$P = P_0 \, e^{i\omega(t - x/U_{Ph.})} \, e^{-\alpha x}$$

$$A + B \underset{k_D}{\overset{k_R}{\rightleftharpoons}} AB$$

$$1/\tau = k_R \, (\bar{C}_A + \bar{C}_B) + k_D$$

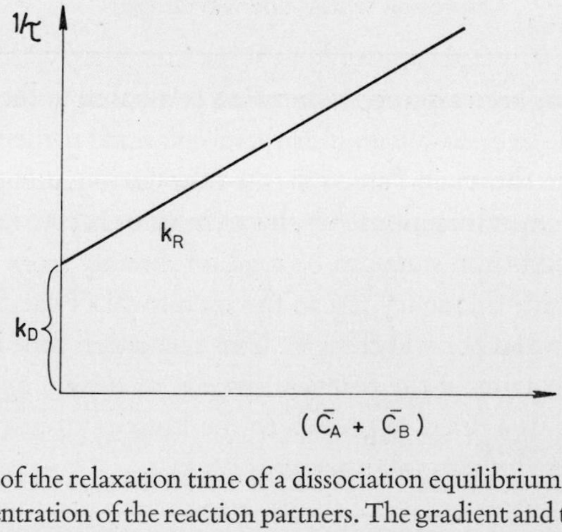

$1/\tau$

k_R

k_D

$(\bar{C}_A + \bar{C}_B)$

Fig. 3. Reciprocal of the relaxation time of a dissociation equilibrium as a function of the equilibrium concentration of the reaction partners. The gradient and the intercept on the ordinate give the rate constants for recombination and dissociation.

soon found: the hydrolysis equilibrium of ammonia in aqueous solution. A predictive calculation of the absorption and relaxation time (assuming diffusion-controlled recombination of NH_4^+ and OH^-) gave full agreement with the experimental data[15,16] (Fig. 4.).

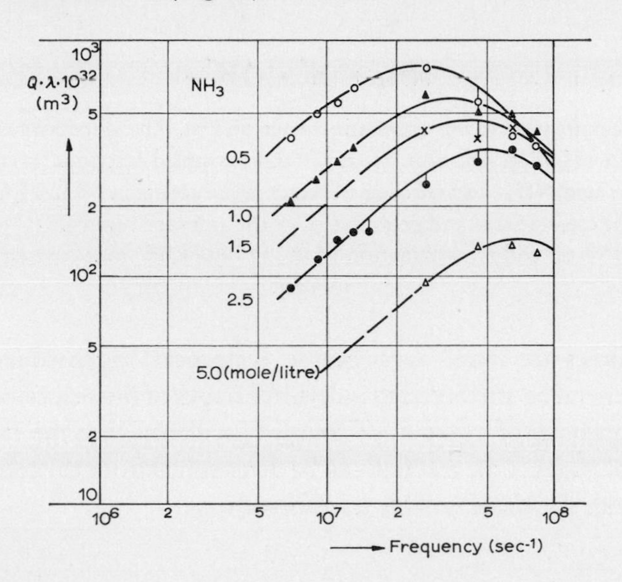

Fig. 4. Absorption of sound as a result of chemical relaxation in a system consisting of NH_3 in aqueous solution. $Q\cdot\lambda$ (absorption cross-section × wavelength) as a function of frequency (Q corresponds to the energy absorption coefficient 2α related to the total number of NH_3 molecules per unit volume). Parameter, molar concentration of NH_3.

At the same time it was also possible to solve a problem which had interested chemists for a long time. It was shown that in alkaline solution the reaction does not proceed *via* dissociation of the cationic acid (*cf.* the reaction scheme given in Fig. 5). In the reverse reaction, the dissociation of H_2O is accelerated by about ten orders of magnitude by the vicinity of NH_3:

$$NH_3\cdot H_2O \rightarrow NH_4^+ + OH^- \quad k = 6\cdot 10^5\ \text{sec}^{-1} \qquad \text{(ref. 15)}$$

$$H_2O\cdot H_2O \rightarrow H_3O^+ + OH^- \quad k = 2.5\cdot 10^{-5}\ \text{sec}^{-1} \qquad \text{(ref. 17)}$$

The situation was much more complicated in the case of magnesium sulphate. Here it was first necessary to assume a scheme of stages for the stepwise formation of an aquo complex between the magnesium and sulphate ions (see Section 8a, p. 193). For the reaction rates of the different stages we get a system of linked differential equations from which, after reduction to linear form, the

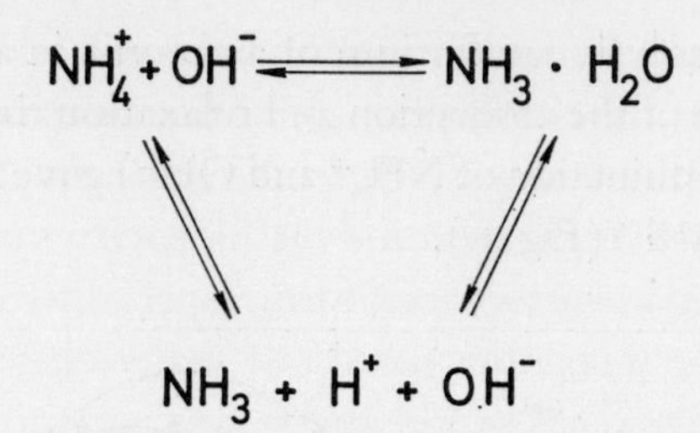

Fig. 5. Equilibrium state in the ammonia–water system. The direction of the reaction depends on the pH of the solution. There are three coupled reactions: (1) neutralization of the cationic acid NH_4^+ and splitting of water in the complex $NH_3 \cdot H_2O$; (2) deprotonation of the cationic acid and protonation of the anhydro base NH_3; (3) dissociation of the solvent water and recombination of the H^+ and OH^- ions originating from the solvent.

relaxation times are found as (negative, reciprocal) eigenvalues[18,19]. They cannot in general be attributed to individual stages of the reaction, as with the normal frequencies of a system of coupled oscillators, but the rate constants of individual stages of the reaction can be determined from the relaxation time spectrum with the aid of suitable transformations.

4. *Historical Review*

The principle of the method described above is quite well-known to physicists. Albert Einstein[20] had already shown in 1916 that relaxation effects appear in a dissociating gas subjected to the periodic temperature variations of a sound wave, and that these effects result in a dispersion of the velocity of sound. At about the same time, Walter Nernst and his associates attempted to detect this effect experimentally in the $2\ NO_2 \rightleftharpoons N_2O_4$ system[21]. However, these measurements were unsuccessful because the techniques of sound transmission were still insufficiently developed. Later, *i.e.* at about the beginning of the thirties, and following investigations by K. F. Herzfeld[22] and H. O. Kneser (ref. 23), the interest of physicists and physical chemists turned to the relaxation effects which result from a lag in the establishment of equilibrium in «internal» temperature, *i.e.* to lags in energy transfers between the different degrees of freedom (translation, rotation, and vibration). It was only after these investigations that the «chemical» relaxation effects were detected and analysed. At about the same time, John Lamb and his school were studying the absorption of sound by rotation isomers[24,25]. The detection of chemical relaxation effects in gases by means of sound absorption measurements was eventually achieved for the classic example of the dissociation of N_2O_4, but only after a delay of about 40 years[26]. In the meantime, the theory of the dispersion of sound based on the thermodynamics of irreversible processes[27,28] had also been developed in very general form, especially by J. Meixner[29,30].

Today we know the relaxation spectra of many chemical reactions, but very few of them derive from sound absorption measurements. The sound absorption method is generally too insensitive for studying chemical relaxation effects. One of the reasons for this is the high background absorption of the solvent. Since this increases with the square of the frequency, very different methods are required to cover a wide frequency range. Thus, at 100 kHz the amplitude of sound in pure water falls by about 37% (1/e) over a distance of 4 km, while at 100 MHz the same percentage reduction takes place over a distance of only 4 mm.

At low frequencies (< 100 kHz) resonance or reverberation techniques are generally used, and at very high frequencies (> 10 MHz) mainly pulse-echo methods are preferred. Several methods are available in the intermediate frequency range: interferometric methods and direct measurement of the damping from observations on the amplitude of the sound from the refraction of light waves in the acoustic field (after Debye and Sears). A review of the various techniques of measurement can be found in refs. 31 and 32. Because of the long wavelengths at low frequencies, large volumes of liquid and hence relatively large amounts of substances are required at such frequencies. At high frequencies it is necessary to work with very high concentrations because of the strong background absorption of the solvent. Even if it has by now by and large proved to be possible to adapt various sound absorption methods to the requirements of chemical relaxation studies[33] – *e.g.* by considerably reducing the liquid volume while at the same time increasing the sensitivity by using difference methods – the application of this method, particularly to the more interesting multi-stage reaction mechanisms of biochemistry, remains very limited.

5. *New Paths*

Our first thoughts were along the lines of developing new relaxation methods in which the pressure of the sound waves was replaced by another variable, possibly electric field strength. It is of course impossible to carry out such measurements in travelling electromagnetic waves, for the wavelengths are larger than those of sound waves of comparable frequency by a factor of 10^5. On the other hand, it should be relatively easy to determine dielectric dispersion and absorption directly by means of capacitance measurements using tuned circuits. On examining the literature, we found that although there was here a highly developed measurement technique and a great wealth of experimental data, it related exclusively to the relaxation of dipole orientation and not to chemical effects. The reason for this can be seen immediately from Fig. 6: the equilibrium constant for a chemical change of polar molecules is a quadratic function of the electric field. Small amplitudes are generally used in determining the dielectric constant (ε) or the energy loss ($\tan \delta$), so that the chemical increment does not appear in the linear field terms measured. It was but a short step from recognizing this to going on to develop an appropriate non-linear method[34], even if this left a whole series of technical problems to be solved. In a resonant circuit of high efficiency (with a sharp resonance line), a strong steady field is superimposed on a low-amplitude alternating field. The associated chemical reaction leads to a broadening of, and shift in, the resonance line, which can be measured as a function of the field strength ($\sim E^2$), concentration (corresponding to a second-order reaction) and fre-

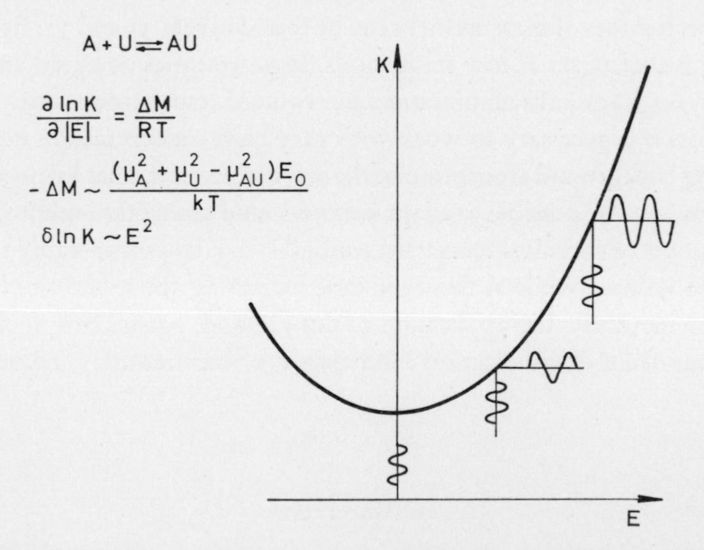

Fig. 6. Dependence of the equilibrium constant K of a chemical change on the electric field. (Example: the base pairing reaction adenine [A] + uracil [U].) In dilute solution the «reaction moment» ΔM varies as shown with the dipole moments of the individual reaction partners. E, electric field strength; kT, Boltzmann factor.

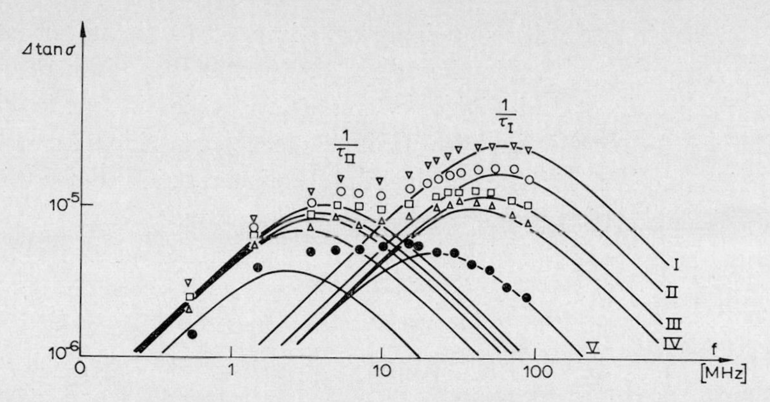

Fig. 7. Dielectric absorption in strong fields as a result of chemical relaxation. (Increment in the loss angle Δ tanδ as a function of frequency.) The two maxima are the result of a relaxation of the coupled chemical reactions indicated. A, ε-caprolactam; B, 2-amino-pyrimidine, in cyclohexane, 22°C, 200 kV/cm.

$$\begin{array}{ccc} 1/\tau_I & & 1/\tau_{II} \\ A_2 \underset{k_{21}}{\overset{k_{12}}{\rightleftharpoons}} A + A + B & & \underset{k_{31}}{\overset{k_{13}}{\rightleftharpoons}} AB + A \end{array}$$

	c_{OA}	c_{OB}
I	0.18	0.01512
II	0.151	0.008
III	0.123	0.008
IV	0.079	0.008
V	0.0226	0.0167

quency ($\omega\tau/1+\omega^2\tau^2$). In contradistinction to the simple orientation relaxation of dipoles (which generally occurs in the microwave range), the relaxation time of the chemical change is concentration-dependent in accordance with the order of the reaction. This enables the chemical effects to be distinguished from other relaxation effects relatively easily—once they have been made accessible to measurement in the first place. The main factor responsible for the technical difficulties of such measurements is that the effect is very small. Field strengths of up to $3 \cdot 10^5$ V/cm (the breakdown limit) were required, and tan δ (the loss angle) had to be accurate to fractions of 10^{-6}. My colleagues Klaus Bergmann[35] and Leo De Maeyer[36] developed a method of this accuracy. Fig. 7 shows some absorption curves (tan δ as a function of the frequency of the alternating field) obtained by Julian Suarez[37]. The curves are for two coupled association reactions of hydrogen-bridge-forming substances in non-polar solvents. Unfortunately, the method cannot be applied directly to polar solvents (because their conductivity is too high), but pulse methods are available in this case, as will be shown later in Section 6 (p. 184).

The dielectric method has attained great importance because it enables direct investigation of the kinetics of hydrogen bonding. This method thus enables the kinetics of the individual stages of base pairing in nucleic acids to be followed directly[38]. With long-chain macromolecules it can happen that orientation requires a longer time than the chemical reaction, and in this case the chemical relaxation appears in the linear terms as well. Gerhard Schwarz used a method like this (which does not require a strong steady field to be superimposed) to investigate the kinetics of structural changes in polypeptides[39].

6. *Things «Get Even Simpler»*

Periodic relaxation methods have added greatly to our knowledge of the kinetics of chemical reactions, particularly in the time range from microseconds to nanoseconds, but their use has always been confined to specific individual cases. In many cases the contribution of a chemical change to the thermodynamic parameters of the solution as a whole (*e.g.* its compressibility or dielectric permittivity) is so small that precise measurement is impossible. This is particularly so when we are concerned with multi-stage reactions, in which the relaxation spectrum could give detailed information on the intermediate stages and hence on the mechanism of the reaction. Here the approach to adopt was to follow the chemical change directly by means of specific properties of the reaction partners. This principle has now become familiar to us; in its simplest form, it can be formulated as follows:

The equilibrium constant must be rapidly changed by a constant (small) amount and the establishment of the new equilibrium followed immediately.

We then obtain a system of homogeneous linear differential equations for the reaction rates. The solutions are exponential functions with real, negative arguments.

However, it is not always simple to carry out in practice what appears to be simple in mathematical terms or in its physical principle.

Although we have now learned how to produce such step-like disturbances (with steep rising and descending branches), we initially attempted to use sinusoidal single pulses. As the disturbing parameter for investigating electrolytic dissociation equilibria we used the electric field strength. M. Wien[40] had already shown at the beginning of the thirties that binary electrolytes exhibit increased dissociation in strong fields. Lars Onsager[41] gave a complete theo-

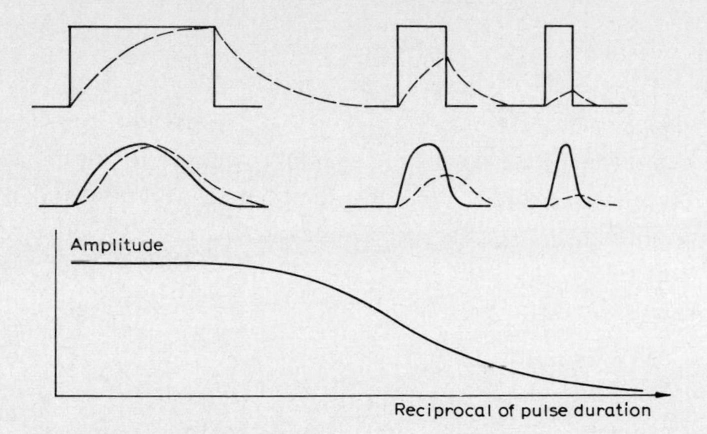

Fig. 8. Relaxation behaviour following pulse-like disturbance of equilibrium.

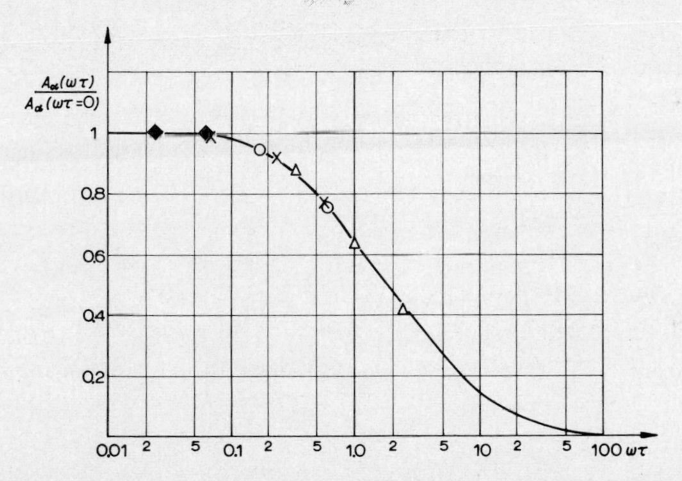

Fig. 9. Dispersion of the dissociation field effect in the systems ammonia and acetic acid in aqueous solution. The (normalized) change in the amplitude of the shift at equilibrium (*cf.* Fig. 8) is plotted as a function of the reciprocal of the pulse duration (relative to the relaxation time). (ω is the angular frequency of the critically damped sinusoidal pulse; *cf.* ref. 42.) t, 20°C.

		C(moles/litre)	*(MHz)*
NH_3	●	$7 \cdot 10^{-3}$	0.44, 1.45
	×	$7 \cdot 10^{-4}$	0.44, 1.45, 3.60
	○	$2 \cdot 10^{-4}$	0.44, 1.45
	Δ	$8 \cdot 10^{-5}$	0.44, 1.45, 3.60
CH_3COOH	×	$3 \cdot 10^{-4}$	0.44, 1.45, 3.60
	○	$1 \cdot 10^{-4}$	0.44, 1.45

Fig.10. 200 kV high-voltage apparatus, pulse circuit, and impedance bridge with which the first chemical relaxation measurements using the dissociation field effect were carried out in 1953/1954 (*cf.* Fig.9).

retical interpretation of this «dissociation field effect». Fig. 8 shows how the dispersion of this effect can be measured with the aid of short-duration field pulses, and hence used to determine the relaxation time for the chemical equilibrium state. The amplitude of the shiftin equilibrium—measured as a change in conductivity—was determined directly with the help of a null method specially developed for this purpose in collaboration with Josef Schoen[42]. Fig. 9 shows the dispersion of the amplitude of the field effect as a function of pulse duration (or the frequency of a strongly damped harmonic vibration). The high-voltage apparatus used at that time is shown in Fig. 10.

A fairly large number of individual measurements with different pulse lengths were required to obtain a dispersion curve, but in contrast the relaxation time could be measured in a single experiment using a rectangular pulse. This advantage appeared to us decisive in determining the rate of a neutralization reaction. In this case it is possible to obtain a measurable disturbance in the equilibrium only by starting with «very pure» water, in which 10^{-7} moles of H^+ and OH^- ions are in equilibrium with about 55 moles of H_2O. A change in the equilibrium concentration of the ions can easily be followed by means of the electrical conductivity provided no contaminating ions get into the solution. But this is precisely what happens when the highly purified water is subjected to too many high-voltage pulses. A *single* rectangular pulse – produced by a double spark circuit[17] – made it possible for the first time to measure the chemical relaxation of the dissociation of H_2O and hence the neutralization kinetics. At the same time, a stationary field method made possible the direct measurement of dissociation rates (*e.g.* in ice crystals)[43], thus enabling the kinetic parameters of the transport and neutralization of proton charges in hydrogen-bridge systems to be finally established.

7. *On the Way to «Technical Perfection»*

Today we use step-type disturbances, which are the most suitable for studying complex, multi-stage reaction mechanisms (see Fig. 11), almost exclusively. This procedure enables the relaxation spectrum to be followed directly by means of discrete steps (with logarithmic time scale). Fig. 12 shows the relaxation spectrum of a biochemical reaction. With a closely packed sequence of steps (a relaxation continuum) we obtain clearly defined mean values, either from the integral or from the initial gradient of the relaxation curve (mean values of τ and $1/\tau$ respectively[44]).

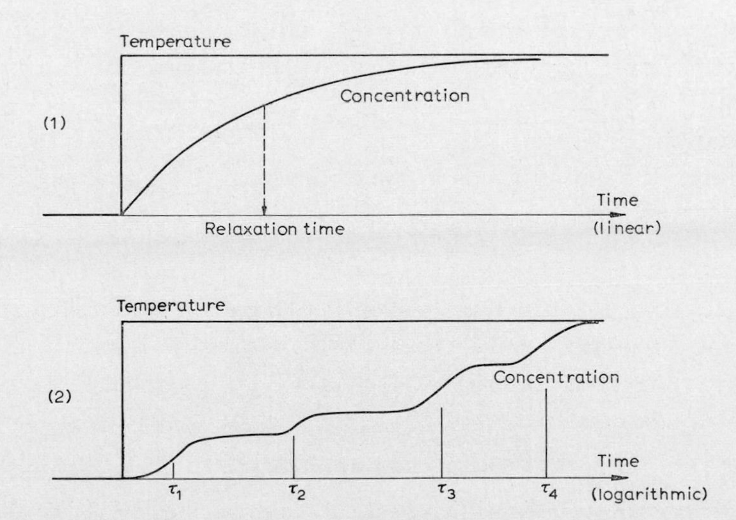

Fig. 11. Relaxation resulting from the disturbance of an equilibrium by a temperature jump. (1) A single relaxation process on a linear time scale. (2) Relaxation spectrum on a logarithmic time scale.

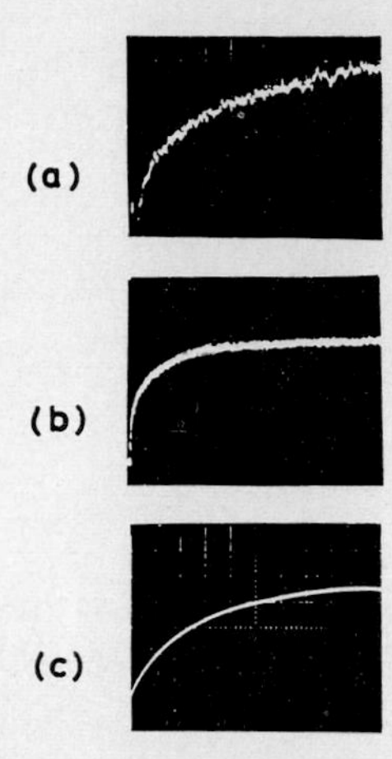

Fig. 12. Oscillogram of a relaxation spectrum with 3 time constants (the system glyceraldehyde-3-phosphate dehydrogenase + β-NAD). The spectrum describes the reaction mechanism of an allosteric enzyme. The relaxation measurements yielded important information on the nature of allosteric control (*cf.* refs. 66–68, see also Figs. 22 and 23). pH, 8.5; 40°C; $D_0 = 6 \cdot 10^{-4}$ (M). (a) 0.2 msec/cm, $1/\tau_1 = 7000$ sec^{-1}. (b) 1.0 msec/cm, $1/\tau_2 = 690$ sec^{-1}. (c) 500 msec/cm, $1/\tau_3 = 0.2$ sec^{-1}.

The time resolution depends on the steepness of the step and also on the signal-to-noise ratio that can be attained. Fig. 13 shows as example the field effect of oxyhaemoglobin, measured by Georg Ilgenfritz[45] with a time resolution of about 50 nsec. In this case, the rectangular field pulse was obtained in the form of a travelling wave by discharging a high-tension cable across two spark gaps (Fig. 14)[46].

Pressure waves with very steep ascending and descending branches can also be produced in a shock tube on a similar principle. The shock tube is now a standard instrument for studying fast reactions in gas kinetics[47]. Fig. 15 shows

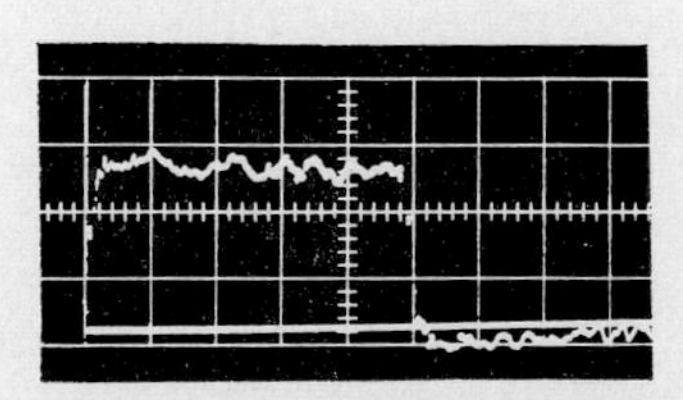

Fig. 13. Oscillogram of the dissociation field effect in oxyhaemoglobin (time scale 10^{-6} sec/cm, relaxation time ~ 50 nsec). The measurement was made by the method described in Fig. 14. System: oxyhaemoglobin (sheep), conc. 10^{-4} *M*, pH 8.9, $\lambda_{obs.}$ 577 mμ, field strength 75 kV/cm, time constant $\leqslant 10^{-7}$ sec.

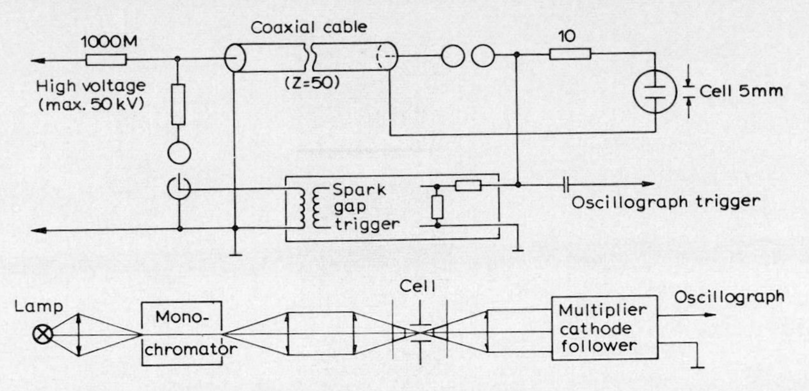

Fig. 14. Schematic representation of the field jump method. Travelling electric wave method for studying chemical relaxation. The (rectangular) electric pulse is produced by discharging a coaxial high-tension cable (time of travel about 3 μsec). The cable is connected through one spark gap (on the left in the figure) with a resistance matched to the impedance of the cable and through the other spark gap (above, right) to the (high-resistance) measuring cell. Depending on which of the two spark gaps is triggered first, it is possible to produce pulse durations that are fractions or multiples of the time of travel (in the former case the energy is dissipated in the matched resistance, and in the latter multiple reflection occurs at the measuring cell). The field effect is recorded spectrophotometrically. (The apparatus was constructed by Georg Ilgenfritz, Dissertation, Göttingen, 1966.)

the principle of a shock tube for liquids developed by Alexander Jost for studying fast reactions in solutions[48]. Simple pressure jump methods have been given by Hans Strehlow[48], among others.

However, particular mention should be made here of temperature jump methods. Very simple in principle, these methods are especially wide-ranging in their applicability to the study of fast reactions.

There are two ways of producing temperature jumps: firstly by adiabatic compression or dilatation, and secondly by heating by means of electrical impulses in electrolyte systems. The first method is not suitable for aqueous solutions, because of the maximum in the density of water at 4°C. Electrical heating can be achieved in two ways: (*1*) simply with a current impulse in solutions having a finite electrolytic conductivity; and (*2*) with microwave impulses in the X-band range (dispersion of H_2O orientation) for any conductivity. Strong fields (~ 100 kV/cm) are required in both cases. In the case of microwave heating, this means radar impulses with powers of the order of megawatts. The chemical change is best followed by optical methods (spectrophotometry, fluorimetry, polarimetry). After initial difficulties (inho-

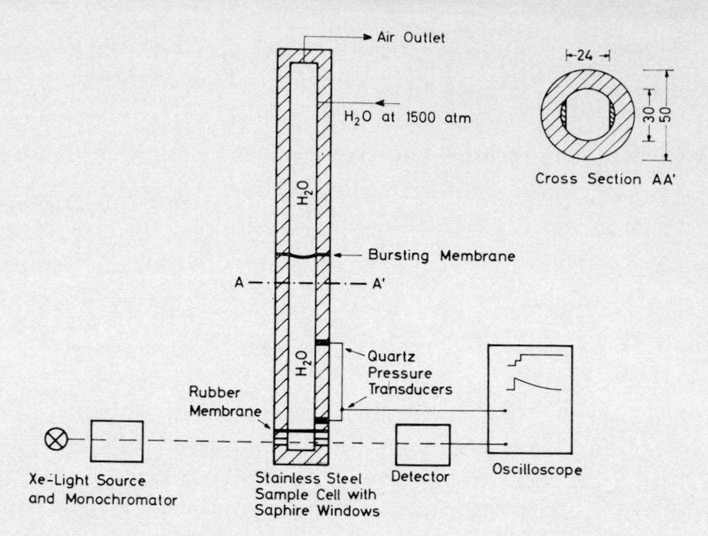

Fig. 15. Mechanical pressure wave method for studying chemical relaxation effects in liquids. The method is a mechanical analogue of the electrical impulse method shown in Fig. 14 (although in this case the time of travel is limited by the velocity of sound which is about 5 orders of magnitude slower). The thickwalled shock tube is completely filled with liquid. The shock wave originates when the metal membrane bursts following the production of a pressure of about 1000–1500 atm in the upper part of the tube. The measurement is made using the reflected shock wave (time of travel from the measurement chamber to the top end of the tube and back: about 10^{-2} sec; steepness of ascending and descending branches $< 10^{-6}$ sec). The solution in the measurement chamber (at the bottom end of the tube) is separated from the liquid filling the tube by a plastic membrane «transparent» to the pressure wave. The chemical reaction is followed spectrophotometrically. The apparatus was constructed by Alexander Jost, Dissertation, Göttingen, 1966, ref. 48.

mogeneous heating in «electric lenses», cavitation resulting from pressure waves, «cross-talk» of the high-voltage impulse on the electrical measurement equipment, unfavourable signal-to-noise ratio in the short time range) had been overcome, the temperature jump method was eventually developed into a standard procedure, which now has an extremely wide range of applications extending from inorganic to biological chemistry. This is particularly due to the development work of Leo de Maeyer[50], Georg Czerlinski[51], Hartmut Diebler[52], Gordon Hammes[53], Roland Rabl[54], and others. Fig. 16 shows a T-jump apparatus developed by Leo de Maeyer and now available commercially. Fig. 17 is a schematic representation of the circuit of a microwave T-jump apparatus developed by Roland Rabl and Leo de Maeyer[54].

Two important improvements considerably extend the possible range of applications: (*1*) A flow arrangement is used instead of a static measurement cell, so that it is also possible to carry out relaxation measurements on systems which are not in equilibrium (but have reached a stationary state). This improvement is of decisive importance in investigating many biochemical reaction mechanisms. (*2*) The stationary-state reaction mixture flowing through the observation capillary can be heated periodically with the aid of repeated microwave impulses. A «time sampling» procedure can be used to obtain an average over many individual measurements, thus considerably improving the signal-to-noise ratio. The desicive limitation on previous methods in their application to chemical reaction systems concerned the sensitivity of recording rather than the time range involved. The greater the precision and sensitivity of our measurements, the more reaction stages become accessible to direct analysis (*cf.* also ref. 55).

Present-day methods of measurement[32] cover without a gap the time range between fractions of a nanosecond and several seconds, and thus bridge

Fig. 16. Photograph of a standard present-day temperature jump instrument, developed by Leo de Maeyer and built by Messanlagen Studiengesellschaft, Göttingen. The lower part contains the high-tension generator, impulse circuit, lamps and multiplier power source and amplifier. The upper part consists of a spectrophotometer with the T-jump cell set up in its beam. This part contains interchangeable units which can also be combined to make a fluorimeter or polarimeter.

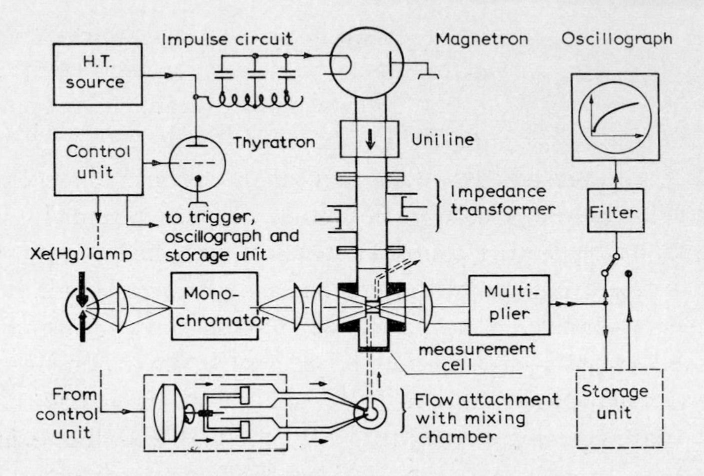

Fig. 17. Schematic circuit diagram of a microwave temperature jump apparatus. The temperature jumps are produced by microwave impulses from a magnetron (frequency 9.35 GHz, power 0.6 MW, pulse duration 0.5–3.0 μsec, temperature jump 3–15°C). The optical measurement cell, which is matched to the impedance of the hollow conductor, is either a static microcell (volume 30 μl) for simple T-jump studies or a capillary for relaxation studies on flowing reaction mixtures in a stationary state. In the latter case, the temperature jump can be repeated with a frequency < 300 Hz. Measurements by spectrophotometry. (Developed by Carl-Roland Rabl, Dissertation, in preparation.)

the gap between the time ranges of classical kinetics and molecular spectroscopy (*cf.* Fig. 18). It has therefore been possible to elucidate the individual steps in many of the reactions previously considered to be «immeasurably fast». Some examples of this will be discussed in the following sections.

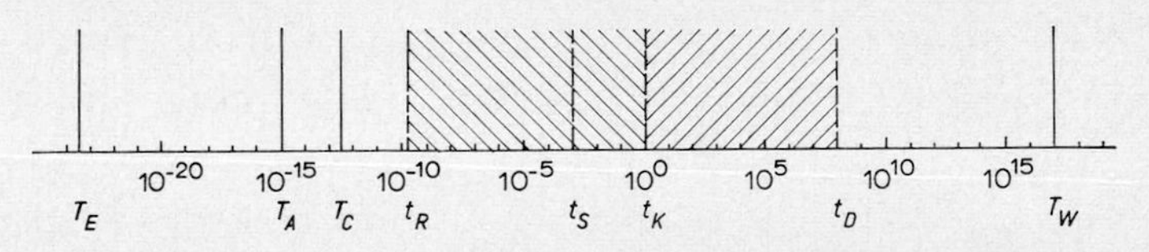

Fig. 18. The time scale of the chemist (see also ref. 56). Hatched part, from left to right: proton transfer, electron transfer, formation and rupture of H bridges, isomerization, substitution in coordination compounds, secondary structure changes in proteins, enzymatic changes, elementary processes in biology. T_E, elementary time (dimension of the nucleus/velocity of light); T_A, time constant for processes in the atomic shell (excitation); T_C, time limit for chemical changes ($\sim h/kT$); t_R, present lower time limit for direct measurements of reaction rate; t_S, time limit for flow methods; t_K, classical time limit for reaction rate measurements; t_D, time to obtain a doctor's degree; T_W, age of the world.

8. Applications

(*a*) *For the inorganic chemist : a «periodic system» of reaction rates*

We have already learned some of the applications of relaxation spectrometry from individual examples. Charge neutralization in the reaction $H^+ + OH^- \rightleftharpoons H_2O$ does in fact take place almost «instantaneously». Every encounter between the solvated ions results in combination ($k = 1.4 \cdot 10^{11}$ mole$^{-1} \cdot$ sec^{-1} at 25°C). In this process the proton «tunnels» through a whole chain of hydrogen bonds, as has been shown directly by measurements on ice cyrstals[43]. Are all reactions involving a neutralization of charges then diffusion-controlled?

Let us first have a look at inorganic chemistry. Here we find many reactions in which a positively charged metal ion combines with a negatively charged ligand to form a neutral complex. It is precisely this kind of reaction that is frequently described in the literature as «immeasurably fast». The formation of an aquo complex between Mg^{2+} and SO_4^{2-}, mentioned at the beginning of this lecture, is a typical example of this class of reaction.

A metal ion in aqueous solution is surrounded by one or more shells of coordinated water molecules. If it is to combine with another ion of opposite charge, the latter must penetrate the hydration shells, substituting successive water molecules in the different shells. Since the water molecules in the inner coordination shell are bound most strongly, their substitution will be the slowest step of the process. Relaxation studies on very widely differing metal ions have confirmed this assumption. The mechanism of the stepwise substitution reveals itself in a relaxation spectrum with several time constants[57]. The chemist is primarily interested in substitution in the inner coordination shell. The specific properties of the metal ion, such as charge, radius, coordination number, and electronic structure should be expressed directly at this stage.

Fig. 19 summarizes the measured values of the rate constants for substitution in the inner coordination shell. With a few exceptions, these values are typical of the metal ion alone, *i.e.* they are practically independent of the nature of the substituting ligand. (A more detailed compilation of measurements can be found in ref. 58 and a discussion of the mechanisms involved in ref. 59). Metal ions with an electronic configuration similar to that of the noble gases exhibit the expected dependence on charge and radius. The smaller the radius and the higher the charge the more strongly are the H_2O molecules to be substituted bound, and hence the more slowly does the substitution take place. An in-

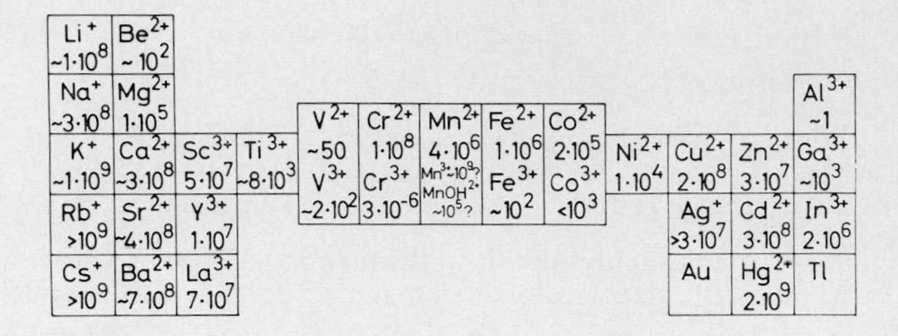

Ce^{3+}	Pr^{3+}	Nd^{3+}	Pm^{3+}	Sm^{3+}	Eu^{3+}	Gd^{3+}	Tb^{3+}	Dy^{3+}	Ho^{3+}	Er^{3+}	Tm^{3+}	Yb^{3+}	Lu^{3+}
$9\cdot10^{7}$	$9\cdot10^{7}$	$9\cdot10^{7}$		$9\cdot10^{7}$	$8\cdot10^{7}$	$5\cdot10^{7}$	$3\cdot10^{7}$	$2\cdot10^{7}$	$1\cdot10^{7}$	$1\cdot10^{7}$	$1\cdot10^{7}$	$1\cdot10^{7}$	$1\cdot10^{7}$

Fig. 19. Characteristic rate constants (in sec^{-1}) for H_2O substitution in the formation of metal complexes. Most of the values are specific to the metal ion and relatively independent of the nature of the ligand (*cf.* refs. 57–59), which also discuss in particular detail the exceptions to this rule). The rate constants for the alkali metal ions and alkaline earth metal ions are essentially determined by radius and charge. Among the alkaline earth metal ions–and especially among the rare earths–the coordination number has an additional specific effect. The ions of the transition metals also reflect the specific properties of their electronic structure (ligand field stabilization and the Jahn–Teller effect.)

crease in coordination number makes the coordination shell more labile and therefore accelerates the reaction (*cf.* the lanthamides). Specific effects of electronic structure are revealed only in the case of the transition metals (filling of the d shell). V^{2+} and Ni^{2+} ions have strikingly slow rates of substitution, as a result of particularly strong ligand field stabilization in the transition state[59], whereas Cr^{2+} and Cu^{2+} ions are extremely labile to substitution because of distortion of the octahedral structure in consequence of the Jahn–Teller effect[59]. Fig. 19 shows that most of the rate constants for substitution lie in the range from 10^3 to 10^9 sec^{-1}. The mechanism of many inorganic reactions was thus inaccessible to experimental elucidation until the introduction of relaxation methods.

(*b*) *For the organic chemist: unused possibilities*

The method proved even more successful in measuring the rates of protolytic reactions. The kinetics of proton transfer have been studied exhaustively with reference to a large number of organic acids and bases[60]. These studies have resulted in the elucidation of the reaction mechanism of acid–base catalysis and enabled the Bronsted relations to be extended and generalized[61].

We shall now consider one example: the keto–enol transformation in heterocyclic compounds, *e.g.* barbituric acid[62].

The anion of this acid (enolate, E^-) can acquire a proton either at the $O^{(-)}$ atom or at the $C^{(-)}$ atom. The first reaction results in the enol (EH) and the second in the ketone (KH). The reaction scheme is as shown in Fig. 21. Both reactions take place so quickly that they cannot be followed by classical methods. When barbituric acid ($pK \approx 4.0$) is dissolved in water, the ionization equilibrium (conductivity) is established «at once». Relaxation measure-

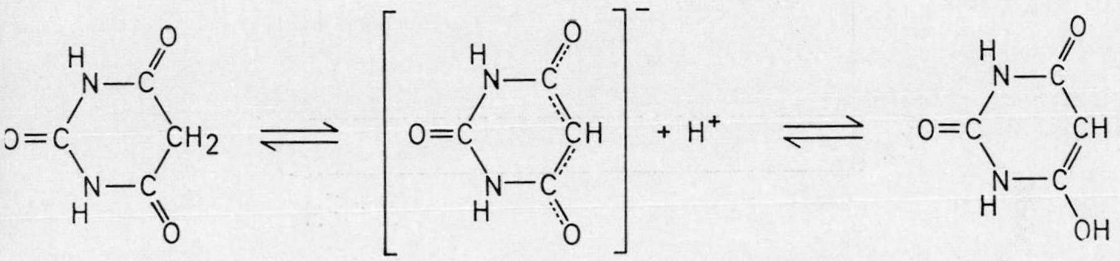

Fig. 20. The keto–enol transformation in heterocyclic compounds.

ments using the electric field method showed that enol formation, like most protolytic recombination processes, is entirely a diffusion-controlled reaction ($k \approx 10^{10}$ mole$^{-1} \cdot$ sec^{-1}). Direct determination of the rate of this step is difficult when there is only a small amount of enol in the presence of a large excess of ketone. If we measure the relaxation time in such a system, then at low concentrations of H^+ and E^- we find a reaction of second order, namely the formation of KH from E^- and H^+. The second step, $E^- + H^+ \rightleftharpoons EH$, is negligible in this case. However, at high concentrations of E^- and H^+ we eventually reach a state in which the enol concentration becomes very much higher than the enolate concentration. What we then find is practically only the conversion of KH into EH (with $E^- + H^+$ as a stationary intermediate state), *i.e.* a first-order reaction (which can, however, be base-catalysed by E^-). The transition from a second-order type of reaction ($1/\tau$ increases linearly with $[E^- + H^+]$) to a first-order type of reaction ($1/\tau = \text{constant}$) is shown in Fig. 21. Relaxation measurements by the temperature jump confirmed that this shape of curve applied to barbituric acid[62].

It was found that in the case of rapid keto–enol transformations this method enables small amounts of enol to be determined quantitatively in the presence of a large excess of ketone. The gradient of the linear part of the curve in

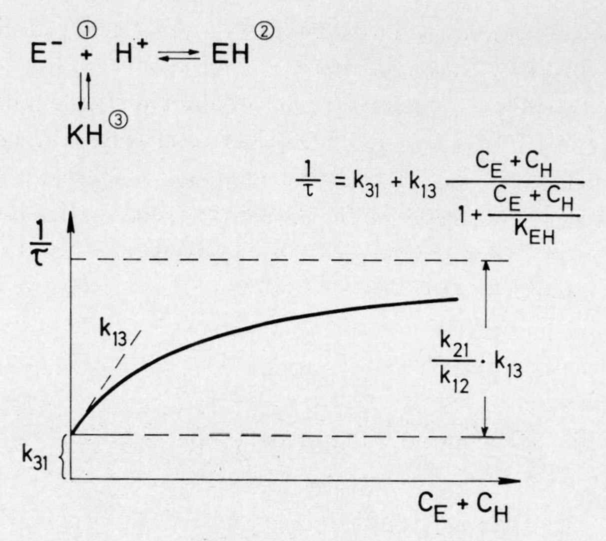

Fig. 21. Reciprocal of the relaxation time of a keto–enol equilibrium. The reaction scheme gives a relaxation spectrum with two time constants. The establishment of the enol equilibrium is very rapid compared to the establishment of the keto equilibrium. The shorter relaxation time is therefore of the form shown in Fig. 3: $1/\tau = k_{12}(\bar{c}_E + \bar{c}_H) + k_{21}$. The longer relaxation time is also of this form (see Fig. 3) as long as $\bar{c}_E$ and $\bar{c}_H$ are small compared to K_{EH} (there is then so little EH present that the keto equilibrium may establish itself independently). However, at high concentrations of E^- and H^+ the quickly established enol equilibrium is always coupled with the keto reaction. $1/\tau$ ultimately tends to a constant limiting value–behaviour characteristic of a first-order reaction (namely EH ⇌ KH). (In the case of base catalysis by E^-, the limiting value can be found by extrapolation.) This method permits the determination of all the kinetic constants and also a very precise determination of both equilibrium constants (and hence also of the keto/enol ratio, even when one form is present in large excess and *both* reactions are taking place quickly).

Fig. 21 and its intercept on the ordinate give the kinetic constants of the keto reaction. The plateau value contains the equilibrium constant of the enol. Other methods generally fail when the dissociation constant of the enol becomes very much larger than that of the ketone and the equilibrium between ketone and enol is rapidly established.

(*c*) *For the biochemist: «intelligent molecules»*

The main field of application of relaxation spectrometry is now in biology. Many methods of measurement have been developed solely for the purpose of studying complex sequences of biological rections or single specific elementary steps in such reactions. The slight disturbance to which the system is subjected enables us to «listen in» on the natural course of processes without going beyond the relatively narrow limits imposed by the «conditions of life».

Naturally enough, it was the enzymes, the key substances in the whole mass and energy balance of the cell, that first attracted our interest. Where previously only the gross, overall rates of enzymatic catalysis had been accessible to the biochemist, it was now possible to come to grips with the «fine structure» on the reaction mechanisms.

This opens up an entirely new world: a world of well-«planned», economically-functioning molecular «machines». The molecules of the inorganic chemist can say only «yes» or «no» – by reacting or not reacting. They may also occasionally exchange a «perhaps» by temporarily entering into a «non-specific» interaction, but this achieves little, for they «forget» everything as soon as they are parted. The molecules of the biochemist are quite different; they can «read», «program», «control», «correlate» different functions – and even «learn». Here is an example.

For a long time biochemists were uncertain how to interpret the fact that certain enzymes do not bind their substrates in accordance with the law of mass action (as any proper molecule should), but exhibit a cooperative behaviour, even though the individual binding groups are so far apart from each other that there is no possibility of a direct interaction between them.

In this case, it is evident that the affinity (or «attraction») of the enzyme for its substrate is initially only moderate, *i.e.* the first molecules of the substrate are bound with only a relatively low affinity but they succeed in arousing the «attraction» of the enzyme. This increases with the supply of substrate molecules, so that finally – close to saturation point – all substrate molecules are bound with high affinity. The interaction of haemoglobin with oxygen is such a case. In the lung, when there is a large supply of O_2, there is a complete (cooperative) saturation, but if the partial pressure of O_2 drops below a certain threshold value the haemoglobin suddenly becomes disinterested and once more gives up all its oxygen. If the saturation of the enzyme with substrate is measured as a function of substrate concentration, a sigmoid curve is obtained instead of the hyperbola expected from simple application of the law of mass action[63].

Monod, Wyman and Changeux[64] have proposed a model to explain this cooperative interaction. Their starting point is that an enzyme consisting of several subunits can occur in two different spatial conformations – one with affinity for the substrate or activator (or of high catalytic efficiency) and one with low affinity (or low catalytic efficiency). In Fig. 22, the two different conformations are indicated schematically by squares and circles; their frequency in the population at the various stages is indicated by the boldness of type used for the lines. In a given conformation all the subunits bind with the same affinity; the sites of binding are so widely separated that they have no direct influence on each other. The arrangement of the subunits is symmetrical. It is assumed that all the subunits are present in the same form, and that this can change only according to an «all-or-none» law. Hybrids are excluded, for the transformation of a single subunit would have a marked effect on the pattern of the interactions between the subunits. Such a model can be described by three parameters and can be made to fit the measured sigmoid binding curves perfectly. These three parameters describe the binding of the substrate by both forms and their isomerization. The Monod model makes a number of assumptions whose justification can be tested experimentally. D. E. Koshland *et al.*[65] put forward an alternative model, which also enabled the binding curves to be reproduced exactly with the aid of 3 parameters. One such alternative is the «induced-fit» mechanism, shown in the diagonal of Fig. 22, in which the change in the structure of a subunit is always associated with the uptake of substrate or activator by that subunit. A decision between the different models can be made with the aid of relaxation measurements. The individual stages of the reaction can be analysed in detail in the relaxation time spectrum, and inappropriate models can be eliminated[66].

My colleague Kasper Kirschner was able to show that the mechanism postulated by Monod, Wyman and Changeux applies for the enzyme glyceraldehyde-phosphate dehydrogenase, a key enzyme in glycolysis[67]. It was possible to give all the values for the rate and equilibrium constants of the individual stages[68]. The relaxation spectrum obtained has already been shown in Fig. 12. This mechanism dies not apply so simply in the case of haemoglobin, and there are deviations in the direction of the alternative mechanism proposed by Koshland *et al.*[69]. It has been found that the two models represent the two possible limiting cases of a single, more general reaction scheme (Fig. 22), and that it is possible to state the conditions under which the system will approximate to one case or the other[66].

These investigations have proved to be very important, for, as Fig. 23 shows, such a mechanism can explain properties which we meet nowhere else on the plane of molecules and which we have only come to know through the man-made circuit and control elements of electronics and transistor technology. It has been known for a long time that such properties occur in biology,

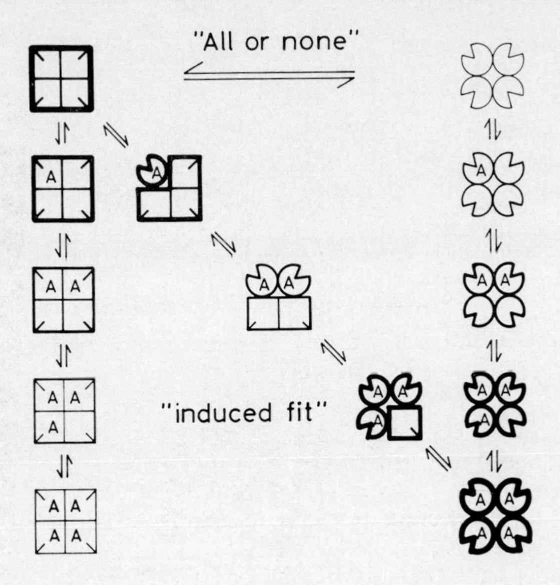

Fig. 22. Reaction scheme of allosteric control in enzymes. The enzyme consists of four identical subunits. The squares and circles indicate two different conformations, only one of which (circle) is capable of transforming the substrate, *i.e.* exhibits catalytic activity; this is indicated by «opening» a site of binding which is «closed» in the square form. The change in conformation is «regulated» by the binding of an activator A. Two alternative mechanisms are shown here: (1) The change in conformation of both forms, which have different affinities for A, always takes place cooperatively according to an «all-or-none» law, but with the square form being preferred in the absence of A and the circular form preferred when A is present at saturation concentration (Monod *et al.*[64]). (2) The change in conformation of the subunits takes place independently of each other, but must first be «induced» in the subunit concerned by the binding of A. A can thus be bound only in one conformation. The induced change in conformation affects the behaviour of adjacent units and thus also brings about cooperative binding (Koshland *et al.*[65]). The two alternatives are naturally only idealized limiting cases of a more general reaction scheme[66].

but it is a new discovery, confirmed by quantitative analysis of molecular mechanisms, that such properties are possessed by single – indeed programmed – molecules, and are not the consequence of a complex of mutually coupled reactions.

Many enzymes have been analysed in this way in various laboratories in recent years[70–72]. Here I should like in particular to mention the work of Britton Chance[73] and his school and that of H. T. Witt and his school[74] on the mechanism of photosynthesis[72].

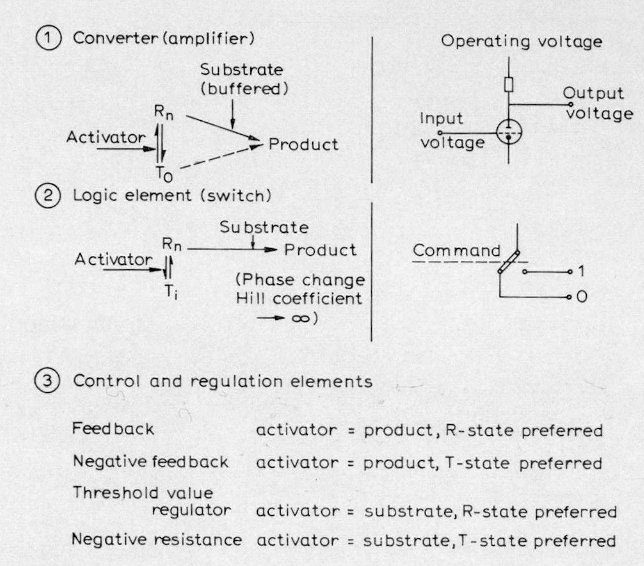

Fig. 23. Comparison of an allosteric enzyme (after Fig. 22) with an electronic circuit element. The activator has the role of a control lattice. Conversion of the substrate into a reaction product corresponds to the current to be regulated. T and R indicate the different conformations of the enzyme (*e.g.* T = squares, R = circles in Fig. 22). Enzyme activity is restricted to the R-form, *i.e.* there is no conversion of the substrate in the T-form (see also ref. 66).

Other classes of biological macromolecules have also been investigated in detail, for example nucleic acids and lipids. The dynamics of code translation in nucleic acids[66,75,76] are of particular interest. Here again we encounter extremely fast reactions. The lifetime of a base pair is measured in fractions of a microsecond. The reading of a code sequence makes use of these reaction steps. Replication, with many correlated individual steps per code unit (such as reading, linking, transporting), takes place in fractions of a millisecond.

9. Where Now

We are just beginning to understand how molecular reaction systems have found a way to «organize themselves». We know that processes of this nature ultimately led to the life cycle, and that (for the time being?) Man with his central nervous system, *i.e.* his memory, his mind, and his soul, stands at the end of this development and feels compelled to understand this development. For this purpose he must penetrate into the smallest units of time and space, which also requires new ideas to make these familiar concepts from physics of service in understanding what has, right into our century, appeared to be beyond the confines of space and time.

10. Epilogue

This description has perhaps failed to do justice to some things that were essential to the development of the main idea. I remember with gratitude those who taught and encouraged me: Arnold Eucken, Ewald Wicke, Karl Friedrich Bonhoeffer and Carl Wagner, who showed me the way and self-lessly encouraged and helped my work. Karl Friedrich Bonhoeffer who turned my path from physical chemistry to biology. Much that I have described is based on the fundamental work of Lars Onsager, Josef Meixner, and many others I was unable to mention. Much was achieved by named and unnamed colleagues and associates, to two of whom I should like to give special mention as representing the others: Konrad Tamm and Leo de Maeyer.

1. A. Eucken, *Lehrbuch der chemischen Physik*, Vol. II/2, Akad. Verlagsges., Leipzig, 1949, p. 1135.
2. P. Langevin, *Ann. Chim. Phys.*, 28 (1903) 433.
3. M. von Smoluchowsky, *Physik. Z.*, 17 (1916), 557, 585.
4. A. Einstein, *Ann. Physik.*, 17 (1905) 549, 19 (1906) 289, 371.
5. L. Onsager, *J. Chem. Phys.*, 2 (1934) 599.
6. P. Debye, *Trans. Electrochem. Soc.*, 82 (1942) 265.
7. H. Hartridge and F. J. W. Roughton, *Proc. Roy. Soc. (London), Ser. A*, 104 (1923) 376.
8. B. Chance, in A. Weisberger (Ed.), *Technique of Organic Chemistry*, Vol. 8, Part II, Interscience, New York, 1963, p. 728.
9. L. Liebermann, *Phys. Rev.*, 76 (1949) 1520.
10 O. B. Wilson and R. W. Leonhard, *J. Acoust. Soc. Am.*, 26 (1954) 223.
11. L. Hall, *J. Acoust. Soc. Am.*, 24 (1952) 704.
12. (a) W. Nernst, *Z. Elektrochem.*, 33 (1927) 428; (b) N. Bjerrum, *Dansk. Mat. Fys. Medd.*, 7 (1926) 9.
13. K. Tamm and G. Kurtze, *Nature*, 168 (1951) 346; *Acustica*, 3 (1953) 33.
14. M. Eigen, G. Kurtze and K. Tamm, *Z. Elektrochem.*, 57 (1953) 103.

15. M. Eigen, *Z. Physik. Chem. (Frankfurt)*, 1 (1954) 176.
16. K. Tamm, G. Kurtze and R. Kaiser, *Acustica*, 4 (1954) 380.
17. M. Eigen and L. de Maeyer, *Z. Elektrochem.*, 59 (1955) 986.
18. M. Eigen, *Discussions Faraday Soc.*, 24 (1957) 25.
19. M. Eigen and K. Tamm, *Z. Elektrochem.*, 66 (1962) 107.
20. A. Einstein, *Sitz. ber. Preuss. Akad. Wiss., Physik.-math. Kl.*, (1920) 380.
21. W. Nernst, see F. Keutel, *Dissertation*, Berlin, 1910; E. Grüneisen and E. Goens, *Ann. Physik*, 72 (1923) 193.
22. K. F. Herzfeld and F. O. Rice, *Phys. Rev.*, 31 (1928) 691; see also G. W. Pierce, *Proc. Am. Acad. Arts Sci.*, 60 (1925) 271.
23. H. O. Kneser, *Ann. Phys.*, 11 (1931) 761, 777.
24. J. Lamb and J. Sherwood, *Trans. Faraday Soc.*, 51 (1955) 1674.
25. R. O. Davies and J. Lamb, *Quart. Rev. (London)*, 11, No. 2 (1957) 134.
26. H. J. Bauer, H. O. Kneser and E. Sittig, *Acustica*, 9 (1959) 181; G. Sessler, *Acustica*, 10 (1960) 44.
27. L. Onsager, *Phys. Rev.*, 37 (1931) 405; 38 (1931) 2265.
28. I. Prigogine, *Etude Thermodynamique des Phénomènes Irreversibles*, Dunod, Paris, 1947; S. R. de Groot and P. Mazur, *Non-equilibrium Thermodynamics*, North-Holland, Amsterdam, 1962.
29. J. Meixner, *Ann. Phys.*, 43 (1943) 470.
30. J. Meixner, *Kolloid-Z.*, 134 (1953) 3.
31. K. Tamm, *Z. Elektrochem.*, 64 (1960) 73.
32. M. Eigen and L. de Maeyer, in A. Weissberger (Ed.), *Technique of Organic Chemistry*, Vol. 8, Part II, Interscience, New York, 1963, p. 895.
33. F. Eggers, *Acustica*, in the press.
34. K. Bergmann, M. Eigen and L. de Maeyer, *Ber. Bunsenges. Physik. Chem.*, 67 (1963) 819.
35. K. Bergmann, *Ber. Bunsenges. Physik. Chem.*, 67 (1963) 826.
36. L. de Maeyer, *Methods in Enzymology*, Academic Press, New York, 1968.
37. J. Suarez, *Dissertation*, T. H. Braunschweig, 1967; L. de Maeyer, M. Eigen and J. Suarez, *J. Am. Chem. Soc.*, in the press.
38. M. Eigen and T. Funck, in preparation.
39. G. Schwarz and J. Seelig, *Biopolymers*, in the press.
40. M. Wien and J. Schiele, *Z. Physik*, 32 (1931) 545.
41. L. Onsager, *J. Chem. Phys.*, 2 (1934) 599.
42. M. Eigen and J. Schoen, *Z. Elektrochem.*, 59 (1955) 483.
43. M. Eigen and L. de Maeyer, *Proc. Roy. Soc. (London), Ser. A*, 247 (1958) 505; see also M. Eigen, L. de Maeyer and H. Ch. Spatz, *Ber. Bunsenges. Phys. Chem.*, 68 (1964) 19.
44. G. Schwarz, *Rev. Mod. Phys.*, 40 (1968) 206.
45. G. Ilgenfritz, in preparation.
46. G. Ilgenfritz and L. de Maeyer, in preparation.
47. E. F. Greene and J. P. Toennies, *Chemical Reaction in Shock Waves*, Arnold, London, 1964.
48. A. Jost, *Ber. Bunsenges. Physik. Chem.*, 70 (1966) 1057.
49. H. Strehlow and H. Wendt, *Inorg. Chem.*, 2 (1963) 6.

50. L. de Maeyer, lecture at the 1968 Spring Meeting of the Optical Society of America, 13–16 March 1968, Washington.
51. G. Czerlinski and M. Eigen, *Z. Elektrochem.*, 63 (1959) 652.
52. H. Diebler, *Dissertation*, Göttingen, 1960.
53. G. G. Hammes and J. I. Steinfeld, *J. Am. Chem. Soc.*, 84 (1962) 4639.
54. R. Rabl, *Dissertation*, in preparation.
55. M. Eigen, in S. Claesson (Ed.), *Fast Reactions and Primary Processes in Chemical Kinetics*, Nobel Symposium No. 5, Almqvist and Wiksell, Stockholm, 1967, p. 477.
56. M. Eigen, *Jahrb. Max-Planck-Ges. Förderung Wiss.*, (1966) 40.
57. M. Eigen, *Z. Elektrochem.*, 64 (1960) 115; H. Diebler and M. Eigen, *Proc. 9th Intern. Conf. Coordination Chem.*, Verlag Helv. Chim. Acta, Basel, 1966, p. 360.
58. M. Eigen and R. G. Wilkins, *Advan. Chem. Ser.*, 49 (1965) 55.
59. M. Eigen, *Pure Appl. Chem.*, 6 (1963) 97.
60. M. Eigen, W. Kruse, G. Maass and L. de Maeyer, *Progr. Reaction Kinetics*, 2 (1964) 285.
61. M. Eigen, *Angew. Chem.*, 75 (1963) 589; *Angew. Chem. Intern. Edn.*, 3 (1964) 1.
62. M. Eigen, G. Ilgenfritz and W. Kruse, *Chem. Ber.*, 98 (1965) 1623.
63. G. S. Adair, *J. Biol. Chem.*, 63 (1925) 529.
64. J. Monod, J. Wyman and P. Changeaux, *J. Mol. Biol.*, 12 (1965) 88.
65. D. E. Koshland, G. Nemethy and D. Filmer, *Biochemistry*, 5 (1966) 365.
66. M. Eigen, see ref. 55, p. 333.
67. K. Kirschner, M. Eigen, R. Bittman and B. Voigt, *Proc. Natl. Acad. Sci.*, 56 (1966) 1661.
68. M. Eigen, G. Ilgenfritz and K. Kirschner, in preparation; K. Kirschner, *Current Topics of Microbiol.*, 44 (1968) in the press.
69. T. M. Schuster, G. Ilgenfritz and M. Eigen, in preparation.
70. M. Eigen and G. G. Hammes, *Advan. Enzymol.*, 25 (1964) 1.
71. P. Fasella and G. G. Hammes, *Biochemistry*, 6 (1967) 1798; J. E. Erman and G. G. Hammes, *J. Am. Chem. Soc.*, 88 (1966) 5607, 5614.
72. M. Eigen, *New Looks and Outlooks on Physical Enzymology*, in the press.
73. B. Chance, see ref. 55.
74. H. T. Witt, see ref. 55.
75. M. Eigen, in F. O. Schmitt *et al.* (Eds.), *The Neurosciences, a Study Program*, Rockefeller University Press, New York, 1967, p. 130.
76. M. Eigen and D. Pörschke, in preparation.

Biography

Manfred Eigen was born in Bochum on 9 May 1927, the son of the chamber musician Ernst Eigen and his wife Hedwig, née Feld. He received his schooling at the Bochum humanistic Gymnasium.

In the autumn of 1945 he commenced the physics and chemistry course at the Georg-August University in Göttingen and obtained his doctorate in natural science in 1951. He wrote his dissertation on the specific heat of heavy water and aqueous electrolyte solutions under the guidance of Arnold Eucken. After two years as an assistant lecturer at the physical chemistry department of the university under Ewald Wicke, he transferred to the Max-Planck-Institut für physikalische Chemie, which had moved to Göttingen under the Directorship of Karl Friedrich Bonhoeffer. The influence of Bonhoeffer, who provided him with magnificent working conditions at the Institut, is reflected in his later work in the field of biophysical chemistry.

Eigen began his work on the problem of fast ionic reactions in solution in the period 1951–1953, encouraged to do so by the ultrasound absorption measurements carried out by his colleagues Konrad Tamm and Walter Kurtze. During the following years he developed a series of measuring techniques involving times down to the order of a nanosecond. He developed many of these techniques with Leo de Maeyer, who joined him in the autumn of 1954, and with whom he is still collaborating closely at the Göttingen Max-Planck-Institut. The Max-Planck-Gesellschaft appointed Eigen a Scientific Member in 1957 and head in 1964. In 1967 he was elected Managing Director of the Institute for a period of three years. At the same time he was appointed to the Scientific Council of the German Federal Republic.

Eigen's scientific development is reflected in the close on 100 papers he has published. The subject matter of these works ranges from the thermodynamic properties of water and aqueous solutions, and the theory of electrolytes, through thermal conductivity and sound absorption, to fast ionic reactions.

In the years 1953–1963 followed the description of a series of novel measuring techniques used for the study of very fast reaction in the range from one second to one nanosecond. The gap between the region of classical reaction kinetics and spectroscopy was thus closed. Eigen was particularly interested in proton reactions: together with De Maeyer he was the first to determine the neutralization rate and found the anomalous conduction characteristics of protons in ice crystals. The development of the theory of relaxation of multi-stage processes was followed by studies on metal complex reactions, in which the fast reactions of a large number of metal ions were investigated in relation to their position in the periodic table. Around 1960 the emphasis in his work shifted towards physical-organic chemistry. The individual steps of a series of reaction mechanisms were elucidated, and a general theory of acid-base catalysis was verified experimentally.

At the same time, however, his attention turned also to biochemical questions, which now claimed his chief interest. These questions ranged from hydrogen bridges of nucleic acids, through the dynamics of code transfer, to enzymes and lipid membranes. Biological control and regulation processes, and the problem of the storage of information in the central nervous system also occupy his attention. Practically every year he travels together with his friend and colleague Leo de Maeyer to Boston to discuss topics of common interest with American neurologists, biochemists, and biophysicists.

Eigen holds the following honours and distinctions: Bodenstein prize of the Deutsche Bunsengesellschaft, 1956; Otto-Hahn Prize for Chemistry and Physics, 1962; Kirkwood Medal (American Chemical Society), 1963; Harrison Howe Award (American Chemical Society), 1965; Andrew D. White Professor at large at Cornell University, Ithaca, N.Y., 1965; Honorary Professor at the Technische Hochschule, Braunschweig, 1965; Foreign Honorary Member of the American Academy of Arts and Sciences, 1964; Member of the «Leopoldina», Deutsche Akademie der Naturforscher in Halle, 1964; Member of the Göttingen Akademie der Wissenschaften, 1965; Honorary Member of the American Association of Biological Chemists, 1966; Honorary degree of doctor of science at Harvard University, U.S.A., 1966; Honorary degree of doctor of science at Washington University, U.S.A., 1966; Foreign Associate of the National Academy of Sciences, Washington, U.S.A. 1966; Honorary degree of doctor of science, University of Chicago, U.S.A., 1966; Carus Medal of the Deutsche Akademie der Naturforscher «Leopoldina», Halle, 1967; Linus Pauling Medal of the American Chemical Society, 1967.

Manfred Eigen is married to Elfriede, née Müller. They have two children, Gerald (born 1952) and Angela (born 1960). In his free time he is a keen amateur musician. His favorite holiday pastime is mountaineering.

R. G. W. Norrish

Some fast reactions in gases studied by flash photolysis and kinetic spectroscopy

Nobel Lecture, December 11, 1967

Realisation that free radicals and atoms take part in chemical reactions has focussed attention on the processes of photo-chemistry which are not only paramount in the geochemistry of the upper atomosphere but are also basic to many reactions of organic chemistry involving free radicals and the triplet state; this realisation also has led to the development of gas lasers, and to the exploration in detail of the intimate anatomy of reactions of pyrolysis, combustion and explosion.

Classical photochemistry emerged in 1908 with the understanding by Stark of the distinction between the primary and secondary photochemical processes, of which the former is the immediate result of the absorption of a light quantum by a molecule or atom and the latter the subsequent «dark» reactions initiated by the products of the former[1]. Into this simple pattern it has been possible to fit the whole gamut of photochemical phenomena – fluorescence, phosphorescence, photolytic and photosynthetic processes, photocatalytic and photosensitised reactions. Determination of quantum yields led to the distinction of endoactinic and exoactinic reactions; the former being endothermic in character draw their energy requirement from the absorbed quantum and rarely exceed an overall quantum yield of 2, the latter, being exothermic release their «pent up» energy by photochemical initiation and are usually of the nature of chain reactions, with high quantum yields, and sometimes explosive characteristics. For example, the dissociation of hydrogen iodide into its elements is 2200 cal endothermic and its quantum yield is limited to two[2] while the synthesis of hydrogen chloride from its elements is exothermic to the extent of 22,000 cal and may have a quantum yield[3] as high as 10^6.

It was indeed the study of these two reactions that first led to the conclusion that the primary reaction may involve photolysis of the reactant into atoms (and later free radicals). In the former case we have

$$HI + h\nu = H + I$$

in the latter

$$Cl_2 + h\nu = Cl + Cl$$

followed by the well known H_2-Cl_2 chain reaction. We owe much to Bodenstein, Warburg and Nernst[4] by whose early work the reality of the participation of atoms in chemical reactions was made apparent and the concept of the chain reaction established. Following this, the reactions of H atoms generated by an electric discharge through hydrogen gas were established by R. W. Wood[5] and by Bonhoeffer[6], and the production of free alkyl radicals by the pyrolysis of metal alkyls proved unequivocably by Paneth[7].

Simultaneously the growth of the study of the band spectra of gaseous molecular species in particular by Frank[8] and V. Henri[9] clarified the quantum mechanisms of the processes of thermal dissociation, photo dissociation and predissociation, indicating the production of free radicals and atoms in both ground and electronically excited states. It may justly be claimed that from the marriage of photokinetics with spectroscopy there resulted a new insight into the mechanism of chemical reactions; the part played by atoms, free radicals and excited species as transient intermediates became abundantly apparent. The reactions of these transients however, which together make up the overall process of conversion of reactants to final products are so fast that they can neither be observed nor isolated by classical means, and their nature and participation could until recently only be deduced from the circumstantial evidence of reaction kinetics, quantum yields, and the spectroscopic characteristics of the reactants.

It therefore became of importance if further progress was to be made, to endeavour to obtain objective evidence of the presence of short lived transients both in thermal and photochemical reactions. Using continuous sources of the highest attainable intensity (*e.g.* a 10 kW high pressure mercury arc) the author and his collaborators in 1946 attempted to obtain evidence by spectroscopic means of a stationary concentration of intermediates in such reactions as the photolysis and photo oxidation of ketene without success. In no case could any absorption spectrum which could be attributed to reacting transients be observed in the reacting system and it became apparent that their reactivity was so great that no sufficient stationary concentration for detection by the means then available could be achieved.

It was the realisation that enormously greater «instantaneous» light intensities could be obtained from a powerful light flash than from a conventional light source, and that such a flash need not be of greater duration than the half life of the elusive transients that led Porter and me to study the results of applying such flashes to suitable responsive photochemical systems[10]. Using an electric discharge from a condenser bank through inert gas, dissipating about 10 000 joules it was immediately found that the resulting light flashes of about 2–3 milliseconds duration were able to create large measures of photodecomposition in reactants such as nitrogen peroxide, chlorine, ketene, acetone and diacetyl, amounting to 100% in some cases. It was obvious that momentarily there must be very high concentrations of free radicals or atoms in such reacting systems which by suitable means should be detectable by absorption spectroscopy. This was first achieved by Porter[11] who using a second less powerful flash triggered mechanically by the method of Oldenberg[12] at specific short intervals after the first was able to observe the complete dissociation of chlorine by the disappearance of the Cl_2 absorption spectrum and its return over a period of milliseconds as the atoms recombined.

The modern method of flash photolysis developed from this uses an electronic technique by which the first flash (photoflash) is caused photo-electrically to trigger the second flash (specflash) at specific short intervals measured in microseconds and milliseconds[13]. The photoflash is generated by discharging a capacity of the order of 40 μF at 10 kV through an inert gas such as krypton or xenon contained in a quartz tube generally 50 cm in length and 1 cm in diameter. The reaction vessel is a quartz tube of similar dimension with plane quartz end plates, lying close to and parallel to the photoflash tube. The specflash lamp consisting of a quartz capillary tube about 10 cm in length is placed «end on» to the reaction vessel, and has a plane quartz end plate so that by means of a lens and limiting stop, a beam of light can pass longitudinally through the reaction vessel to a suitable spectrometer to register the absorption spectrum of the reacting system at any specific interval after the photoflash (see Fig. 1). The discharge is made as before through inert gas.

The energy dissipated by the discharge of a condenser is given in joules by the relationship, $E = 1/2\ CV^2$, where the capacity is measured in microfarads, and the potential difference in kilovolts. For a given energy the duration of the flash is the shorter the smaller C and the greater V; the self inductance of the circuit must be kept as low as possible. For the photoflash, a convenient energy

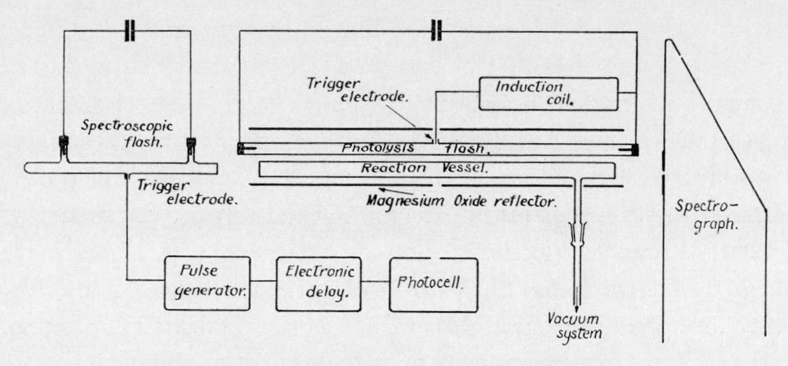

Fig. 1. Diagram of flash photolysis apparatus.

dissipation is 2 000 J derived from the discharge of 40 μF at 10 kV. The half life of the light flash is about 10 μsec. For the specflash a discharge of 100 J is generally used, obtained by discharging 2 μF at 10 kV; its half life is of the order of 2 μsec. The pressure of gas in both lamps is of the order 5–10 cm Hg. The reaction vessel may be double walled for the introduction of gaseous or liquid colour filters in the annular space. Both it and the photolysis lamp are surrounded by a tubular reflector coated on the inside with magnesium oxide, and when necessary the whole can be mounted in a tubular electric furnace. A general description of the apparatus which throughout our work has had several minor modifications is given in detail by Norrish, Porter and Thrush (ref. 13); the technique at present in use, represents a compromise between all the factors affecting its operation. Improvements have been effected by using highly transparent «spectrosil» quartz which transmits down to 1 600 Å, end plates of lithium fluoride, and vacuum spectrographs for detection of transients whose absorption spectra lie in the far ultraviolet. Of great importance for the future is the reduction in the periods of the photoflash and specflash to achieve greater time resolution, and the development of highly transparent materials for construction of apparatus suitable for shorter wave photolysis than is at present available.

It may readily be calculated that the «instantaneous» dissipation of only 1 joule of energy (*i.e.* about 0.05% of the total output of the photoflash) by 150 ml of gas at 1.0 mm pressure will raise the temperature of the reactant by about 5 000°C for there is no time for cooling during the short period of the flash. Thus the early results of flash decomposition are more properly regarded as flash pyrolysis than flash photolysis, and unless steps are taken to neutralise this rise in temperature by the dilution of the system by the addition of a large excess of inert gas, we cannot expect to study the photochemical effects divorced from thermal complications. This however is readily done: by the introduction of inert gas at pressures of 100 to 500 times that of the reactant, the temperature rise can be kept below 10°C, which for practical purposes may be regarded as isothermal, while for reactions in solution of course, there is no problem. On the other hand we may take advantage of flash heating in undiluted systems to administer an adiabatic shock which for many purposes is superior and certainly simpler than the technique of shock wave kinetics. This arises from the fact that by flash heating the whole system is instantaneously and nearly homogeneously heated to high temperatures, making possible the detection of the transient products of pyrolysis and growth and decay of intermediates in chain reactions leading to explosion in suitable systems. Indeed, it is the homogeneity of the explosive processes which makes it possible to observe in absorption the unexcited radicals taking part; we have in fact in a reaction vessel 0.5 m in length a «flame front» virtually 0.5 m thick which is very different from the thin element propagating an explosive wave. This is important because it makes possible for the first time the observation of the reactions of unexcited species leading to and taking part in explosion as well as the electronically excited species to which we were limited in the past.

Thus there are two ways in which we can employ the techniques of kinetic spectroscopy and flash photolysis, the *isothermal* method and the *adiabatic*. The field of their application is almost unlimited; I must content myself with general remarks, and three specific examples.

The first objective of flash photolysis, namely to observe the growth and decay of radical species by kinetic spectroscopy has been achieved; following the first demonstration of the dissociation of chlorine, the spectrum of the ClO radical was first seen in absorption on flashing a mixture of chlorine and oxygen. Its origin was ascribed to the almost complete dissociation of chlorine, and to the reaction to be expected from the chlorine atom in an atmosphere of oxygen. It was possible to show that the sequence of reactions

$$Cl_2 + h\nu = Cl + Cl$$

$$Cl + O_2 = ClOO$$

$$ClOO + Cl = ClO + ClO$$

$$ClO + ClO = Cl_2 + O_2$$

in which the final state of the system is the same as the first provides a complete basis for explaining the reaction.

The study of this reaction[14] constituted an early success in the application of flash photolysis to chemical kinetics and will be described in some detail today by my colleague Professor Porter.

Dr. Husain has collected references to some sixty simple free radicals and atoms which have been discerned and characterised in absorption, either by isothermal flash photolysis or by adiabatic flash pyrolysis and explosion. Prominent among them are CH, CH_2, CH_3, NH, NH_2, OH, HCO, HNO, CN, CS, ClO, BrO, IO, NCl, NCl_2, PH, PH_2, PO, PN, SH, SO, SiO, TeO, TeH, W, Te, Sn, Hg, Fe, Mn and also highly vibrating states of several molecular species, such as O_2. The collection of this information is the first step towards identifying the nature of radical reactions observed by kinetic spectroscopy. To illustrate this we shall now consider two examples of the application of the isothermal technique, the first involving the primary photolysis of nitrosyl halides, the second the secondary reactions associated with the photolysis of nitrogen dioxide, chlorine dioxide and ozone.

Vibrational Excitation by Primary Reaction

The flash photolysis of nitrosyl halides – Vibrational relaxation

The sequence of spectra shown in Fig. 2 show the course of the photolytic dissociation of nitrosyl chloride, typical also of nitrosyl bromide, which takes place in the region of 2600 Å (Basco and Norrish[15]). A study of a large number of plates showed that the primary product, NO is highly vibrationally excited in the ground state comprising all levels from $v''=11$ to $v''=0$. All these were observed in absorption in the β, γ, δ and ε spectra of NO; the rotational temperature of the molecule was, however, unaffected. By using NO as a light filter surrounding the reaction vessel it was proved that these excited species do not have their origin in the secondary excitation of NO molecules and after consideration of all possibilities it was concluded that they are in fact the product of the primary photolysis of the halide, NOX:

$$NOX + h\nu = NO^{*} + X$$

It was found that the relative «instantaneous» population of the higher levels of nitric oxide increased as the halide pressure decreased and that *at first* the level $v''=1$ was barely detectable. The decay of the higher excited levels was however extremely rapid and increased with the pressure of the halide yielding ultimately the level $v''=1$ which accumulated and was virtually the only excited level detectable after the photoflash. It was in fact established that

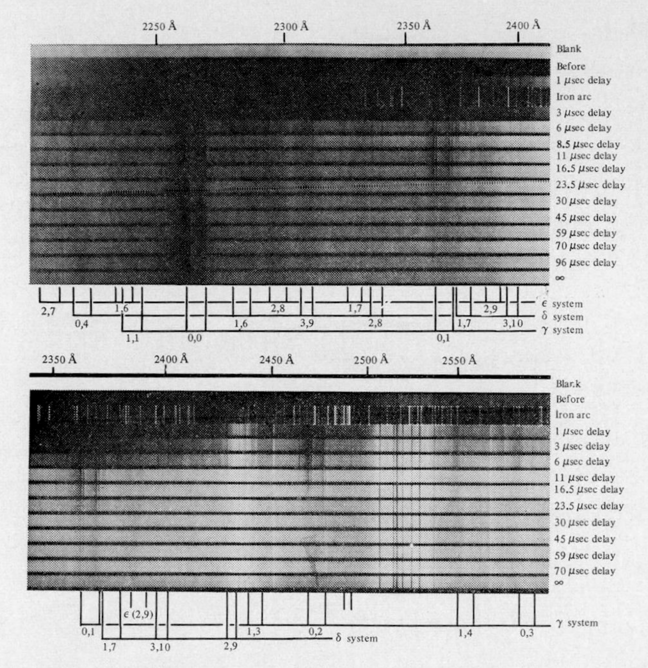

Fig.2. Vibrationally excited NO produced in the flash photolysis of NOCl. Upper picture: pressure of NOCl, 1.0 mm Hg; pressure of N_2, 375 mm Hg. Lower picture: pressure of NOCl, 2.0 mm Hg; pressure of N_2, 420 mm Hg. Flash energy, 1600 J. (Basco and Norrish[15])

the rate of decay is determined by the pressure of the unchanged nitrosyl halide, and that on the other hand, the effect of inert gases was not detectable.

The rapidity of decay of NO^* and the specific effect of the parent NOX suggests that near-resonance transfer processes are operating in deactivation as indeed is confirmed by the fact that the vibration frequencies of NO in the range of levels $v = 11$ to $v = 1$ lie between 1900 and 1600 cm^{-1} while for both NOCl and NOBr the frequency associated with the NO bond was found by Burns and Bernstein[16] to be 1800 cm^{-1}.

At this point, however, there arises an apparent anomaly. The observation of Pearse and Gaydon[17] showed that the first levels in the ground state of NO can be populated by fluorescence as shown diagrammatically in Fig. 3. This fluorescence which consists of the banded $v = 0$ progression $A^2\Sigma^+ \leftarrow X^2\Pi$ of NO was also seen by Basco, Callear and Norrish[18] using the flash technique; yet by the same means they were unable to observe levels higher than $v = 1$ in absorption, with the exception of $v = 2$ *very* faintly, Fig. 4. It might be postulated that the higher levels are populated very weakly relatively to the first,

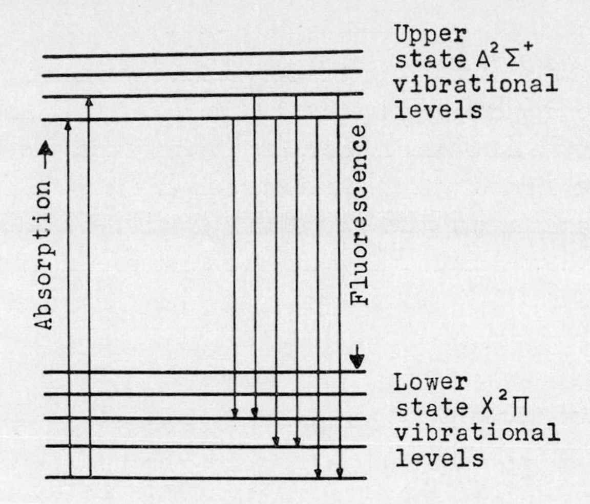

Fig. 3. Diagrammatic representation of population of vibrational levels of NO in the ground state by fluorescence.

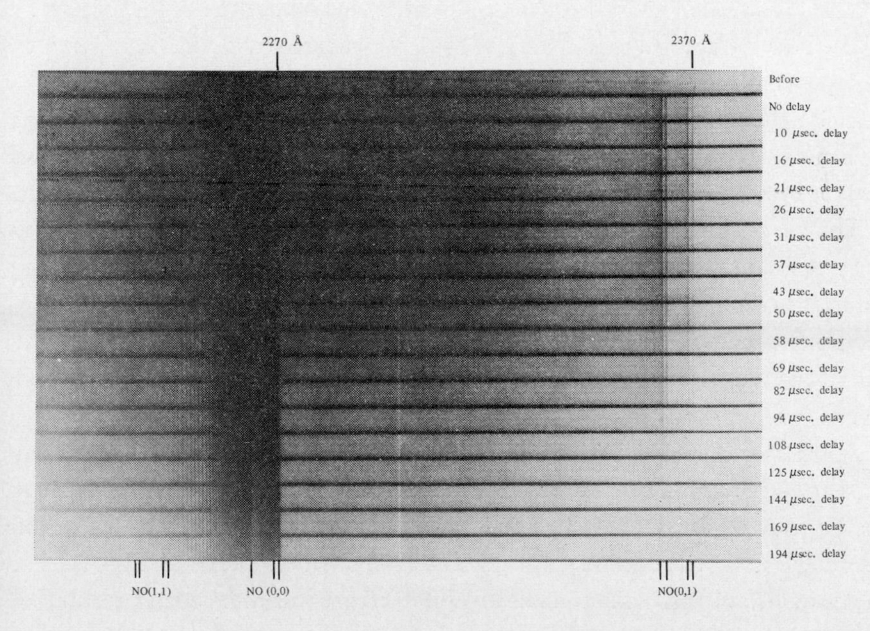

Fig. 4. NO $^2\Pi$ ($v=1$) produced by flash fluorescence of NO, showing decay. Pressure of NO, 5 mm Hg; pressure of N_2, 600 mm Hg; flash energy, 1600 J. (Basco, Callear and Norrish[18])

but this is not so; Pearse and Gaydon from a measurement of the intensities of the fluorescent bands found the first five levels to be populated almost equally. Herein lies the problem: why is only the level $v=1$ seen by kinetic spectroscopy and why do the higher levels $v=2$, 3, 4 and 5 decay too rapidly to be observed in absorption when the same and higher levels derived from the photolysis of the nitrosyl halides are readily detected and their decay, albeit rapid, easily followed in times measured in microseconds?

The solution to this apparent anomaly may be achieved by means of the two following hypotheses[19]:

(*1*) The most favourable resonant collisions are between closely associated levels of the vibrating species, *e.g.*

$$NO_{v=n} + NO_{v=(n-2)} \rightarrow 2NO_{v=(n-1)}$$

and owing to change in frequency of levels due to anharmonicity, the most favourable of all will be obtained when the frequency levels differ by 2 as above.

(*2*) At the instant of production from the nitrosyl halide the $NO^{\star}$ is formed in very high vibrating states–say $v=12$, 11, or 10.

The vibrational energy of $v=11$ is 55 kcal, and since the bond strength of NO–Cl is 38 kcal there is plenty of energy available from the light quantum (say 98 kcal for 2800 Å) for this to occur. The same applies for NOBr. In consequence there is a gap between $v=10$ (say) and $v=0$ and in the absence of other deactivating species (inert gases ineffective) the high vibrational levels cannot be relaxed. This of course is an ideal conception; lower levels will be built up by collisional deactivation by species such as the nitrosyl halides as we have seen, but it will be a relatively slow process compared with self deactivation. As the lower levels are populated so will resonant self-quenching increase, but there will always be an irregular distribution which will cause a retardation, and further, since high levels are continually fed in by the flash, the irregular distribution will be preserved and all levels will be observed during its operation.

In contrast, when the first five vibrational levels of the ground state of NO are populated by fluorescence they are populated as we have seen above, nearly equally; thus the highly efficient process of self-quenching described in hypothesis (*1*) can take place as shown in Fig. 5, and all levels are deactivated to $v=1$ when the resonant process must of necessity stop. The collapse of the pattern is so rapid that only the first level is seen to be overpopulated, and this can only be deactivated slowly by the inefficient process of collisional con-

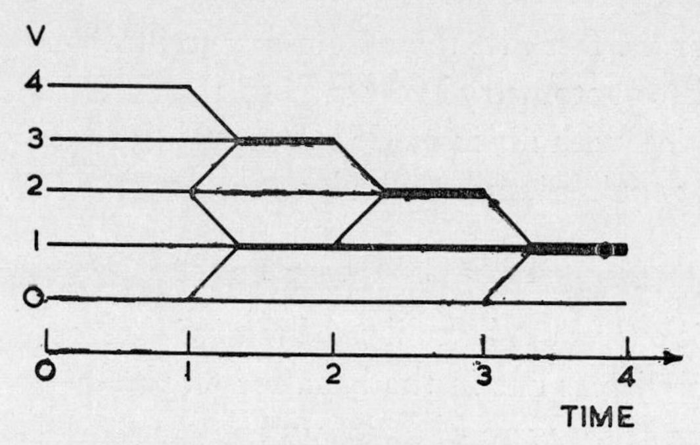

Fig. 5. Diagrammatic representation of relaxation of vibrational energy of NO by self quenching.

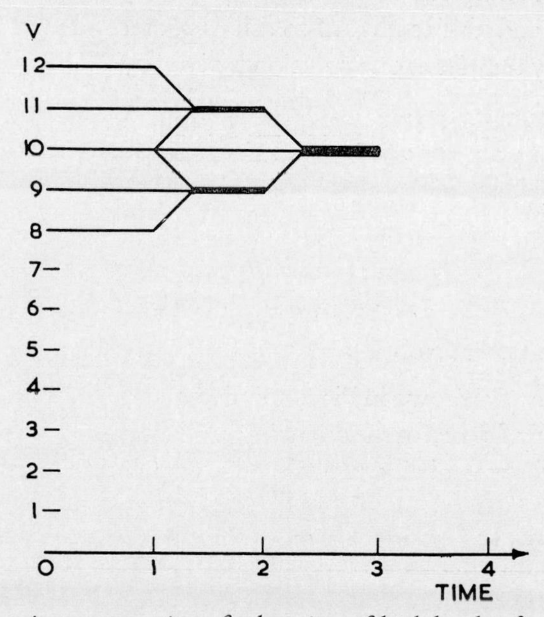

Fig. 6. Diagrammatic representation of relaxation of high levels of vibrational energy of NO, restricted by isolation. (Ideal)

version to translational energy. If, however, we have a gap in the vibrational distribution or a series of irregularities in the sequence of population of the pattern of vibrational levels as with NO* derived from nitrosyl halides the resonant deactivation must be brought to a halt, or slowed down, shown for an ideal case in Fig.6. Overpopulation of all higher levels is observed.

The overpopulation of the $NO_{v=1}$ level in the ground state by fluorescence (Fig. 4) makes possible the quantitative study of the relaxation reaction[18]

$$NO_{v=1} + M = NO_{v=0} + M'$$

This arises from the fact that the absolute concentration of $NO_{v=1}$ can be measured by plate photometry because the (0,1) band is visible spectroscopically in absorption with nitric oxide at atmospheric pressure, and since its concentration at equilibrium is given by

$$[NO^*] = [NO]\, e^{-h\nu/kT}$$

the photometric curves can be calibrated to give absolute concentrations by choosing one particular line in the band for measurement. In this way the curves shown in Fig. 7 were obtained; when plotted logarithmically they give good straight lines indicating first order decay from which the unimolecular constant k_3 can be obtained. $1/k_3$ is the mean life time τ of the excited species, and if this suffers Z collisions per second then P_{1-0} the probability of energy transfer at one collision is given by

$$P_{1-0} = \frac{1}{\tau Z} = \frac{k_3}{Z}$$

k_3 can be split into two terms depending on relaxation by NO, and by any added gas M. Thus

$$k_3 = k_4(NO) + k_5(M)$$

and k_4 and k_5 may be calculated from the various values of k_3 derived from the curves of the type shown in Fig. 7 for nitrogen. The data shown in Table 1 show preliminary figures for the quencing probabilities of various added gases, the high value for water being probably due to chemical reaction.

Table 1

Molecule	NO	CO	H_2O	CO_2	N_2	Kr
P_{1-0}	$3.55 \cdot 10^{-4}$	$0.25 \cdot 10^{-4}$	$7 \cdot 10^{-3}$	$1.7 \cdot 10^{-4}$	$4 \cdot 10^{-7}$	zero

Further studies[18] of relaxation by CO indicated unmistakably that the process occurs by resonant transfer of vibration

$$NO_{v=1} + CO_{v=0} = NO_{v=0} + CO_{v=1}$$

The concentration of CO^* was measured by photometering the unresolved band of the fourth positive $A^1\Pi \rightarrow X^1\Sigma^+$ system which is visible in the spectrum of CO at atmospheric pressure and so can be used to measure in absolute terms the vibrational exchange between NO and CO shown in Fig. 8.

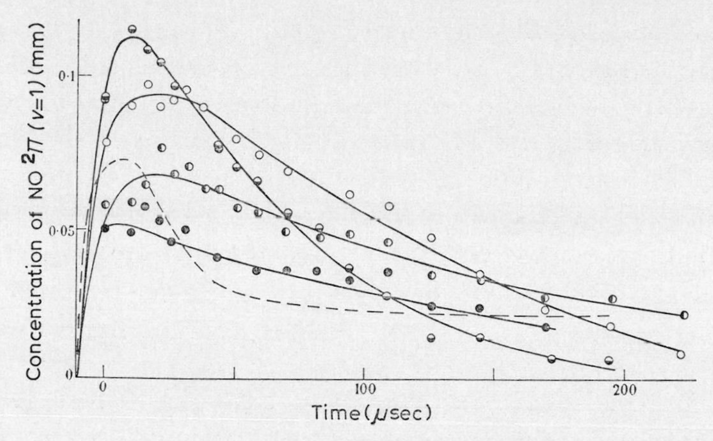

Fig. 7. Rise and decay of NO $^2\Pi$ ($v = 1$) measured by plate photometry. ○, 2 mm NO + 600 mm N_2·(1600 J). ●, 2 mm NO + 220 mm N_2·(1600 J). ◐, 1 mm NO + 600 mm N_2·(1600 J). ◒, 5 mm NO + 600 mm N_2·(1600 J). --, 50 mm NO + 467 mm N_2·(900 J). (Basco, Callear and Norrish[18])

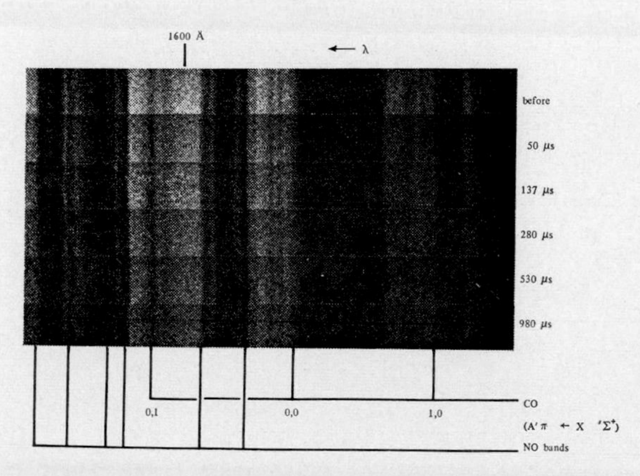

Fig. 8. Production of CO ($v = 1$), ground state, by resonance with NO ($v = 1$), ground state. NO pressure, 5 mm; CO pressure, 100 mm; N_2 pressure, 650 mm. (Basco, Callear and Norrish[20])

Studies of the photolysis of $(CN)_2$, CNBr and CNI generically represented as CNR by kinetic spectroscopy yield results similar to those described for NOBr and NOCl (Basco *et al.*[20]). These substances absorb at the short end of the quartz ultraviolet below 2300 Å and on flashing in the presence of inert gas yield vibrationally excited CN radicals up to $v = 6$ which are observed spectroscopically in absorption in the $\Delta v = 0 \pm 1$ and -2 sequences of the violet ($B^2\Sigma - X^2\Sigma$) system at 3 590, 3 883, 4216 and 4660 Å. Decay sequences with time of CN^* indicated the preferential production of CN^* in the higher excited vibrational states and their decay by collision with CNR as with the analogous nitrosyl halide reactions, but owing to the very high extinction coefficient of the CN radical itself there was also detected, using colour filters, a high secondary population of CN^* resulting from absorption of light in the region 3 500–4 500 Å by CN far outside the photolytic wave lengths of CNR. The process

$$CN \cdot X^2\Sigma(v=0, 1, 2\ldots) \rightleftharpoons CN \cdot B^2\Sigma\ (v=0, 1, 2\ldots)$$

is indicated involving many reversible excitations during the flash, the reverse reaction taking place either by fluorescence or collision; but in the end only $v=1$ persists as before.

Vibrational Excitation by Secondary Reactions

The reactions of oxygen atoms

The photolysis of nitrogen dioxide, chlorine dioxide and ozone studied by the techniques of classical photochemistry were all concluded to proceed by similar mechanisms, involving the primary generation of oxygen atoms, as follows[21]:

Nitrogen dioxide:

$$NO_2 + h\nu = NO + O$$
$$O + NO_2 = NO + O_2$$

Chlorine dioxide:

$$ClO_2 + h\nu = ClO + O$$
$$ClO_2 + O = ClO + O_2$$
$$ClO + ClO = Cl_2 + O_2$$

Ozone:

$$O_3 + h\nu = O_2 + O$$
$$O + O_3 = O_2 + O_2^*$$
$$O_2^* + O_3 = O_2 + O_2 + O$$
$$O_2^* + M = O_2 + M'$$

The quantum yield of the first two reactions in the near ultraviolet is of the order 2. In the case of O_3 it was measured as up to 8 in the region of 2000–

2 500 Å, but limited to 2 when photolysis occurs at the red end of the spectrum. Thus a chain reaction is indicated in the former case, which owing to the inherent simplicity of the system must be propagated as shown by excited oxygen molecules, considered by the earlier workers to be an electronically excited species. On studying these reactions by isothermal kinetic spectroscopy, we found that not only is the kinetic scheme of reactions shown above confirmed, but that in addition highly vibrating oxygen molecules in the ground state, cold rotationally and translationally, are produced in each case. Thus the reactions

$$NO_2 + O = NO + O_2^*$$

$$ClO_2 + O = ClO + O_2^*$$

$$O_3 + O = O_2 + O_2^*$$

were indicated[22,23]. With NO_2 vibrational levels up to $v = 11$ were observed; with ClO_2 levels up to $v = 8$, and with O_3 levels up to $v = 17$–20. In each case more than half the exothermic energy of reaction appeared unequilibrated as vibrational energy of the oxygen molecule observed in absorption in the Schumann–Runge spectrum. Fig. 9 shows the flash photolysis of ClO_2 in which after flashing, the transient spectrum of the ClO radical is seen together with the absorption by highly vibrating oxygen molecules. The production of excited O_2^* is seen more clearly in Figs. 10 and 11 resulting from the photolysis of NO_2 and O_3 respectively.

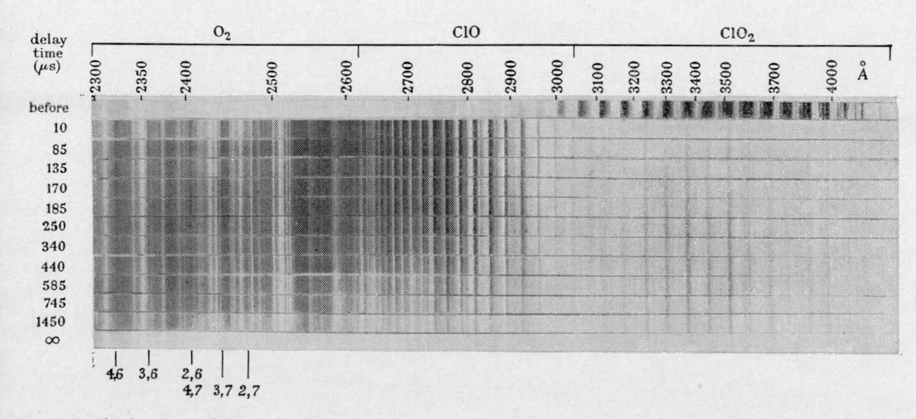

Fig. 9. Flash photolysis of ClO_2. ClO_2 pressure, 0.5 mm Hg; N_2 pressure, 580 mm Hg. Flash energy, 320 J, showing ClO and vibrationally excited O_2 (latter seen with difficulty owing to low dispersion). (Lipscomb, Norrish and Thrush[22])

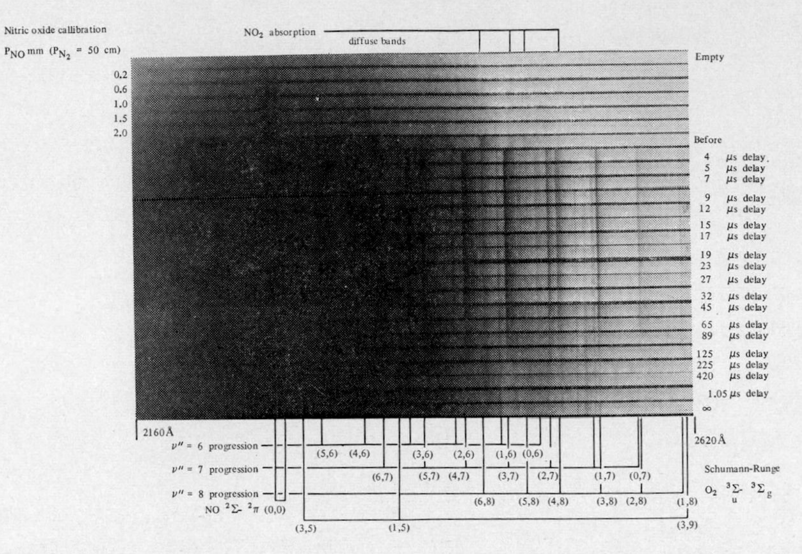

Fig. 10. Decay of vibrationally excited O_2 resulting from the flash photolysis of NO_2 under isothermal conditions. NO_2 pressure, 2 mm Hg; N_2 pressure, 500 mm Hg. Flash energy, 2025 J. (Husain and Norrish[34])

These results led McGrath and Norrish[24] to the tentative generalisation that when an atom reacts with a polyatomic molecule, a large proportion of the exothermic energy of reaction is preferentially located initially as vibration in the newly formed bond, *i.e.*

$$A + BCD = AB^{\star} + CD$$

Qualitatively this seems reasonable since the main interaction must be visualised as between A and B, while the elimination of CD could well occur without much appreciable change in the interatomic distance between the parts C and D. The generalisation has now been widely confirmed. McGrath and Norrish[24] have shown by flash photolysis that the reactions

$$Cl + O_3 = ClO^{\star} + O_2$$

$$Br + O_3 = BrO^{\star} + O_2$$

yield highly vibrating ClO and BrO with up to six quanta of vibration, while likewise, the reactions of ^1D oxygen atoms derived from ozone on reacting with a wide range of hydrides yield vibrationally excited OH[26], *e.g.*

$$O + H_2O = OH^* + OH$$

$$O + NH_3 = OH^* + NH_2$$

$$O + H_2 = OH^* + H$$

Other examples such as the reactions of hydrogen atoms observed by McKinley, Garvin and Boudart[27] and Cashion and Polyani[28],

$$H + O_3 = OH^* + O_2$$

$$H + Cl_2 = HCl^* + Cl$$

$$H + Br_2 = HBr^* + Br$$

further confirm the correctness of our generalisation which invites detailed quantitative study, and must probably await greater time resolution in our technique before it can be achieved. For example, we cannot yet be sure whether the vibrationally excited products are produced *ab initio* in their highest vibrating state and relax subsequently, or whether a complete spectrum of vibrating states results directly as part of the reaction mechanism.

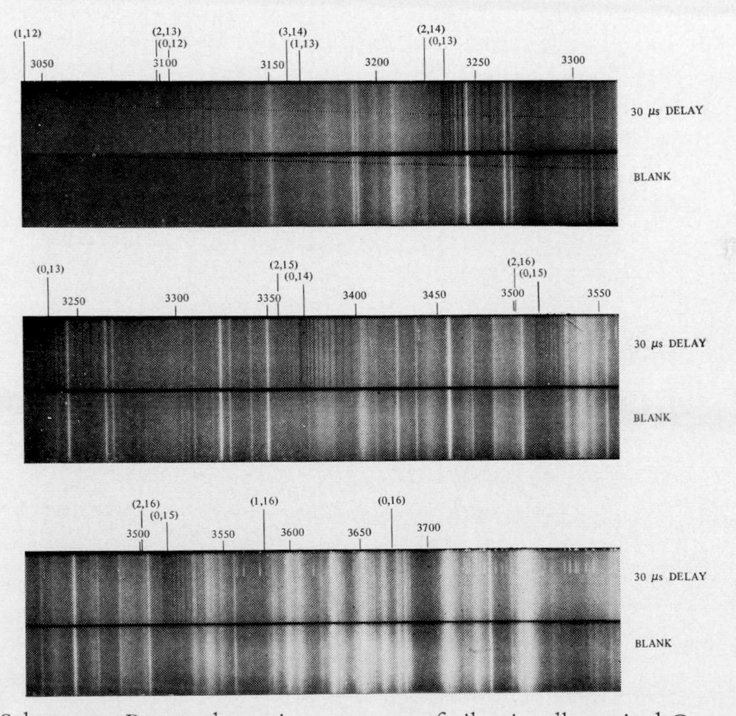

Fig. 11. Schumann–Runge absorption spectrum of vibrationally excited O_2 produced by flash photolysis of ozone. O_3 pressure, 20 mm Hg; N_2 pressure, 800 mm Hg. O_3/N_2 ratio = 1/40. Flash energy, 2000 J. (McGrath and Norrish[23])

The Photolysis of Ozone

The photolysis of ozone was first discerned as a chain reaction by Heidt and Forbes[21] and confirmed for pure ozone by Norrish and Wayne[29] who observed quantum yields up to 16 in the ultraviolet. The nature of the excited oxygen functioning as chain carrier would now appear to be identified as the vibrating molecule with more than 17 quanta of vibration. For the propagation of the chain the endothermic reaction

$$O_3 + O_2^{\star} = O_2 + O_2 + O$$

requires 69 kcal, and this is supplied precisely by a molecule vibrating with more than 17 quanta. All those vibrating with less are visible by flash photolysis and decay by normal relaxation processes. Those with more react so rapidly with ozone molecules that they are not seen, except that they may be faintly discerned up to $v = 20$ as a consequence of competition between reaction and collisional deactivation. This conclusion is based upon the deduction that the oxygen atom is generated in the first electronically excited state, ^1D, lying 45 kcal above the ground state and that the chain reaction is propagated uniquely by ^1D oxygen atoms, because no chain reaction follows photolysis

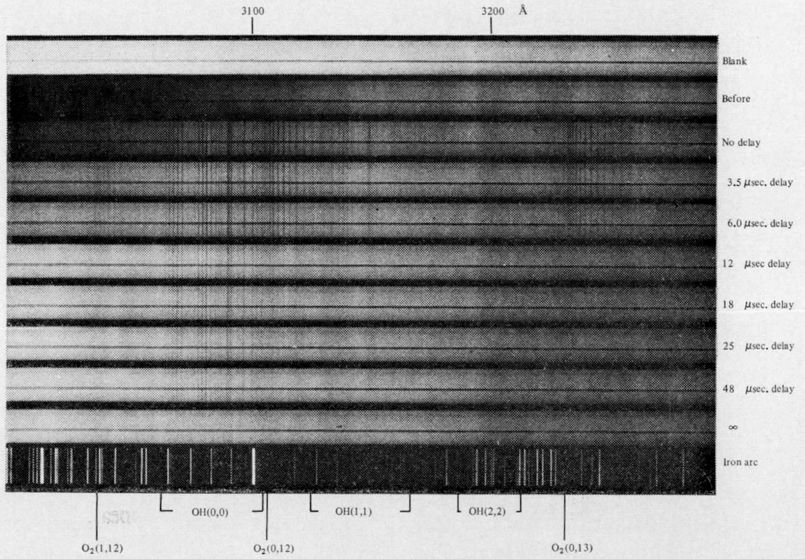

Fig. 12. The flash photolysis of water/ozone mixtures. Production of excited hydroxyl by reaction of $O(^1D)$ with water vapour. Pressure of ozone, 6 mm Hg; pressure of water vapour, 4 mm Hg; pressure of nitrogen, 200 mm Hg. Flash energy, 1600 J. (Basco and Norrish[23])

by «orange» light where the magnitude of the quantum is only sufficient for the generation of ^3P oxygen atoms.

The chemical proof that the oxygen atoms generated by the photolysis of ozone in ultraviolet light are in the ^1D state lies in the fact that when small quantities of water vapour are added to the system the spectrum of vibrationally excited O_2 molecules is progressively suppressed and replaced by the absorption spectrum of OH as seen in Fig. 12. This is to be correlated with the observation of Forbes and Heidt[39] that in «damp» ozone the quantum yield is increased to values as high as 130, as compared with their maximum value of 8 for dry ozone, and in the light of our observation it may be concluded that an entirely new mechanism of chain propagation is substituted as a consequence of the successful competition of water with ozone for the oxygen atom, *i.e.*:

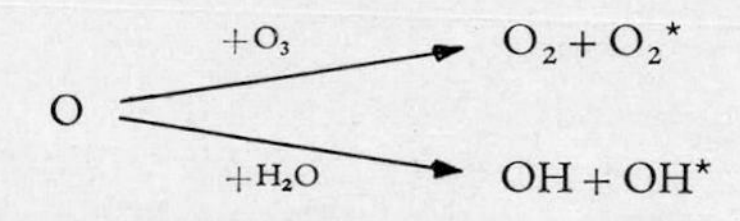

This however can only take place if the O atom is excited to the ^1D state for the reaction of $O(^3P)$ with water is endothermic; we have

$$O(^3P) + H_2O = 2\,OH - 11\,\text{kcal}$$

$$O(^1D) + H_2O = 2\,OH + 34\,\text{kcal}$$

In the presence of water, the chain reaction may be written

$$O_3 + h\nu = O_2 + O(^1D)$$

$$O(^1D) + H_2O = OH + OH^*$$

$$\left.\begin{array}{r} OH + O_3 = HO_2 + O_2 \\ HO_2 + O_3 = OH + 2\,O_2 \end{array}\right\} \text{propagation}$$

followed by chain ending by intercombination of radicals. This scheme satisfies the kinetic findings of Forbes and Heidt; it explains the appearance of the OH radical and demands the formation of the excited O atom. The reaction of O atoms with other hydrides referred to above is also equally dependent on the photolytic generation of $O(^1D)$ in the ultraviolet. It is significant that water has no effect on ozone photolysis in «orange» light where only ^3P oxygen atoms can be generated. The quantum yield remains unchanged at 2 in accord with the simple scheme[31]

$$O_3 + h\nu = O_2 + O(^3P)$$
$$O(^3P) + O_3 = O_2 + O_2$$

analogous to the photolysis of NO_2 and ClO_2.

It was shown by McGrath and Norrish[32] that the rate of decomposition of ozone by the secondary reactions subsequent to the flash is strongly affected by the addition of inert gases. Starting with 2.94 mm of O_3 and diluting with added gas to give a mixture ratio of $O_3/M = 1:163$, the rate of disappearance of O_3 was determined by photometering the O_3 absorption in a series of spectra such as those shown in Fig. 13. In Fig. 14 are seen three typical curves showing ozone decay. From these curves could be measured the efficiences of third bodies M in the back reaction

$$O(^1D) + O_2 + M = O_3 + M'$$

When M is O_2 the ratio of O_2 to O_3 is 163 : 1 so it is hardly surprising that the above reaction predominates over the reaction

$$O(^1D) + O_3 = O_2 + O_2^*$$

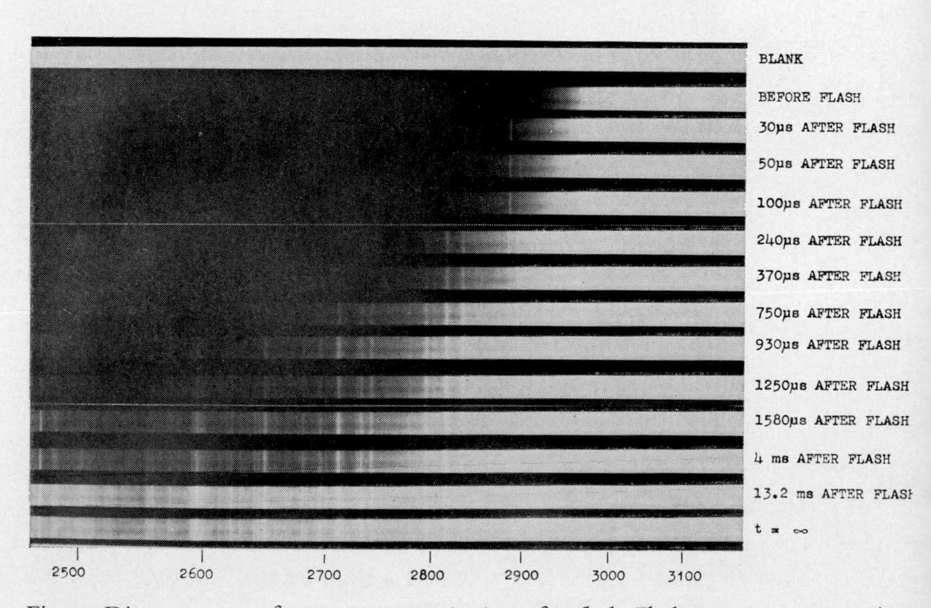

Fig. 13. Disappearance of ozone spectrum in time after flash. Flash energy, 1280 J. O_3/N_2 mixture, ratio 1 : 163. Ozone pressure, 2.93 mm Hg.

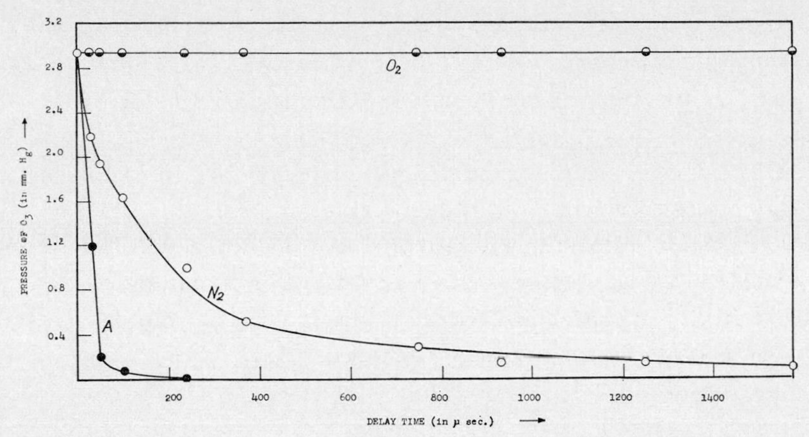

Fig. 14. Typical ozone decay curves for O_3/N_2, O_3/A and O_3/O_2 mixtures. Mixture ratio in all cases 1:163.

to such an extent as to reverse all O_3 decomposition. With other added gases the relative efficiences of the molecules M for the three body recombination were determined as He = 1, A = 1, SF_6 = 1.5, CO_2 = 14, N_2 = 16, N_2O = 17.

The gases divide into two groups: (*1*) the inert gases and SF_6 and (*2*) N_2, CO_2 and N_2O. Group (*1*) exhibiting low efficiency are spherically symmetrical and chemically inert. Group (*2*) are much more efficient. It is possible with group (*2*) that some form of chemical affinity is operative in forming intermediate transition species, and that a more facile energy transfer is possible due to readily stimulated vibrational modes. Further work along these lines may well prove rewarding.

Application of the Adiabatic Method

The study of explosive processes exemplified by the oxidation of hydrides

The gaseous oxidation of hydrides, including hydrocarbons occurs by exothermic processes which have the characteristics of chain reactions, that is to say they proceed by initiation, propagation, multiplication and extinction of reacting centres. The reactions are said to be autocatalytic and if the conditions are such that multiplication of propagating centres exceeds extinction, the process may develop to explosion. These conditions depend on the parameters of temperature, total pressure, relative concentrations of reactants, catalytic activity of the surface in initiating or terminating reaction chains, the geometry of the reaction vessel, and the activity of added catalysts and inhibitors. The variation of these parameters gives rise to sharp limits of explosion, and by judicious kinetic experiment the separate effect of each can be isolated and defined by keeping all but the one under examination constant.

The development of the slow reaction and the incidence of ignition are subject to an induction or incubation period during which autocatalysis occurs (initially exponentially) to a steady state or to explosive reaction. This autocatalysis is dependent on the magnitude of the «net branching factor», which is the result of the interplay of the physical parameters leading to multiplication and extinction of reaction centres. If in the notation of Semenov, f represents the sum of the reactions leading to multiplication and g the sum of those leading to extinction

$$(f-g) = \Phi$$

the net branching factor, which may obviously be positive or negative and the development of the reaction velocity (v) in time (t) is given by

$$v = A\,e^{\Phi t}$$

where the pre-exponential term A varies only slowly and in a much less dramatic way then Φ, with changing kinetic conditions. When Φ is negative from the beginning a finite and small stationary reaction velocity is imposed. When Φ is positive, rapid and exponential development of velocity to explosion may occur. This is the case with the reaction of hydrogen with oxygen which shows sharp explosive limits dependent on the parameters listed above. There exist cases, however, where Φ starting positive, may give rise to exponential development of the reaction in a big way, but owing to consumption of reactants or varying catalytic factors may become negative during the course of reaction which, as it were, starting hopefully towards explosive build up is finally quenched to a stationary state and subsequent decline, by the failure of the net branching factor to remain positive. Such reactions are termed degenerate explosions by Semenov. They are distinguished by having a small but positive initial value of Φ and depend for branching on the reaction of a «precariously stable» intermediate which builds up as the reaction proceeds and which can be detected by kinetic and analytical observation. The recognition of degenerately branched-chain reactions represents the culminating triumph in Semenov's interpretation of branching-chain reactions and in particular provides a pattern for the understanding of hydrocarbon oxidation[33].

But while giving us the overall pattern of reaction, neither the experimental methods nor the mathematical conceptions were capable of exposing the intimate nature of the precise reactions involved. These were deduced in some instances from circumstantial evidence, with the gradual realisation that atoms and free radicals are more often than not concerned in the chain processes.

It has remained for flash photolysis and kinetic spectroscopy not only to confirm and amplify the general conclusions of the classical studies of chain reactions, but also to provide objective proof of the nature and reactions of the transient participating species. For this purpose we use the *adiabatic* method taking advantage of the free radicals produced by pyrolysis and photolysis for initiation, and flash heating to generate temperatures suitable to sustain the propagation and branching reactions upon which the autocatalytic chain reaction depends. We take for example the case of hydrides.

Since oxygen does not absorb energy from the photolytic flash under the condition imposed by the limitations of the transparency of quartz, it is fortunate therefore that many hydrides absorb sufficiently to provide the necessary pyrolysis for initiation. This is true for hydrogen sulphide, hydrogen telluride, ammonia, hydrazine, and phosphine, all of which photolyse isothermally and pyrolyse adiabatically by eliminating a hydrogen atom:

$$XH_n + h\nu = XH_{n-1} + H$$

The growth and decay of the free radical XH_{n-1} so generated can be followed by kinetic spectroscopy. Under pyrolytic conditions however when the concentration of free radicals generated may be high, the above reactions may be followed by further elimination of hydrogen from the free radical, *e.g.*

$$XH_{n-1} + XH_{n-1} = XH_n + XH_{n-2}$$

This is true for example for ammonia (Husain and Norrish[34]), which under isothermal conditions gives only NH_2 but under adiabatic conditions yields NH radicals as well. We observe the same result with PH_3 (Norrish and Oldershaw[35]), H_2S (Norrish and Zeelenburg[36]), and H_2Te (Norrish and Osborne[37]), the last two yielding HS and S, and HTe and Te respectively even under isothermal conditions as shown for example in Fig. 15.

The pyrolytic reactions under our conditions are generally limited in extent, but on the addition of oxygen in sufficient quantity oxidation proceeds to explosion, unless the system is partially cooled by the addition of an inert diluent. Sufficient excess fuel or oxygen has the same effect. Under such conditions the oxidation proceeds by a quenched-chain reaction, and is much more limited in extent.

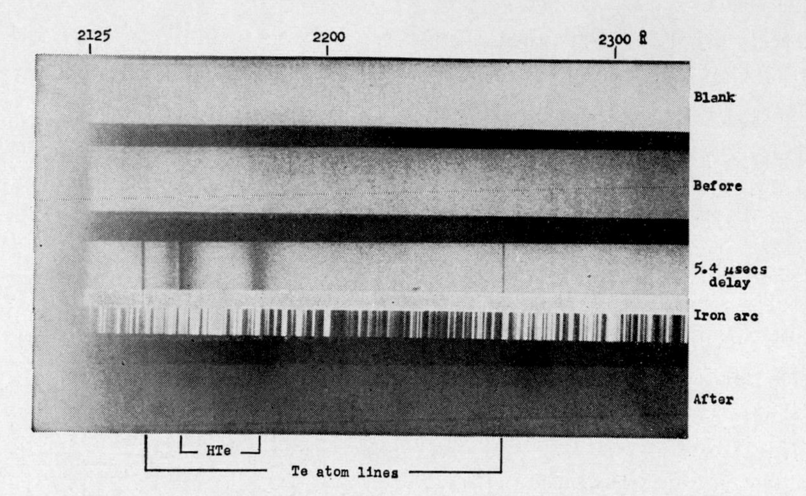

Fig. 15. Flash photolysis of tellurium hydride. Pressure of TeH_2, 0.25 mm Hg; pressure of N_2, 250 mm Hg. Flash energy, 2 500 J. (Norrish and Osborne[37])

The development of reaction from initiation to explosion involves an incubation period of less than a millisecond, and in oxygen rich mixtures the onset of ignition is marked by a copious burst of hydroxyl radicals. It is clear that in all cases studied the hydroxyl radical acts as a chain carrier.

Hydrocarbons on the other hand do not in general absorb light transmitted by quartz (with the exception of highly unsaturated compounds) and the fuel oxygen mixture therefore does not respond to the flash. To initiate explosive reaction it is necessary to add a small quantity of sensitizer such as chlorine, nitrogen peroxide or alkyl nitrite. These, by absorbing strongly, raise the temperature of the system, and simultaneously photolyse and pyrolyse to give free atoms or radicals which act as initiators. Nitrogen peroxide for example absorbs strongly throughout the spectrum and yields oxygen atoms which give ready initiation[38]

$$NO_2 \rightarrow NO + O$$

In Fig. 16 is shown a sequence of absorption spectra illustrating the explosion of a mixture of $2\,H_2 + O_2$ sensitized by nitrogen peroxide[39]. The growth and decay of the OH radical is seen in the (0,0) and (0,1) bands of the transition $^2\Sigma^+ - ^2\Pi$. This and the earlier study of the reaction of oxygen atoms with hydrogen by Norrish and Porter[38] go far to confirming the scheme of oxidation of hydrogen proposed by Lewis and Von Elbe[40] of which the following are some constituent reactions

$$\left.\begin{aligned} OH + H_2 &= H_2O + H \\ H + O_2 &= OH + O \\ O + H_2 &= OH + H \end{aligned}\right\} \text{Propagation and branching}$$

$$\left.\begin{aligned} OH + \text{surface} &= \text{products} \\ H + O_2 + M &= HO_2 + M' \end{aligned}\right\} \text{termination}$$

The explosion of hydrocarbons sensitized by amyl nitrate was studied by Erhardt and Norrish[41] and is illustrated in Figs. 17 and 18 which show the effect of adding tetraethyl lead to a mixture of hexane and oxygen. The first shows the ignition in the absence of the addendum with the rapid disappearance of the spectrum of the sensitizer on flashing, followed by an incubation period of 875 μsec to the onset of explosion as marked by the sudden growth of the OH radical. The second shows the ignition under identical conditions in the presence of the addendum. It is seen that the incubation period is increased some three-fold to *ca.* 2600 μsec, while during the growth to explosion the spectrum of gaseous lead oxide is strongly developed. At the point of ignition the PbO spectrum disappears completely and is replaced by the

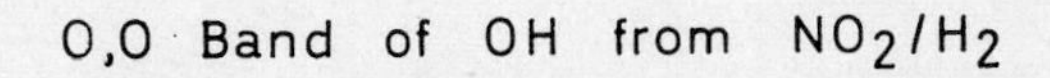

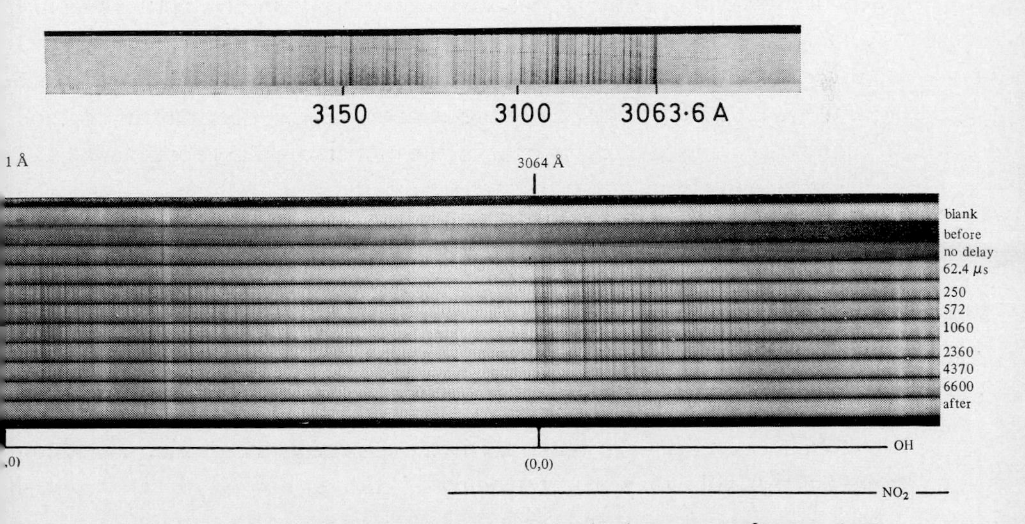

Fig. 16. (a) Photo-reaction of $NO_2 + H_2$ giving OH radical. Pressure of H_2, 2 mm Hg; pressure of NO_2, 2 mm Hg. No delay. (Norrish and Porter[38]) (b) Flash photolysis of NO_2 (2 mm)+H_2 (20 mm)+O_2 (10 mm)+N_2 (15 mm) showing the formation and decay of OH during a typical explosion. Flash energy, 3300 J.

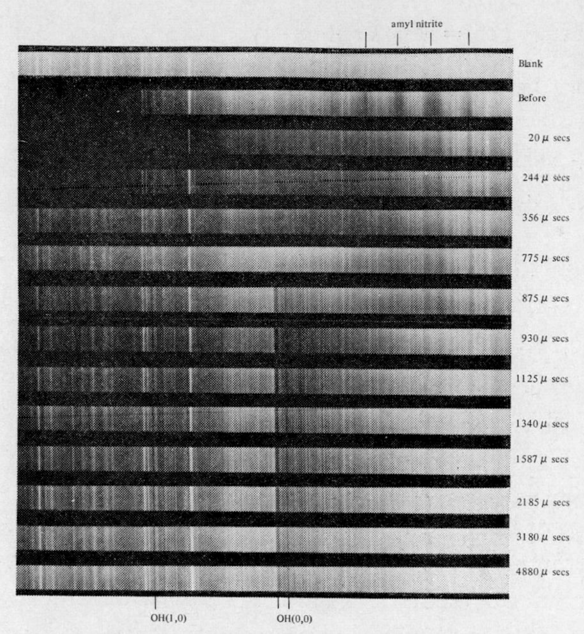

Fig. 17. Spectra *vs.* time. Explosion of hexane and oxygen sensitized by amyl nitrite. Pressure of C_6H_{14}, 2 mm Hg; pressure of $C_5H_{11}ONO$, 2 mm Hg; pressure of O_2, 32.5 mm Hg. Flash energy, 2000 J. (Erhard and Norrish[41])

resonance spectrum of lead. Both the OH and the Pb spectra are very faintly visible before ignition. These and other experiments in which ignition was observed photoelectrically by the sudden growth of OH emission, throw light upon the mechanism of antiknock in the internal combustion engine which we conclude to be dependent on the moderating effect of Pb and PbO on the development of the autocatalytic growth to explosion.

Knock has been proved by Miller[42] and Male [43] to be due to the homogeneous detonation of the residual charge in the cylinder at the end of the ignition stroke. It is believed to be due to the generation of centers of autoignition (peroxides, aldehydes, etc.) due to adiabatic rise of temperature, which replaces the smooth explosion wave generated by the spark ignition. We have concluded that the tetraethyl lead clearly must operate in the gas phase and suggest that (*1*) it may remove the centers of autoignition by reduction – *e.g.* peroxides may be removed by

$$R{\cdot}OOH + Pb(C_2H_5)_4 \rightarrow PbO + \text{products}$$

and (*2*) moderate the liberation of energy as the reaction develops to explosion by the following reactions

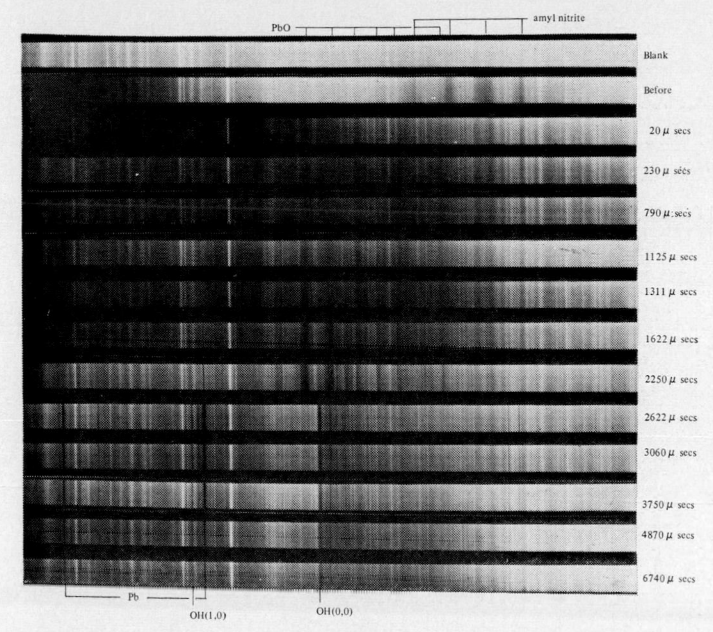

Fig. 18. Spectra *vs.* time. Effect of tetraethyl lead on the hexane explosion. Pressure of C_6H_{14}, 2 mm Hg; pressure of $C_5H_{11}ONO$, 2 mm Hg; pressure of O_2, 32.5 mm Hg; pressure of tetraethyl lead, 0.2 mm Hg. Flash energy, 2000 J. (Erhard and Norrish[41])

$$OH + Pb \rightarrow PbOH \xrightarrow{+OH} Pb(OH)_2 \xrightarrow{-H_2O} PbO$$

$$PbO + R \rightarrow RO + Pb$$

Thus by alternate oxidation and reduction from lead to lead oxide and back again the atomic lead and lead oxide can intervene in chain propagation by the removal of OH, and so by shortening the chains retard their development. With the onset of explosion, the PbO is instantaneously decomposed to atomic lead, which as the system cools is finally deposited on the surface of the reaction vessel.

The question as to whether moderation of the explosive process occurs in the gas phase, or by chain ending on heterogeneous particles of lead or lead oxide («smoke») would appear to be answered by these results, since no «smoke» is observed during the course of the reaction, which is seen to be completely homogeneous. In contrast the addition of tetraethyl tin which has no anti-knock action is accompanied by the copious formation of smoke. There is no sign of the production of gaseous SnO during the incubation period and no effect whatsoever on the said incubation period and the reactions leading to ignition[44]. This is due to the lower volatility of SnO.

Many other studies of the affects of addenda on explosive reactions of hydrocarbons have been made by Callear and Norrish[44] with interesting results which cannot be discussed here. Reactions of this kind provide a plentiful source of free radicals and atoms derived from the addenda in high temperature reactions.

The growth and decay of free radicals as we pass through ignition is shown for the combustion of acetylene sensitized by NO_2 in Figs. 19 and 20 (Norrish, Porter and Thrush[45]). The curves were obtained by plate photometry of the various radical spectra seen in absorption at increasing intervals after initiation.

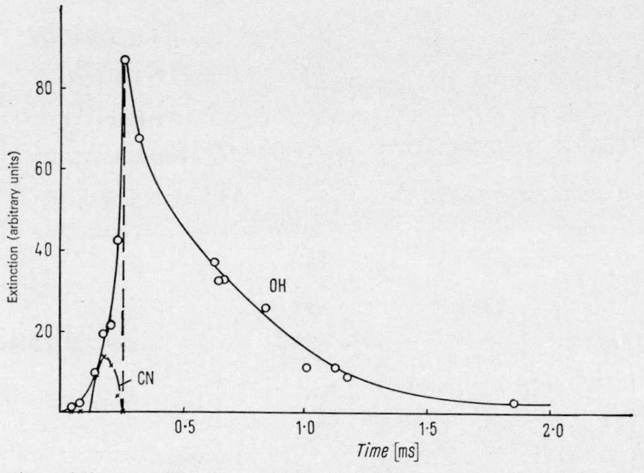

Fig. 19. Growth and decay of OH and CN radicals in oxygen-rich mixture of acetylene and oxygen (NO_2 counting as O_2). Pressure of C_2H_2, 10 mm Hg; pressure of O_2, 10 mm Hg; pressure of NO_2, 1.5 mm Hgl (Norrish, Porter and Thrush[45])

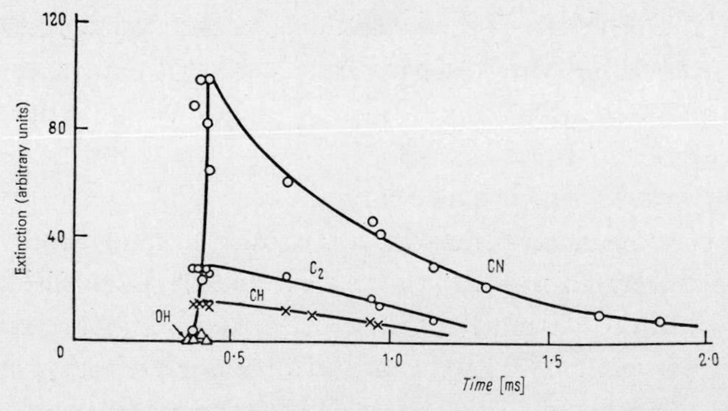

Fig. 20. Growth and decay of radicals in fuel-rich mixture of acetylene and oxygen. Ordinates of curves not comparable since extinction coefficients of radicals unknown. Pressure of C_2H_2, 13 mm Hg; pressure of O_2, 10 mm Hg; pressure of NO_2, 1.5 mm Hg. (Norrish, Porter and Thrush[45])

They are also typical of curves obtained for ethylene and methane[46] and indicate the growth and decay of the observed radicals with time, though they cannot be compared in terms of absolute concentration since the extinction coefficients of the radicals are at present unknown. The combustion of acetylene and ethylene were shown by Bone[47] to depend on the stoichiometric equations

$$C_2H_2 + O_2 = 2CO + H_2$$
$$C_2H_4 + O_2 = 2CO + 2H_2$$

according to which there is an apparent preferential burning of carbon. With oxygen in excess, water is formed, while in fuel-rich mixtures free carbon in the form of smoke is produced. These two conditions are very sharply distinguished on either side of the fuel–oxygen ratio of 1:1. This classical result is very clearly confirmed by the curves shown in Figs. 19 and 20. In the former we have an oxygen-rich system and the formation of water is indicated by the copious display of OH, in the latter – the fuel-rich system – the OH is barely in evidence and the precursors of free carbon are observed in the CH, C_2, and C_3 radicals. The change from one type of display to the other takes place extremely sharply at the fuel: oxygen ratio of 1 : 1, nitrogen peroxide for this purpose being counted as oxygen.

The CN radical which is strongly in evidence in fuel-rich systems is derived from the sensitizer. It has been shown[48] that during the induction period of about 0.5 msec the temperature rises exponentially, slowly at first and very sharply at the end. With the sudden appearance of the free radicals the explosive reaction is complete: we are witnessing in fact the afterburning of hydrogen in oxygen-rich mixtures, and the after-cracking of the fuel in fuel-rich mixtures. The only radical which can be seen during the induction period before ignition is the OH radical which grows in concentration as the reaction develops.

Further experiments with fuel-rich mixtures[46] indicated the growth and decay of a precursor of free carbon which followed closely the growth and decay of the carbon radicals C_2, C_3 and CH. It was possible to deduce the extinction coefficient of this carbon precursor at 3 700 Å and to show that its high value is characteristic of aromatic polynuclear hydrocarbons. It may be suggested[49] that in the high temperature of the flame (> 3 000°C) cracking of some of the excess fuel occurs to yield free carbon atoms, which progressively «crystallise» through C_2 and C_3 to the «aromatic» structure of graphite. The confirmation or otherwise of this view must await further studies of the products of explosion by means of the vacuum spectrograph when we may hope to see the resonance line of carbon in absorption.

A General Mechanism for the Combustion of Hydrides

As I have mentioned above we have noted that the hydroxyl radical is common to the ignition processes of all the hydrides so far examined. Where initiation occurs by the direct photolysis of the hydride yielding an H atom, it may be generated by the reaction

$$H + O_2 = OH + O$$

When initiation is by the photolysis of a sensitizer yielding oxygen atoms as with NO_2, OH may be derived from the reaction

$$O + XH_n = OH + XH_{n-1}$$

During the incubation period the OH is observed gradually to increase, and the instant of ignition is marked in oxygen-rich mixtures by a very sudden and enormous increase in its concentration. By detailed comparison of the oxidation reactions of H_2S, NH_3, PH_3 and hydrocarbons it may be concluded that the pattern of chain propagation is the same in all cases, and represented by the general scheme

$$\left.\begin{aligned} OH + XH_n &= H_2O + XH_{n-1} \\ XH_{n-1} + O_2 &= XH_{n-2}O + OH \end{aligned}\right\} \text{propagation}$$

Branching is dependent on the intermediate and may take place variously by any one of the following reactions

$$XH_{n-2}O + O_2 = XH_{n-2}O_2 + O$$

$$XH_{n-2}O + O_2 = XH_{n-3}O + HO_2$$

$$XH_{n-2}O = XH_{n-3}O + H$$

The first reaction takes place in the autocatalysis of H_2S oxidation, in which SO (seen by kinetic spectroscopy) is the intermediate. The second occurs in the oxidation of methane which yields formaldehyde (readily detectable during the reaction by conventional methods of analysis). The third is exemplified by the oxidation of ammonia in which HNO is concluded to be the origin of chain branching.

Table 2

Hydride	*Uniradical*	*Associated intermediate*	*Reference*
SH_2	SH	SO	36
TeH_2	TeH	TeO	37
NH_3	NH_2	HNO	34
PH_3	PH_2	HPO	35
N_2H_4	N_2H_3	NH_2NO	34
CH_4	CH_3	H_2CO	50
C_2H_4	CH_3	H_2CO	51
B_2H_6	BH_2	HBO	52

Table 2 shows the uniradical and the associated intermediate derived from a series of hydrides which take part in chain propagation and branching, in accordance with our conclusions based on comparative study both by classical kinetic methods and flash photolysis. In cases where the intermediate is moderately stable, as with SO from H_2S and H_2CO from CH_4 the overall oxidation exhibits the slow autocatalysis associated with degenerate branching. In other cases, with extremely unstable intermediates the branching factor may be high, and is reflected in kinetics which show very short incubation periods and sharp transition from very slow reaction to explosion.

All the uniradicals and associated intermediates are seen to be isoelectronic or electronically structurally similar. This and the uniform participation of the OH radical in all the chain-propagation reactions would seem to provide a generalizing hypothesis of value and one which invites further experimental examination.

In connexion with the continued study of the reactions of the OH radical, Horne and Norrish[53] have recently been able to measure quantitatively the kinetics of the reactions

$$OH + C_2H_6 \xrightarrow{k_4} C_2H_5 + H_2O$$

$$(\log_{10}k_4 = (11.1 \pm 0.7) - (3600 \pm 600)/2.303\,RT \text{ l mol}^{-1}\text{ sec}^{-1})$$

and

$$OH + CH_4 \xrightarrow{k_5} CH_3 + H_2O$$

$$(\log_{10}k_5 = 10.7 - 5000/2.303\,RT \text{ l mol}^{-1}\text{ sec}^{-1} \text{ approximately})$$

by kinetic spectroscopy.

The OH radicals were generated by flashing water vapour in highly transparent quartz and comparing their rates of decay in the presence of inert gases and hydrocarbons. Further measurements of this kind with other hydrides will be of value to the continued study of the combustion of hydrocarbons along the lines indicated above. They are also of course of importance in consideration of the reactions involved in the evolution of planetary atmospheres, as are many other reactions studied by kinetic spectroscopy such as the photochemistry of NO, and of ozone and the reactions of the oxygen atom described above.

The examples which I have cited give, I hope, some indication of the breadth of application of methods based on flash photolysis in the study of gas reactions. Other results of importance involve the discovery of new absorption spectra of chlorine and bromine by Briggs and Norrish[54], and the detection of population inversion such as is observed in the study by Donovan and Husain (ref. 55) of spin orbit relaxation of the metastable iodine atom I $(5^2P_{1/2})$ produced in the photolysis of CF_3I

$$I(5^2P_{1/2}) \rightleftharpoons I(5^2P_{3/2})$$

Population inversion in favour of high vibrational levels is as we have seen also observed in NO and CN, and has also been studied effectively by Polanyi and his co-workers[56] for atomic reactions such as

$$H + Cl_2 = HCl^* + Cl$$

$$Cl + HI = HCl^* + I$$

All these reactions form the basis of potential gas laser action and are being effectively studied in this connexion.

The opportunity for the application of the methods of flash photolysis to chemical kinetics, not only in the gas phase, but also to the study of photochemical reactions in solution is very great, and is increasing steadily as improvement in technique gives greater time resolution, and ever increasing accessibility to reactions of the «vacuum ultraviolet».

In conclusion I give thanks to those who have collaborated and contributed to the work described in this lecture, many of whom are continuing to direct and develop it with distinction.

Mr. Chairman, ladies and gentlemen, thank you for your interest and attention.

1. J.Stark, *Z.Physik*, 9 (1908) 889, 894.
2. E. Warburg, *Sitz.Ber. Preuss. Akad.Wiss., Physik.-Math. Kl.*, (1916) 314, (1918) 300.
3. M. Bodenstein and W. Dux, *Z. Physik. Chem. (Leipzig)*, 85 (1913) 297.
4. See R. G. W. Norrish, *Bakerian Lecture*, 1967; *Proc.Roy. Soc. (London), Ser. A*, 301 (1967) 1.
5. R.Wood, *Phil. Mag.*, VI, 44 (1922) 538.
6. K. F. Bonhoeffer, *Z. Physik. Chem. (Leipzig)*, 113 (1924) 199.
7. F. Paneth and W. Hofeditz, *Chem.Ber.*, 62 (1929) 1335. Also F. Paneth and W. Lautsch, *ibid.*, 64 (1931) 2702.
8. J. Franck, *Trans. Faraday Soc.*, 21 (1926) 536.
9. V. Henri, *Compt. Rend.*, 177 (1923) 1037.
10. R. G. W. Norrish and G. Porter, *Nature*, 164 (1949) 658.
11. G. Porter, *Proc. Roy. Soc. (London), Ser. A*, 200 (1950) 284.
12. O. Oldenberg, *J. Chem. Phys.*, 2 (1934) 713; 3 (1935) 266.
13. R. G. W. Norrish, G. Porter and B. A.Thrush, *Proc. Roy. Soc. (London), Ser. A*, 216 (1955) 165.
14. G. Porter and F. J. Wright, *Discussions Faraday Soc.*, 14 (1953) 23.
15. N.Basco and R. G. W. Norrish, *Proc. Roy. Soc. (London), Ser. A*, 268 (1962) 291.
16. W. G. Burns and H. J. Bernstein, *J. Chem. Phys.*, 18 (1950) 1669.
17. R. W. B. Pearse and A. G. Gaydon, *Identification of Molecular Spectra*, Chapman and Hall, London, 1950.
18. (a) N. Basco, A. B.Callear and R. G. W. Norrish, *Proc. Roy. Soc. (London), Ser. A*, 260 (1961) 293; (b) 269 (1962) 180.
19. R. G. W. Norrish, The study of energy transfer in atoms and molecules by photochemical methods, in *The Transference of Energy in Gases*, 12th Solvay Conference, Brussels, 1962, Interscience, New York, 1964, p. 99.
20. N. Basco, J. E.Nicholas, R. G. W. Norrish and W. H. J. Vickers, *Proc. Roy. Soc. (London), Ser. A.*, 272 (1963) 147.
21. R. G. W. Norrish, *J. Chem. Soc.*, (1929) 1158; J. W. T.Spinks and J. M. Porter, *J. Am. Chem. Soc.*, 56 (1934) 264; G. Kistiakowsky, *Z. Physik. Chem. (Leipzig)*, 117 (1925) 337; J. Heidt and G. S. Forbes, *J. Am. Chem. Soc.*, 56 (1934) 2365.
22. F. J. Lipscomb, R. G. W. Norrish and B. A. Thrush, *Proc. Roy. Soc. (London), Ser. A*, 233 (1956) 455.
23. W. D. McGrath and R. G. W. Norrish, *Proc. Roy. Soc. (London), Ser. A*, 142 (1957) 265; N. Basco and R. G. W. Norrish, *ibid.*, 260 (1960) 293; *idem, Discussions Faraday Soc.*, 33 (1962) 99.
24. W. D. McGrath and R. G. W. Norrish, *Z. Physik. Chem. (Frankfurt)*, 15 (1958) 245; *Proc. Roy. Soc. (London), Ser. A*, 254 (1960) 317.
25. N. Basco and R. G. W. Norrish, *Proc. Roy. Soc. (London), Ser. A*, 260 (1961) 293.
26. N. Basco and R. G. W. Norrish, see ref. 24.
27. J. D. McKinley, D. Garvin and M. J. Boudart, *J. Chem. Phys.*, 23 (1955) 784.
28. J. K. Cashion and J. C. Polanyi, *J. Chem. Phys.*, 29 (1958) 455; 30 (1959) 316, 1097; J. C. Polanyi, *ibid.*, 25 (1955) 784; *idem, Chemistry in Britain*, 2 (1966) 151.
29. R. G. W. Norrish and R. P. Wayne, *Proc. Roy. Soc. (London), Ser. A*, 288 (1965) 200, 361.

30. G.S.Forbes and L.J.Heidt, *J.Am.Chem.Soc.*, 56 (1934) 1671.
31. G.Kistiakowski, *Z.Physik.Chem.(Leipzig)*, 117 (1925) 337.
32. W.D.McGrath and R.G.W.Norrish, *Proc.Roy.Soc.(London)*, *Ser.A*, 242 (1957) 265.
33. N.N.Semenov, *Chemical Kinetics and Chain Reactions*, Oxford University Press, London, 1935.
34. D.Husain and R.G.W.Norrish, *Proc.Roy.Soc.(London)*, *Ser.A*, 273 (1963) 145.
35. R.G.W.Norrish and G.A.Oldershaw, *Proc.Roy.Soc.(London)*, *Ser.A*, 262 (1961) 1.
36. R.G.W.Norrish and A.P.Zeelenburg, *Proc.Roy.Soc.(London)*, *Ser.A*, 240 (1957) 293.
37. R.G.W.Norrish and M.Osborne, in preparation.
38. R.G.W.Norrish and G.Porter, *Proc.Roy.Soc.(London)*, *Ser.A*, 210 (1952) 439.
39. J.F.Nicholas and R.G.W.Norrish, *Proc. Roy. Soc. (London)*, *Ser. A*, 309 (1969) 171.
40. B.Lewis and G.von Elbe, *Combustion, Flames and Explosions of Gases*, Academic Press, New York, 1951.
41. K.Erhard and R.G.W.Norrish, *Proc.Roy.Soc.(London)*, *Ser.A*, 234 (1956) 178.
42. S.A.E.Miller, *Quart. Trans.*, 1 (1947) 98.
43. T.Male, *Third Symposium on Flame and Combustion Phenomena*, Williams and Wilkins, Baltimore, 1949, p.271.
44. A.P.Callear and R.G.W.Norrish, *Proc.Roy.Soc.(London)*, *Ser.A*, 259 (1960) 304.
45. R.G.W.Norrish, G.Porter and B.A.Thrush, *Proc.Roy.Soc.(London)*, *Ser.A*, 216 (1953) 165.
46. R.G.W.Norrish, G.Porter and B.A.Thrush, *Proc.Roy.Soc.(London)*, *Ser.A*, 227 (1955) 423.
47. W.A.Bone, *Proc.Roy.Soc.(London)*, *Ser.A*, 137 (1932) 243.
48. R.G.W.Norrish, Plenary Conference, Section Mélanges Gazeux, *16th Congress of Pure and Applied Chemistry*, *Experientia*, Suppl.7 (1958) 87.
49. R.G.W.Norrish, see ref.48, p.97.
50. R.G.W.Norrish, *Rev.Inst.Franc.Pétrole Ann.Combust.Liquides*, 7 (1949) 288.
51. A.Harding and R.G.W.Norrish, *Proc.Roy.Soc.(London)*, *Ser.A*, 212 (1952) 291.
52. M.D.Carabine and R.G.W.Norrish, *Proc.Roy.Soc.(London)*, *Ser.A*, 296 (1967) 1.
53. D.Horne and R.G.W.Norrish, *Nature*, 215 (1967) 1373.
54. A.G.Briggs and R.G.W.Norrish, *Proc.Roy.Soc.(London)*, *Ser.A*, 276 (1963) 57.
55. R.J.Donovan and D.Husain, *Trans.Faraday Soc.*, 62 (1962) 11, 1050.

Biography

Ronald George Wreyford Norrish was born in Cambridge on November 9th, 1897. His father, a native of Crediton, Devonshire, came to Cambridge as a young Pharmacist to open one of the early shops of Boots, the Chemists, and remained there for the rest of his long life.

After spending his early years at the local Board school, Norrish obtained a scholarship to the Perse Grammar School in 1910. He remembers with deep gratitude his early teachers, in particular Rouse, Turnbull and Hersch, who gave dedicated and individual help to promising young scholars. In 1915 he obtained an entrance scholarship to Emmanuel College, Cambridge in Natural Sciences, but left in 1916 with a commission in the Royal Field Artillery for service in France. He was made prisoner of war in March 1918 and spent the rest of the war in Germany, first at Rastatt and later at Graudenz in Poland. Repatriated in 1919, he returned to Emmanuel College where he has remained ever since, first as a student and after 1925 as a Fellow. Norrish's early research was inspired by Eric Redeal (now Sir Eric Redeal) under whose lively supervision he first took up the study of Photochemistry.

In 1925 he was made Demonstrator and in 1930, Humphrey Owen Jones Lecturer in Physical Chemistry in the Department of Chemistry at Cambridge and upon the death of the first Professor of Physical Chemistry, Dr. T.M.Lowry, was appointed to the Professorship in 1937. He occupied the chair until 1965 when he retired as Emeritus Professor of Physical Chemistry in the University.

Norrish has had the good fortune to work with many gifted students and with them has carried out a wide range of research in the fields of Photochemistry and Reaction Kinetics, including Combustion and Polymerisation. As the study of Chemical Kinetics developed, there was a fortunate integration in the various aspects of the study in which his school of work was engaged, as a result of which the importance of Photochemistry and Spectroscopy to Chemical Kinetics in general emerged. All this was sadly brought to a temporary halt in 1940. During the second world war, while still continuing to direct the Department of Physical Chemistry and to teach, Norrish was concerned with a good deal of research work in connection with various ministries and was able to collaborate with his colleagues on various government committees. It was after the war in 1945 when research was recommenced that work was started with the object of observing short lived transients in chemical reactions. In collaboration with his student, now Professor George Porter, this led to the development of Flash Photolysis and Kinetic Spectroscopy which has had considerable influence on the subsequent development of Photochemistry and Reaction Kinetics, and in the hands of workers in many parts of the world is continuing to develop as a powerful technique for the study of all aspects of chemical reaction.

In 1926 Ronald Norrish married Annie Smith who was Lecturer in the Faculty of Education in the University of Wales in Cardiff. They have two daughters and four grandchildren. Much of their time has been spent in travel.

Norrish has served on the Councils of the Chemical Society, the Faraday Society of which he became President in 1953-1955 and on the Council of the Royal Institute of Chemistry of which he was Vice President from 1957 to 1959. He delivered the Liversidge Lecture to the Chemical Society in 1958, the Faraday Memorial Lecture to the Chemical Society in 1965, and the Bakerian Lecture to the Royal Society in 1966. He was President of the British Association Section B (Chemistry) in 1961, and in the same year was made Liveryman of the Worshipful Company of Gunmakers. In 1958 he received the honorary degree of D. de l'U. at the Sorbonne in Paris and also honorary degrees D. Sc. in Leeds and Sheffield in 1965, Liverpool and Lancaster (1968) and British Columbia (1969). He is an honorary member of the Polish Chemical Society and Membre d'honneur of the Societé de Chimie Physique in Paris. He is a foreign member of the Polish and the Bulgarian Academies of Sciences, a corresponding member of the Academy of Sciences in Göttingen and of the Royal Society of Sciences in Liège. He is a honorary member of the Royal Society of Sciences in Uppsala and the New York Academy of Sciences. He has received the Meldola medal of the Royal Institute of Chemistry (1926), the Davy medal of the Royal Society (1958), the Lewis medal of the Combustion Institute (1964), the Faraday medal of the Chemical Society (1965) and their Longstaff medal (1969). He was elected Fellow of the Royal Society in 1936 and is still endeavouring to continue to prosecute his scientific activities in Cambridge.

To mark his retirement in 1965, many of his old friends and younger colleagues now occupying distinguished positions in academic and industrial work in Great Britain and abroad collaborated to publish a work entitled «Photochemistry and Reaction Kinetics». To them and to all others with whom he has worked for over 50 years he is deeply grateful.

GEORGE PORTER

Flash photolysis and some of its applications

Nobel Lecture, December 11, 1967

One of the principal activities of man as scientist and technologist has been the extension of the very limited senses with which he is endowed so as to enable him to observe phenomena with dimensions very different from those he can normally experience. In the realm of the very small, microscopes and microbalances have permitted him to observe things which have smaller extension or mass than he can see or feel. In the dimension of time, without the aid of special techniques, he is limited in his perception to times between about one twentieth of a second (the response time of the eye) and about $2 \cdot 10^9$ seconds (his lifetime). Yet most of the fundamental processes and events, particularly those in the molecular world which we call chemistry, occur in milliseconds or less and it is therefore natural that the chemist should seek methods for the study of events in microtime.

My own work on «the study of extremely fast chemical reactions effected by disturbing the equilibrium by means of very short pulses of energy» was begun in Cambridge twenty years ago. In 1947 I attended a discussion of the Faraday Society on «The Labile Molecule». Although this meeting was entirely concerned with studies of short lived chemical substances, the four hundred pages of printed discussion contain little or no indication of the impending change in experimental approach which was to result from the introduction, during the next few years, of pulsed techniques and the direct spectroscopic observation of these substances. In his introduction to the meeting H. W. Melville referred to the low concentrations of radicals which were normally encountered and said «The direct physical methods of measurement simply cannot reach these magnitudes, far less make accurate measurements in a limited period of time, for example 10^{-3} sec.»

Work on the flash photolysis technique had just begun at this time and details of the method were published two years later[1,2]. Subsequent developments were very rapid, not only in Cambridge but in many other laboratories. By 1954 it was possible for the Faraday Society to hold a discussion on «The study of fast reactions» which was almost entirely devoted to the new techniques introduced during the previous few years. They included, as well as flash photolysis, other new pulse techniques such as the shock wave, the stopped flow method, and the elegant pressure, electric field density and temperature pulse methods described by Manfred Eigen. Together with pulse radiolysis, a sister technique to flash photolysis which was developed around 1960, these methods have made possible the direct study of nearly all fast reactions and transient substances which are of interest in chemistry and, to a large extent, in biology as well.

The various pulse methods which have been developed are complementary to each other, each has its advantages and limitations, and the particular power of the flash photolysis method is the extreme perturbation which is produced, making possible the preparation of large amounts of the transient intermediates and their direct observation by relatively insensitive physical methods. Furthermore, the method is applicable to gases, liquids and solids and to systems of almost any geometry and size, from path lengths of many metres to those of microscopically small specimens.

My original conception of the flash photolysis technique was as follows: the transient intermediates, which were, in the first place, to be gaseous free radicals, would be produced by a flash of visible and ultraviolet light resulting from the discharge of a large condenser bank through an inert gas. The flash would be of energy sufficient to produce measurable overall change and of short duration compared with the lifetime of the intermediates. Calculation showed that an energy of 10 000 J dissipated in a millisecond or less, in lamps of the type which were being developed commercially at that time, would be adequate for most systems. The bank of condensers was given by my friends in the Navy, and, although I am grateful to them for saving us much expense, it consisted of a motley collection of capacitors which, owing to their high inductance, gave a flash of rather longer duration than was desirable. The detection system was to consist of a rapid-scanning spectrometer and much time was wasted in the development of this before I realised that to demand high spectral resolution, time resolution and sensitivity in a period of a few milliseconds was to disregard the principles of information theory. Subsequent applications of flash photolysis, with a few exceptions, have been content to record, from a single flash, either a single spectrum at one time or a small wavelength range at all times. The use of a second flash, operating after a time delay, to record the absorption spectrum of the transients must now seem a very obvious procedure but it was many months before it became obvious to me. The double-flash procedure was an important step forward and is still, in principle, the most soundly based method for the rapid recording of information.

In the first apparatus the delay between the two flashes was introduced by a rotating sector, with two trigger contacts on its circumference, a photograph of which is shown in Fig. 1. The reason for using this, in preference to an electronic delay, was in anticipation of difficulties with scattered light from the high-energy photolysis flash, which could be eliminated by the shutter incorporated in the rotating wheel. As flash durations were reduced it became necessary to resort to purely electronic methods but the apparatus worked well for several years and provided, for the first time, the absorption spectra of many transient substances and a means for their kinetic study.

Fig. 1. Part of the original flash-photolysis apparatus showing the rotating sector with shutter and trigger contacts.

There have been many reviews of the flash-photolysis method[3-5] and in this lecture I should like to illustrate our work by describing four rather different problems: the first two are simple gas-phase reactions which were the earliest to be investigated in detail and which illustrate rather clearly the two principal variations of the flash photolysis technique; the second two examples are studies of the principal types of transient which appear in photochemical reactions, *i.e.* radicals and triplet states, with special reference to aromatic molecules.

The first free radical to be studied in detail by flash-photolysis methods, both spectroscopically and kinetically, was the diatomic radical ClO, and this study provided a proving exercise for the technique. The spectrum was discovered, somewhat accidentally, in the course of a study of the chlorine–carbon monoxide–oxygen system and provided one of the first of many lessons on the limitation of predictions based on conventional studies. This new spectrum, which is shown in Fig. 2, was produced by flash photolysis of mixtures of oxygen and chlorine, in which no photochemical reaction had previously been suspected. Indeed there is no reaction at all if one speaks of times of conventional experiments, since the system returns to its original state in a few milliseconds.

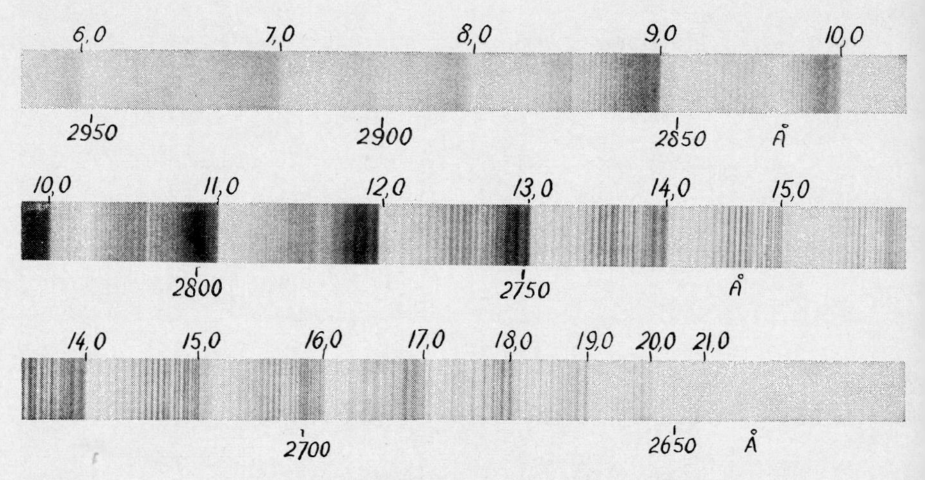

Fig. 2. Absorption spectrum of the ClO radical.

When a new transient species is detected in this way two kinds of information become available. First, analysis of the spectrum itself leads to structural and energetic data about the substance[6]; secondly, studies of the time-resolved spectra provide a measure of its concentration as a function of time and therefore provide kinetic data about the physical and chemical changes which it undergoes[7]. The information which resulted from the early studies of ClO is summarised in Table 1, the kinetic data being obtained from analysis of sequences of spectra of the type shown in Fig. 3.

Kinetic studies, even in this case, would have been more easily and accurately carried out by recording one wavelength only, so that a single flash can provide all the necessary kinetic information. Since flash photolysis of chlorine had already suggested that the amount of decomposition into atoms was very considerable, Norrish and I decided to study halogen atom recombi-

Table 1

Structural and kinetic information about the ClO radical obtained from Figs. 2 and 3

Spectroscopic		
Ground state		Dissociates to Cl ^2P and O ^3P
D''_0	=	22,060 cm^{-1} = 63.0 kcal/mole ± 0.3 %
ω''_0	=	868 cm^{-1}
Upper state		Dissociates to Cl ^2P and O ^1D
D'_0	=	7,010 cm^{-1} = 20.0 kcal/mole ± 1 %
ω'_e	=	557 cm^{-1}
ν_e	=	31,080 cm^{-1}
Kinetic		
$2Cl + O_2$	=	$2ClO$ } k_1
	=	$Cl_2 + O_2$ } k_1
$2Cl + N_2$	=	$Cl_2 + N_2$ k_2
$2ClO$	=	$Cl_2 + O_2$ k_3

$k_1 = 46\,k_2$. $k_3 = 8.6 \cdot 10^7 \exp(0 \pm 650/RT)$ l $mole^{-1}$ sec^{-1}

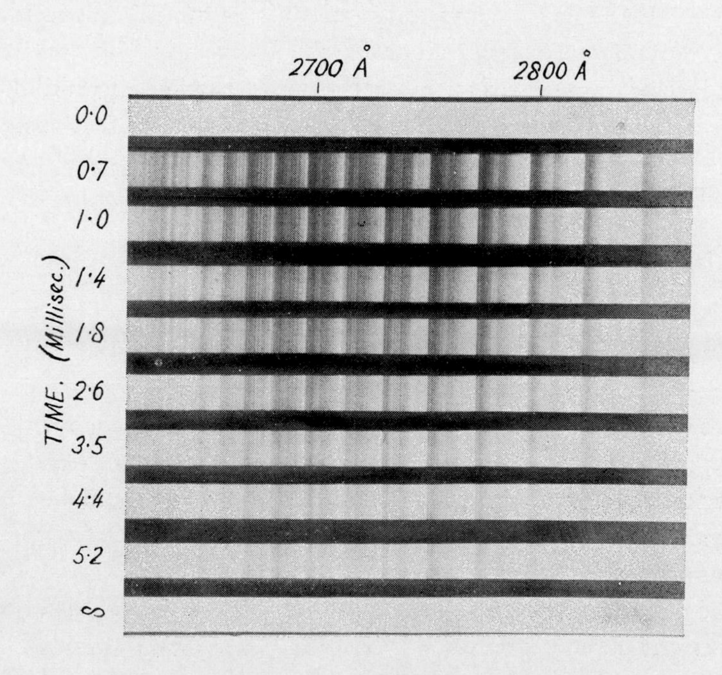

Fig. 3. Sequence of spectra of ClO after flash photolysis of a chlorine–oxygen mixture, showing bimolecular decay.

nation using, for monitoring, a continuous source, a monochromator and photoelectric detector. Iodine is the most convenient gas for these studies and the recombination of iodine atoms has been studied, in several laboratories, in more detail than perhaps any other gaseous reaction.

It was first necessary to show directly that the recombination of atoms, produced photolytically, was indeed a third-order reaction, as had been predicted theoretically for many years. It was satisfying, though not surprising, when this proved to be the case, at least to a first approximation. Oscilloscope traces and second-order plots of the decay of iodine atoms in argon are shown in Figs. 4 and 5, the gradients are proportional to the pressure of argon as they should be for a third-order reaction involving argon as the third body[8].

Studies in a number of laboratories[9-12], revealed interesting complications and problems. There were complex concentration gradients across the reaction vessel, caused by the essentially adiabatic nature of the reaction. When these were eliminated, by working with a high excess of inert gas, further deviations from linearity appeared which were found to be caused by an unexpected very high efficiency of the iodine molecule as third body. There were striking differences in efficiency between different third body molecules and temperature coefficients were negative. Later work[13] showed that the negative temperature coefficients could be expressed in Arrhenius' form as «a negative activation energy» and that the more efficient the third body, the greater the value of this quantity. This is illustrated by the data in Table 2.

All these observations can be interpreted in terms of a mechanism involving intermediate formation of a complex between the iodine atom and a third body

$$\mathrm{I} + \mathrm{M} \overset{K}{\rightleftharpoons} \mathrm{IM}$$

$$\mathrm{IM} + \mathrm{I} \xrightarrow{k_3} \mathrm{I_2} + \mathrm{M}$$
$$(k = k_3 K)$$

the observed negative activation energy being nearly equal to the heat of formation of the complex. The nature of this complex is of interest. In extreme cases, such as that of NO as third body[14] the bonding is undoubtedly chemical. In the more general case of the molecules given in Fig. 2 we have suggested[13,15] that complexes are of the charge-transfer type and indeed, in solution, spectroscopic evidence for charge-transfer complexes of iodine atoms was obtained directly by flash-photolysis studies[16].

These studies of ClO and iodine atom reactions are representative of most of the work using flash photolysis which has followed, although there has

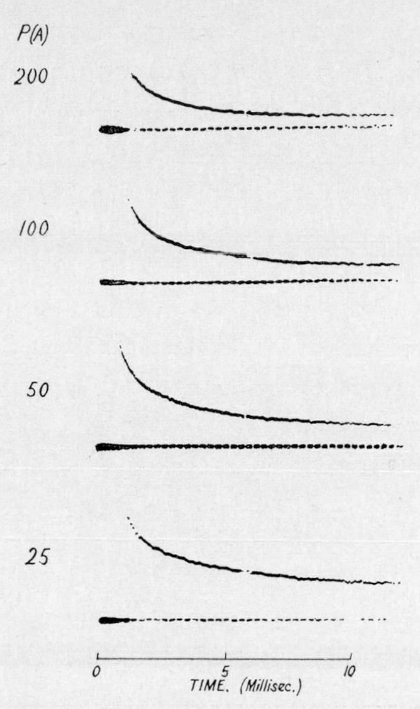

Fig.4. Oscillograph traces of I_2 absorption after flash photolysis of iodine in excess argon. $P(A)$ is argon pressure in mm Hg.

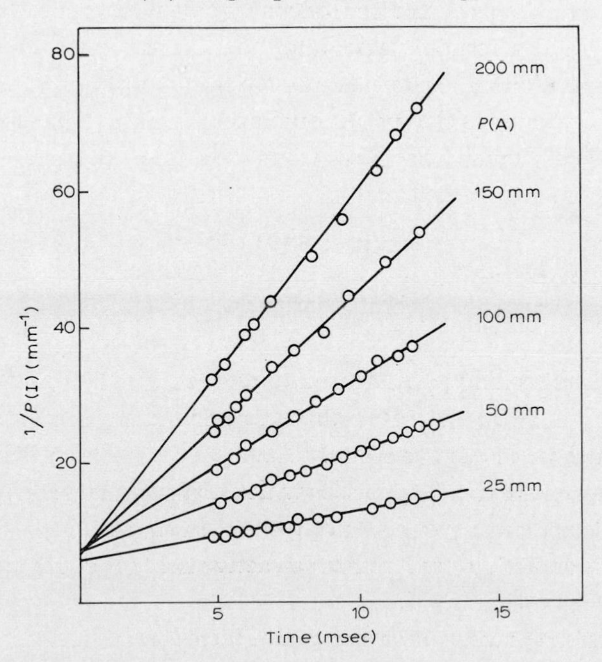

Fig.5. Second-order plots of data from Fig.4. $P(I)$ is iodine atom partial pressure.

Table 2

Termolecular recombination constants (k_{27}) at 27°C and temperature coefficients expressed as negative activation energies (E_a) for the recombination of iodine atoms in the presence of various chaperon molecules M

M	k_{27} (l^2 $mole^{-2}$ sec^{-1}) $\times 10^{-9}$	E_a (kcal)
He	1.5	0.4
A	3.0	1.3
H_2	5.7	1.22
O_2	6.8	1.5
CO_2	13.4	1.75
C_4H_{10}	36	1.65
C_6H_6	80	1.7
CH_3I	160	2.55
$C_6H_5CH_3$	194	2.7
C_2H_5I	262	2.4
$C_6H_3(CH_3)_3$	405	4.1
I_2	1600	4.4

naturally been a trend towards the study of more complex molecules. Some of the most significant spectroscopic work is that which has been carried out by Herzberg and his school on polyatomic radicals containing three or four atoms. When flash photolysis was first introduced, no absorption spectrum of a gaseous polyatomic radical was known and I think the first to be detected must have been that which has been assigned[2] to HS_2. One of our first interests had been the methylene radical, CH_2, and it was with this problem particularly in mind that the flash technique was conceived. It was very gratifying when, in 1959, Herzberg eventually brought methylene into the fold by extending flash spectroscopy into the vacuum ultraviolet region[17].

In addition to the increased complexity of the molecules investigated, the flash-photolysis technique has been increasingly applied to solutions, solids and even to biological systems so that these applications are now more extensive than those in the gas phase. I have been particularly interested, for over ten years, in the transient species which appear upon excitation of larger organic molecules, many of them of interest in organic mechanisms and in biological processes. This will remain an active field for a long time, since it is as large as organic chemistry itself, and I shall devote the rest of this lecture to a brief description of two principal classes of these species, namely the aromatic free radicals and the triplet states of organic molecules.

The Aromatic Free Radicals

The first aromatic free radical, triphenyl methyl, was discovered, quite unexpectedly, by Gomberg[18] in 1900. Many similar resonance-stabilised free radicals, which are stable enough to exist in observable concentrations at equilibrium, were subsequently reported but the direct spectroscopic observation of reactive, shortlived, aromatic free radicals was only achieved relatively recently.

The first detection of the absorption spectra of these species was made possible by the development of low temperature trapping techniques by G.N. Lewis and his school[19] in the early 1940's. This important work attracted little attention, and the method was revived and extended by Norman and Porter[20] who, in 1954, were able to detect a number of simple aromatic free radicals in rigid media, radicals such as benzyl and anilino. This led us to search for similar spectra in flash-photolysis experiments, since the results of low temperature trapping studies are very limited; the spectra are diffuse and no kinetic studies are possible.

Porter and F.J. Wright[21], in 1955, detected a series of aromatic free radicals in the gas phase by flash photolysis of aromatic vapours. For example, spectra in the region of 3 000 Å were attributed to benzyl (from toluene), anilino (from aniline) and phenoxyl (from phenol). These radicals are isoelectronic with each other, and form a type of seven π-electron system which, in aromatic free-radical chemistry, is of comparable importance with the six π-electron benzene ring in normal molecules. The spectra of radicals of this type, as has been shown by Dewar, Longuet-Higgins and Pople[22], result from the interaction of two degenerate configurations and consist of a weak forbidden transition at longer wavelengths and a strong allowed transition at shorter wavelengths. The transition observed near 3 000 Å is the strong, allowed one; the weaker longer wavelength transition has been observed by flash photolysis in the gas phase only relatively recently by Porter and Ward[23] (Fig. 6) though, being from the lowest excited state, it is relatively easily observed in emission[24].

The spectra of benzyl, anilino and phenoxyl in solution are diffuse, but quite characteristic as will be seen from the phenoxyl radical spectra, in various media, observed by Porter and Land[25] and shown in Fig. 7. An interesting complication arises in the anilino radical, where two quite different spectra are observed[26] depending on solvent or, in aqueous solvents, on pH (Fig. 8). The two spectra correspond to the protonated and unprotonated forms, the radical

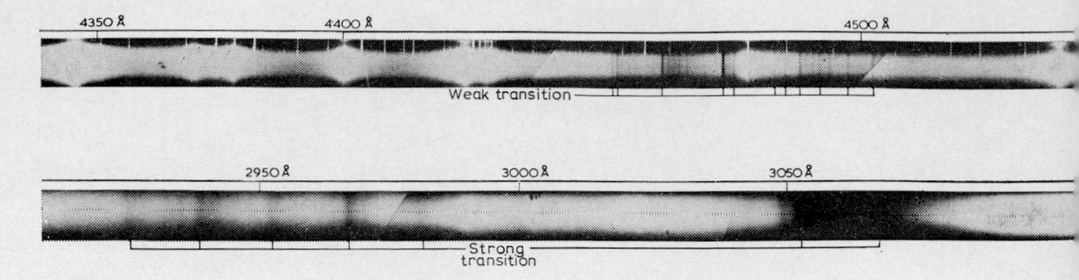

Fig. 6. Absorption spectrum of the benzyl radical in the vapour phase after flash photolysis of toluene. Path length = 8 metres.

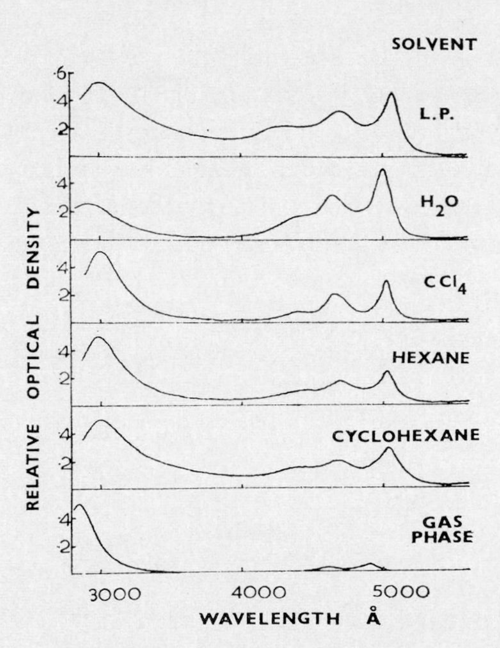

Fig. 7. Absorption spectrum of the phenoxyl radical in the gas phase and in various solvents after flash photolysis of phenol.

ion and the radical, and the equilibrium can be established within the lifetime of the radical making it possible to determine the equilibrium constant of this acid–base equilibrium. The pK of the anilino radical is found to be 7.0.

These seven π-electron radicals are the prototype of many of the most important, resonance-stabilised radicals of organic chemistry. For example, benzyl is the prototype of the Gomberg type radicals such as triphenyl methyl whilst OH substitution in the β position gives ketyl radicals, anilino is the prototype of Würster radical cations whilst phenoxyls, on substitution by OH, become semiquinones. The spectra and physico-chemical properties of

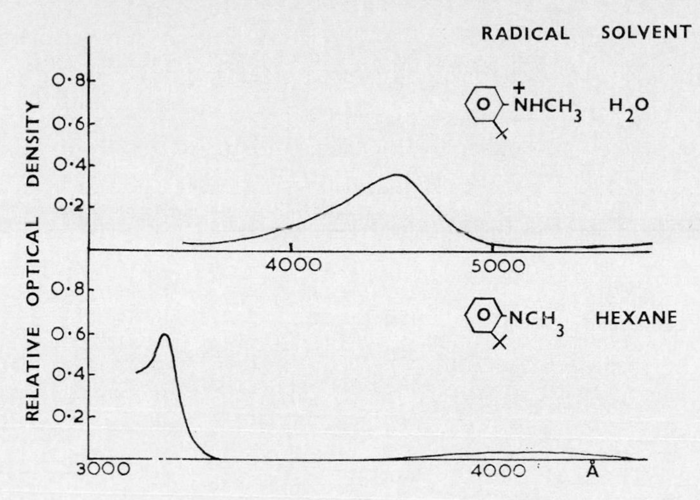

Fig. 8. Absorption spectrum of a substituted anilino radical showing acidic form in water and deprotonated form in hexane solution.

all these radicals are closely similar and provide a large and interesting field of study, of considerable importance both in chemistry and biology. For example the flavins have been shown to behave similarly and to yield semiquinone radicals on flash photolysis[27] and the principal transients observed on flash photolysis of proteins such as ovalbumin are phenoxyl-type radicals derived from tyrosine groups[28].

Flash-photolysis studies carried out more recently by Porter and Ward[29], on aromatic vapours at high resolution have succeeded in detecting many other aromatic free radicals of interest, which are not of the seven π-electron type. The most important of these is phenyl, obtained from benzene and halogeno benzenes and substituted phenyls derived from disubstituted benzene derivatives (Fig. 9).

Phenyl nitrene (or imine), an even electron-number radical analogous to the methylenes, is obtained from flash photolysis of phenyl isocyanate and, by

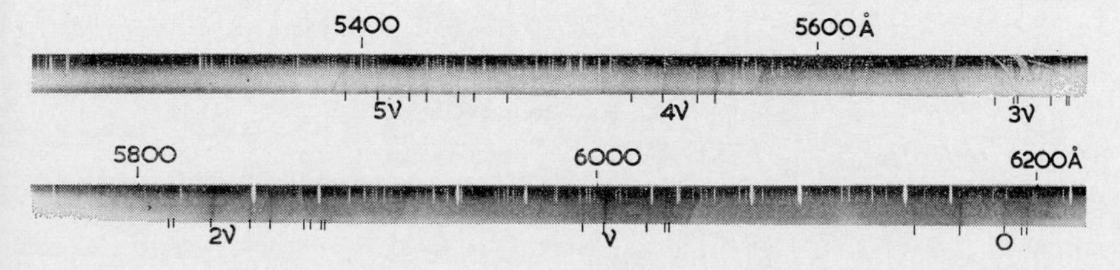

Fig. 9. Spectrum of the *ortho*-fluorophenyl radical after flash photolysis of *ortho*-chlorofluorobenzene.

molecular elimination of HCl followed by tautomeric change, from *ortho*-chloroaniline[30]. One of the most surprising transformations of all is the formation[30,31] of cyclopentadienyl (C_5H_5) radicals from flash photolysis of aniline, phenol, nitrobenzene and many other substituted benzenes, as well as a variety of substituted cyclopentadienyls from disubstituted aromatic compounds (Figs. 10 and 11).

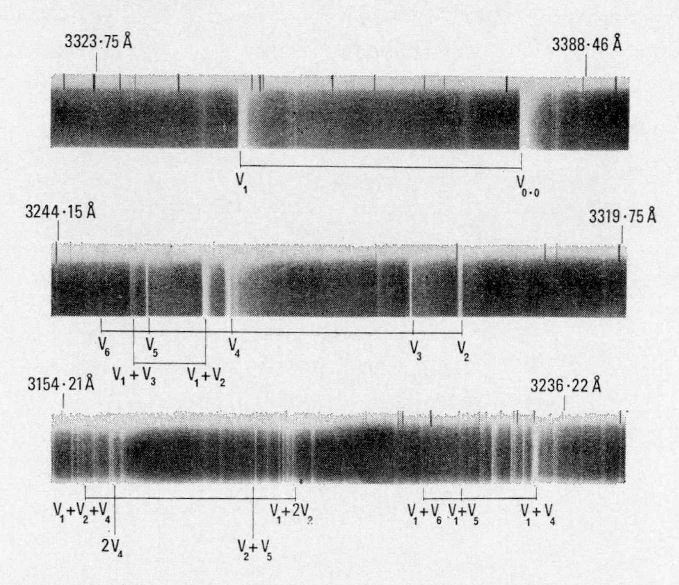

Fig. 10. Absorption spectrum of the cyclopentadienyl radical.

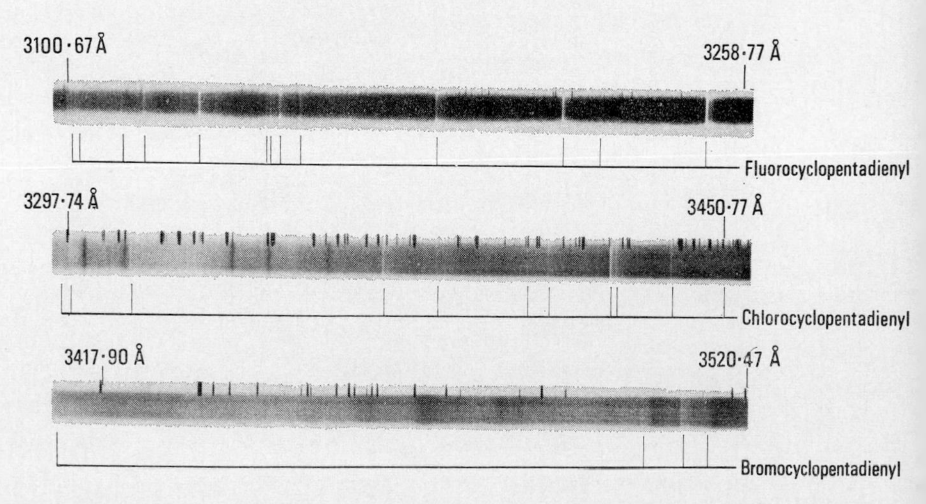

Fig. 11. Absorption spectra of three halogenated cyclopentadienyl radicals.

The assignment of spectra in all these cases is made on the basis of studies of a series of related substituted compounds, and the method is both convenient and reliable in the aromatic series, since so many compounds are available with the possibility of cross checks in most cases. The data on cyclopentadienyl radical formation, Fig. 12, illustrate the method. Although little or no assistance in the identification is possible from the spectra, apart from general similarity and positions of electronic transitions in related radicals, none of the assignments which have been made so far has subsequently had to be revised; a situation which is not always found in the assignment of spectra, even those of much simpler compounds.

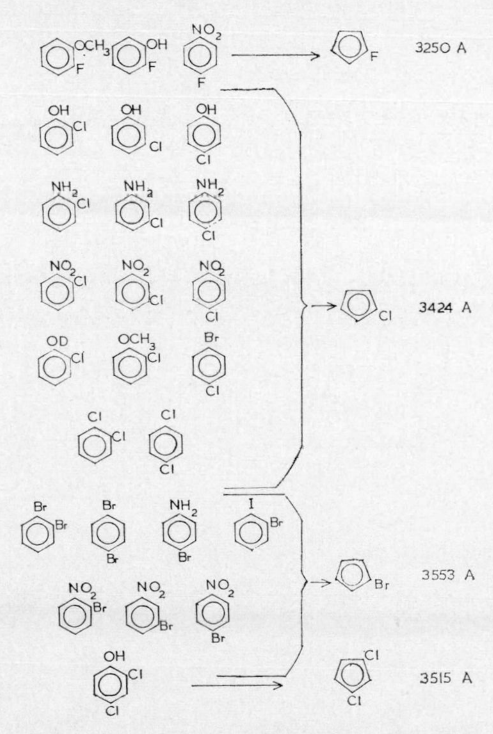

Fig. 12. Observed fission processes of benzene derivatives to form cyclopentadienyl radicals.

The field of aromatic free-radical spectroscopy is relatively new and what has been done already is little more than an indication of what will be done by flash-photolysis techniques in this field in the future. The hundred radicals so far assigned are prototypes of thousands of others which may be observed whenever there is a reason for wanting to study them. Many of the spectra show fine detail which, with the help of computer programmes, should eventually yield information about the radical structures. The chemical problems are almost untouched, and represent a much more extensive series of problems than mere identification. Extinction coefficients are known in only a few cases, usually in solution. Finally there are the intriguing photochemical problems of the primary processes in the excited state by which these often remarkable transformations take place. At the present time it cannot even be taken for granted that the vapour phase dissociations are monophotonic and in the case of cyclopentadienyl it seems probable that a biphotonic process, involving radicals such as phenoxyl and anilino as intermediates, may be operative. In the formation of phenyl, benzyl and similar radicals, biphotonic mechanisms in the gas phase appear less probable and I have tentatively suggested[5] that a mechanism is operative in which radiationless conversion to the ground state is followed by what is essentially a thermal dissociation of a highly vibrationally excited molecule.

The Triplet State

In 1944, Lewis and Kasha[32] showed that the phosphorescence of organic molecules which is observed in rigid media is the emission of light from the lowest excited state of these molecules and that this state is of triplet multiplicity. This work opened up a new realm of spectroscopy and physical investigation which continues with increasing activity today.

The influence of this discovery on chemistry and photochemistry was, at first, very slight. The reason for this was that the chemist does little work in rigid solutions at low temperatures and only under these conditions could the triplet states be observed. The reasons for the absence of phosphorescence in gases and fluid solutions were not altogether clear, though Lewis and Kasha clearly stated their view that the triplet state was formed under these conditions, presumably with much shorter lifetime. If this were the case, it should be possible to detect the triplet state by means of its absorption spectrum in flash-photolysis experiments provided its lifetime was greater than a few microseconds, a question about which there was little information.

In 1952, Windsor and I decided to attempt the observation of triplet absorption spectra of organic molecules by flash photolysis in ordinary fluid solvents at normal temperatures. The scheme of transitions involved in these studies is shown in Fig. 13. Almost immediately the experiments were suc-

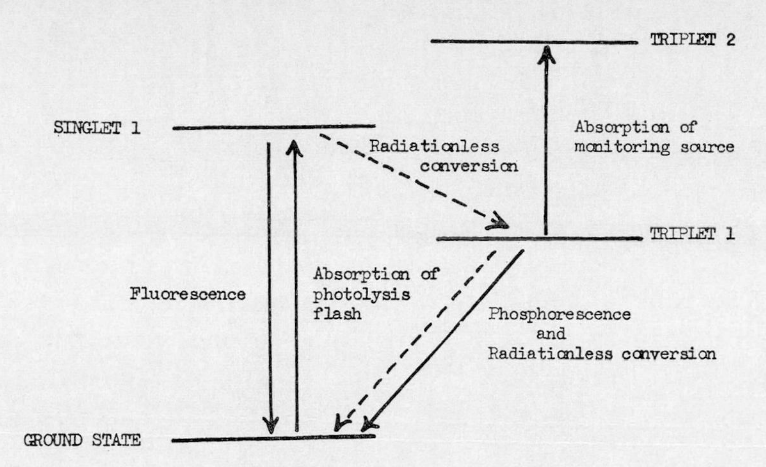

Fig. 13. Transitions involved in flash photolysis studies of the triplet state.

cessful; it transpired that triplet-state lifetimes under these conditions were of the order of a millisecond, ideal for studies by flash photolysis, provided oxygen was excluded. Some of the first flash-photolysis records of triplet states of aromatic molecules in solution[33] are shown in Figs. 14 and 15.

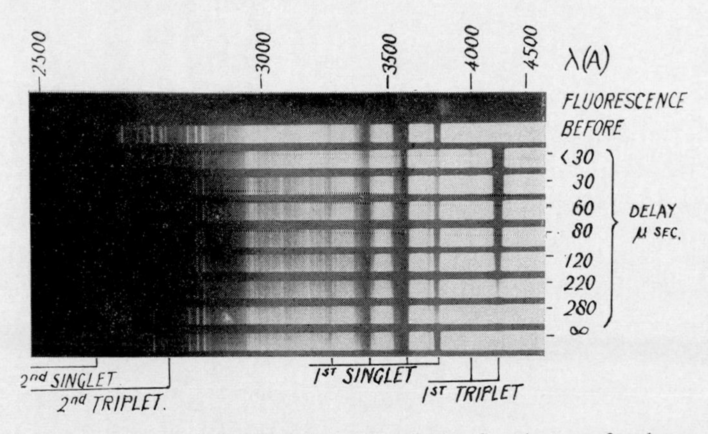

Fig. 14. Absorption spectra following flash photolysis of a solution of anthracene in hexane solution.

Immediately F. J. Wright and I investigated the possibility of detecting triplet states in the gas phase and although the lifetime was shorter the triplet spectra of a number of aromatic hydrocarbons were successfully recorded in 1-m paths of the vapour[34].

It is these studies, perhaps more than any others, which have brought flash photolysis into the chemical laboratory as a routine method of investigation.

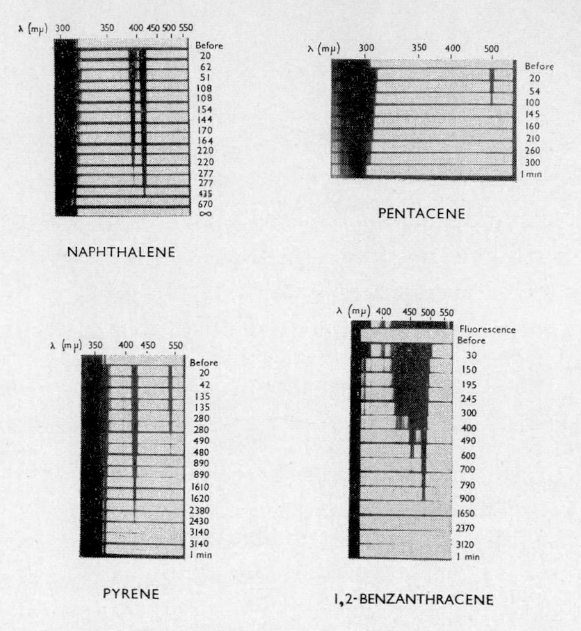

Fig. 15. Sequences of spectra after flash photolysis of four aromatic hydrocarbons in solution showing formation and decay of their triplet states. Delays in microseconds.

Any discussion of mechanism in organic photochemistry immediately involves the triplet state and questions about this state are most directly answered by means of flash photolysis. It is now known that many of the most important photochemical reactions in solution, such as those of ketones and quinones, proceed almost exclusively *via* the triplet state and the properties of this state therefore become of prime importance. Its relatively long lifetime, compared with the time of a flash experiment, has made it possible to study the triplet state almost as readily as the ground state, and in many systems its physical and physico-chemical properties and its chemical reactions are now as well characterised as those of the ground state. The spectrum itself, being a transition between two excited states and usually diffuse, has been of less interest for structural studies than as a means of identification and quantitative estimation of triplet concentrations and, therefore, once identified, most kinetic studies have been carried out at a single wavelength using photoelectric methods.

Studies in our laboratory and others, particularly those of Livingston and of Linschitz, over the last fifteen years, have established the following properties of triplet states in fluid media:

(*1*) The radiationless decay processes which occur in solution are principally first order under normal conditions and due to traces of quenching impurities which are still largely unknown. This apparently trivial and uninteresting process has been the most difficult of all to establish and is still the most unsatisfactory aspect of the work. That the decay in solution is a quite separate process from the true radiationless and radiative conversion which occurs in rigid media is most clearly shown by flash experiments carried out over the whole range of viscosity and temperature[35].

(*2*) At high concentrations of triplet and low quencher concentrations Porter and M. Wright showed that a second-order process of triplet–triplet annihilation is predominant[36]. Parker and Hatchard[37] have shown that part of this process results in the formation of singlet excited states and delayed fluorescence. In the gas phase Porter and West[38] showed that triplet–triplet annihilation is the predominant means of decay.

Physical quenching of the triplet state can be brought about in four principal ways involving the following species: (*a*) Heavy atoms which increase spin-orbit interaction[39]. (*b*) Atoms and molecules with unpaired electrons which can interact with the triplet to form a collision complex *via* which conversion to the ground state can occur without contravention of the spin selection rules[36]. (*c*)Molecules which form charge-transfer complexes with the triplet (ref. 40). (*d*) Molecules with lower electronic states (singlet or triplet), to which energy can be transferred[41].

The last of these processes is of particular interest in photosensitisation and biological systems and it was first observed, under conditions of high concentration in rigid media, by Terenin and Ermolaev[42]. In solution it is readily studied by flash photolysis and, in favourable cases, both the decrease in donor triplet and the increase in acceptor triplet can be followed independently. An early example of this type of transfer studied by Porter and Wilkinson, between phenanthrene and naphthalene, is shown in Fig. 16.

The efficiencies of triplet state formation have recently been studied quantitatively and directly by Bowers and Porter[43] using an optical arrangement of flash photolysis which makes it possible to monitor the light absorbed. At present, in all molecules studied which do not present complexities such as dimer formation, the sum of fluorescence and triplet yields is unity within the precision of the combined measurements (Table 3).

As examples of chemical processes in the triplet state we may mention proton transfer, and electron or hydrogen atom transfer. It is usually easy to arrange, by using buffered solutions, that protonic equilibrium is established

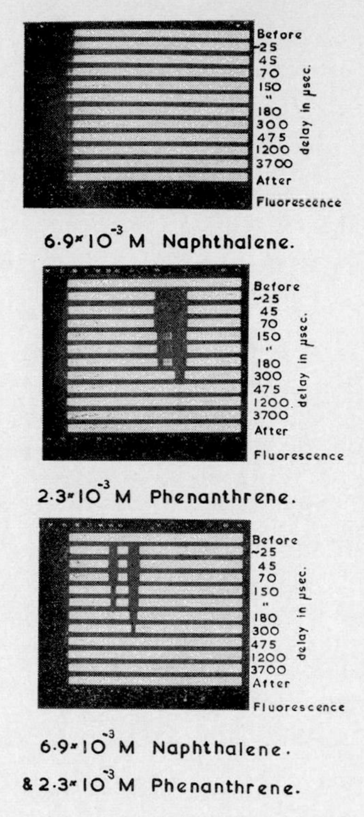

Fig. 16. Energy transfer from triplet phenanthrene to naphthalene. The flash is filtered through a strong solution of naphthalene so absorption is by phenanthrene only. In the bottom sequence, obtained from a solution containing both compounds, the triplet of phenanthrene has been completely quenched and replaced by triplet naphthalene.

Table 3

Fluorescence yield Φ_F, triplet yield Φ_T determined directly by flash photolysis in solution, and the sum of these yields for several organic compounds

Compound	*Solvent*	Φ_F	Φ_T	$\Phi_F+\Phi_T$
Anthracene	Liquid paraffin	0.33	0.58 ± 0.10	0.91
Phenanthrene	3-Methylpentane	0.14	0.70 ± 0.12	0.84
1,2,5,6-Dibenzanthrancene	3-Methylpentane	—	1.03 ± 0.16	1.03
Fluorescein (fl)	Aqueous pH 9	0.92	0.05 ± 0.02	0.97
$flBr_2$	Aqueous pH 9	—	0.49 ± 0.07	—
Eosin ($flBr_4$)	Aqueous pH 9	0.19	0.71 ± 0.10	0.90
Erythrosin (flI_4)	Aqueous pH 9	0.02	1.07 ± 0.13	1.09
Chlorophyll *a*	Ether	0.32	0.64 ± 0.09	0.96
Chlorophyll *b*	Ether	0.12	0.88 ± 0.12	1.00

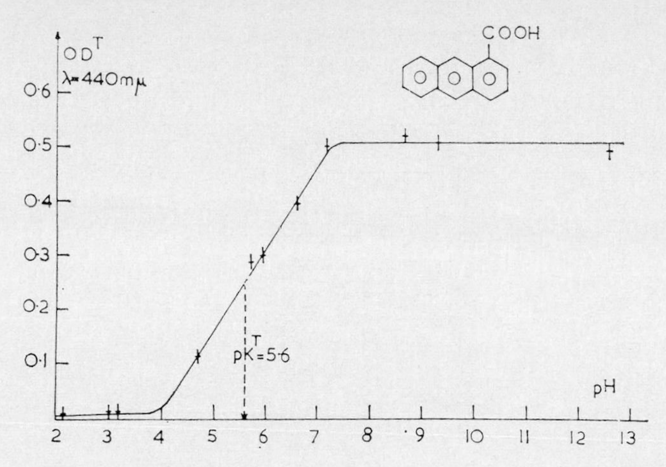

Fig. 17. Optical density plot of triplet anthroic acid *versus* pH used to derive the triplet-state acidity constant of this compound.

during the lifetime of the triplet and, in this case, a titration can be carried out almost as readily as when one determines the p*K* of the ground state, though now the «indicator» is the molecule in its triplet state. It is interesting to compare the results not only with the ground-state properties, but with those of the first excited singlet state which can be determined by fluorescence studies using the methods developed by Forster and Weller[44]. Such studies of triplet states were first carried out in collaboration with Jackson and later with Van der Donckt[45,46]. A typical p*K* plot for the triplet state of anthroic acid is shown in Fig. 17 and a summary of results on p*K*'s in the three states of interest for a variety of molecules is given in Table 4.

Electron and hydrogen atom transfer, particularly from solvent to triplet states of ketones, aldehydes and quinones have been the subject of very extensive investigations in a number of laboratories. My interest in this type of reaction first arose in a rather practical way when I was consulted about a technical problem known as phototendering, in which a dyed fabric, such as cellulose, becomes degraded under illumination in sunlight. The mechanism of this presented few problems; it was merely abstraction, by the excited dye molecule, of a hydrogen atom from the cellulose followed by addition of oxygen to the resulting radicals followed by degradation of the cellulose chain. What was surprising was that dyes, such as the anthraquinones, fell into two classes, one very reactive and one almost completely unreactive, the difference between the two classes being caused by the apparently small effects of substitution[47]. After a long and interesting series of investigations, mostly carried

Table 4

Acidity constants expressed as p*K* values for the ground state (G), the first excited singlet (S_1) and the lowest triplet (T_1) of a number of aromatic molecules

Compound	p$K(G)$	p$K(S_1)$	p$K(T_1)$
2-Naphthol	9.5	3.0	8.1
1-Naphthoic acid	3.7	7.7	3.8
2-Naphthoic acid	4.2	6.6	4.0
Acridine	5.5	10.6	5.6
Quinoline	4.9	(7)	6.0
2-Naphthylamine	4.1	−2	3.3
N,N-Dimethylaniline	4.9	—	2.7
1-Anthroic acid	3.7	6.9	5.6
2-Anthroic acid	4.2	6.6	6.0
9-Anthroic acid	3.0	6.5	4.2
2-Aminoanthracene	3.4	(−4.4)	3.3

out on benzophenone derivatives which show exactly the same phenomena, the matter became quite clear[48 49]. It is always the lowest triplet state which reacts and the electronic structure of this state is, therefore, the prime consideration. Depending on substituents and solvent, this lowest triplet state may be n–π, with electrophilic oxygen and therefore reactive or π–π, with considerable charge-transfer character in the opposite sense to that of the n–π state (CT), and therefore unreactive (Fig. 18).

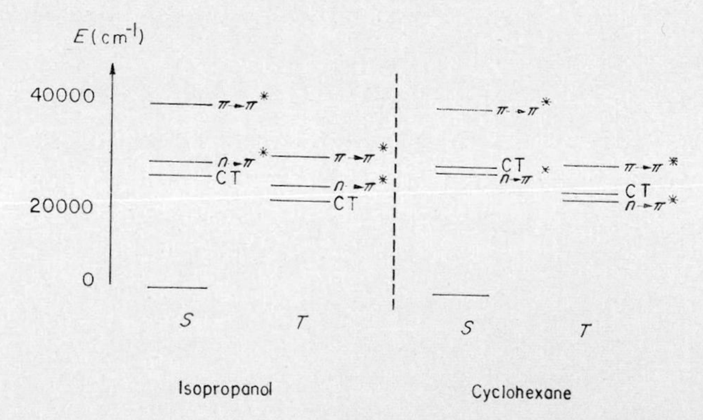

Fig. 18. Energy levels of singlet (S) and triplet (T) states of p-aminobenzophenone in isopropanol and in cyclohexane. In isopropanol the lowest triplet is of charge transfer (CT) type and therefore no reaction occurs; in cyclohexane the lowest triplet is the reactive n–π^* type and hydrogen abstraction occurs from cyclohexane to yield a ketyl radical.

These studies, both of proton and hydrogen atom transfer illustrate how the excited electronic state must be treated as a new species, with its own structure, electron distribution and chemical reactivity and how flash-photolysis techniques make it possible to study these characteristics of the excited triplet almost as readily as those of the ground state. Since each molecule has only one ground state, but several excited states, it is clear that this field of investigation is, in principle, a bigger subject than the whole of conventional chemistry.

At the present time efforts are being made to extend flash photolysis techniques in many ways, but particularly to shorter times. Although gas discharge lamps are unlikely to be much improved beyond the microsecond region, the giant pulsed laser promises to bring nanosecond times into the range of investigation and these, as well as nanosecond sparks coupled with integrating techniques, are now being developed in our laboratory and several others[5]. The nanosecond region will be particularly valuable for the direct study of excited singlet states.

The first flash apparatus resolved times of milliseconds, later ones worked comfortably in microseconds and nanosecond flash photolysis is now possible. This is a very short time. If we carry out an experiment every nanosecond then the results of a few seconds' work are enough to fill all the books and journals of the world. Advances of technique, such as the extension of our chemical experiments into the range of very short times, greatly increases the number of questions we can ask and the number of experiments to be done. To solve a problem is to create new problems, new knowledge immediately reveals new areas of ignorance, and the need for new experiments. At least, in the field of fast reactions, the experiments do not take very long to perform.

In conclusion, I express my deep gratitude to the collaborators who are referred to at the end of this paper and to all the other students and colleagues with whom it has been my good fortune to be associated. I share with them this great honour.

1. R.G.W. Norrish and G. Porter, *Nature*, 164 (1949) 658.
2. G. Porter, *Proc. Roy. Soc.* (*London*), *Ser. A*, 200 (1950) 284.
3. G. Porter, in E. Weissberger (Ed.), *Technique of Organic Chemistry*, Vol. 8, Part 2, Interscience, New York, 1963, p. 1055.

4. G. Porter, in P. G. Ashmore, F. S. Dainton and T. M. Sugden (Eds.), *Photochemistry and Reaction Kinetics*, Cambridge University Press, Cambridge, 1967, p. 93.
5. G. Porter, in Claesson (Ed.), *Nobel Symposium 5—Fast Reactions and Primary Processes in Reaction Kinetics*, Interscience, New York, 1967, p. 141.
6. G. Porter, *Discussions Faraday Soc.*, 9 (1950) 60.
7. G. Porter and F. J. Wright, *Z. Elektrochem.*, 56 (1952) 782; *Discussions Faraday Soc.*, 14 (1953) 23.
8. M. I. Christie, R. G. W. Norrish and G. Porter, *Proc. Roy. Soc. (London), Ser. A*, 216 (1952) 152.
9. K. E. Russell and J. Simons, *Proc. Roy. Soc. (London), Ser. A*, 217 (1953) 271.
10. R. Marshall and N. Davidson, *J. Chem. Phys.*, 21 (1953) 659.
11. R. L. Strong, J. C. W. Chien, P. E. Graf and J. E. Willard, *J. Chem. Phys.*, 26 (1957) 1287.
12. M. I. Christie, A. J. Harrison, R. G. W. Norrish and G. Porter, *Proc. Roy. Soc. (London), Ser. A*, 231 (1955) 446.
13. G. Porter and J. A. Smith, *Nature*, 184 (1959) 446; *Proc. Roy. Soc. (London), Ser. A*, 261 (1961) 28.
14. G. Porter, Z. G. Szabo and M. G. Townsend, *Proc. Roy. Soc. (London), Ser. A*, 270 (1962) 493.
15. G. Porter, *Discussions Faraday Soc.*, 33 (1962) 198.
16. T. A. Gover and G. Porter, *Proc. Roy. Soc. (London), Ser. A*, 262 (1961) 476.
17. G. Herzberg and J. Shoosmith, *Nature*, 183 (1959) 1801.
18. M. Gomberg, *J. Am. Chem. Soc.*, 22 (1900) 757; *Chem. Ber.*, 33 (1900) 3150.
19. G. N. Lewis, D. Lipkin and T. T. Magel, *J. Am. Chem. Soc.*, 66 (1944) 1579.
20. I. Norman and G. Porter, *Nature*, 174 (1954) 508; *Proc. Roy. Soc. (London), Ser. A*, 230 (1955) 399.
21. G. Porter and F. J. Wright, *Trans. Faraday Soc.*, 51 (1955) 1205.
22. M. J. Dewar and H. C. Longuet-Higgins, *Proc. Phys. Soc. (London)*, 67 (1954) 795; H. C. Longuet-Higgins and J. Pople, *ibid.*, 68 (1955) 591.
23. G. Porter and B. Ward, *J. Chim. Phys.*, 61 (1964) 1517.
24. H. Schuler and J. Kusjakow, *Spectrochim. Acta*, 17 (1961) 356.
25. E. J. Land and G. Porter, *Trans. Faraday Soc.*, 57 (1961) 1885.
26. E. J. Land and G. Porter, *Trans. Faraday Soc.*, 59 (1963) 2027.
27. B. Holmström, *Arkiv Kemi*, 22 (1964) 281, 329.
28. L. Grossweiner, *J. Chem. Phys.*, 24 (1956) 1255.
29. G. Porter and B. Ward, *Proc. Roy. Soc. (London), Ser. A*, 287 (1965) 457.
30. G. Porter and B. Ward, *Proc. Roy. Soc. (London)*, in the press.
31. G. Porter and B. Ward, *Proc. Chem. Soc.*, (1964) 288.
32. G. N. Lewis and M. Kasha, *J. Am. Chem. Soc.*, 66 (1944) 2100.
33. G. Porter and M. W. Windsor, *J. Chem. Phys.*, 21 (1953) 2088; *Discussions Faraday Soc.*, 17 (1954) 178; *Proc. Roy. Soc. (London), Ser. A*, 245 (1958) 238.
34. G. Porter and F. J. Wright, *Trans. Faraday Soc.*, 51 (1955) 1205.
35. J. W. Hilpern, G. Porter and L. J. Stief, *Proc. Roy. Soc. (London), Ser. A*, 277 (1964) 437.
36. G. Porter and M. R. Wright, *J. Chim. Phys.*, 55 (1958) 705; *Discussions Faraday Soc.*, 27 (1959) 18.

37. C. A. Parker and C. Hatchard, *Proc. Roy. Soc. (London)*, *Ser. A*, 269 (1962) 574.
38. G. Porter and P. West, *Proc. Soc. Roy. (London)*, *Ser. A*, 279 (1964) 302.
39. A. R. Horrocks, A. Kearvell, K. Tickle and F. Wilkinson, *Trans. Faraday Soc.*, 62 (1966) 3393.
40. H. Linschitz and L. Pekkarinen, *J. Am. Chem. Soc.*, 82 (1960) 2411.
41. G. Porter and F. Wilkinson, *Proc. Roy. Soc. (London)*, *Ser. A*, 264 (1961) 1.
42. A. W. Terenin and V. L. Ermolaev, *Trans. Faraday Soc.*, 52 (1956) 1042.
43. P. G. Bowers and G. Porter, *Proc. Roy. Soc. (London)*, *Ser. A*, 296 (1967) 435; 299 (1967) 348.
44. T. Forster, *Z. Elektrochem.*, 54 (1950) 42, 531; A. Weller, *Progress in Reaction Kinetics*, Vol. 1, Pergamon, Oxford, 1961, p. 187.
45. G. Jackson and G. Porter, *Proc. Roy. Soc. (London)*, *Ser. A*, 260 (1961) 13.
46. E. van der Donckt and G. Porter, *Trans. Faraday Soc.*, 64 (1968) 3215, 3218.
47. N. K. Bridge and G. Porter, *Proc. Roy. Soc. (London)*, *Ser. A*, 244 (1958) 259, 276.
48. A. Beckett and G. Porter, *Trans. Faraday Soc.*, 59 (1963) 2038, 2051.
49. G. Porter and P. Suppan, *Proc. Chem. Soc.*, (1964) 191; *Trans. Faraday Soc.*, 61 (1965) 1664; 62 (1966) 3375.

Biography

George Porter was born in the West Riding of Yorkshire on the 6th December 1920. He married Stella Jean Brooke on the 25th August 1949 and they have two sons, John and Andrew.

His first education was at local primary and grammar schools and in 1938 he went, as Ackroyd Scholar, to Leeds University. His interest in physical chemistry and chemical kinetics grew during his final year there and was inspired to a large extent by the teaching of M. G. Evans. During his final honours year he took a special course in radio physics and became, later in the year, an Officer in the Royal Naval Volunteer Reserve Special Branch, concerned with Radar. The training which he received in electronics and pulse techniques was to prove useful later in suggesting new approaches to chemical problems.

Early in 1945, he went to Cambridge to work as a postgraduate research student with Professor R. G. W. Norrish. His first problem involved the study, by flow techniques, of free radicals produced in gaseous photochemical reactions. The idea of using short pulses of light, of shorter duration than the lifetime of the free radicals, occurred to him about a year later. He began the construction of an apparatus for this purpose in the early summer of 1947 and, together with Norrish, applied this to the study of gaseous free radicals and to combustion. Their collaboration continued until 1954 when Porter left Cambridge.

During 1949 there was an exciting period when the method was applied to a wide variety of gaseous substances. Porter still remembers the first appearance of the absorption spectra of new, transient substances in time resolved sequence, as they gradually appeared under the safelight of a dark room, as one of the most rewarding experiences of his life.

His subsequent work has been mainly concerned with showing how the flash-photolysis method can be extended and applied to many diverse problems of physics, chemistry and biology. He has made contributions to other techniques, particularly that of radical trapping and matrix stabilisation.

After a short period at the British Rayon Research Association, where he applied the new methods to practical problems of dye fading and the photo-tendering of fabrics, he went in 1955, to the University of Sheffield, as Professor of Physical Chemistry, and later as Head of Department and Firth Professor. In 1966 he became Director and Fullerian Professor of Chemistry at the Royal Institution in succession to Sir Lawrence Bragg. He is Director of the Davy Faraday Research Laboratory of the Royal Institution. Here his research group is applying flash photolysis to the problem of photosynthesis and is extending these techniques into the nanosecond region and beyond.

Porter became a fellow of Emmanuel College, Cambridge, in 1952 and an honorary fellow in 1967. He was elected a Fellow of the Royal Society in 1960 and awarded the Davy Medal in 1971. He received the Corday-Morgan Medal of the Chemical Society in 1955 and was Tilden Lecturer of the Chemical Society in 1958 and Liversidge Lecturer in 1969. He has been President of the Chemical Society since 1970. He is Visiting Professor of University College, London, since 1967, and Honorary Professor of the University of Kent at Canterbury since 1966.

Porter holds Honorary D. Sc.'s from the following Universities: 1968, Utah, Salt Lake City (U. S. A.), Sheffield; 1970, East Anglia, Surrey and Durham; 1971, Leeds, Leicester, Heriot-Watt and City University. He is an honorary member of the New York Academy of Sciences (1968) and of the Academy «Leopoldina». He is President of the Comité International de Photobiologie, since 1968. He was Knighted in January 1972.

He is interested in communication between scientists of different disciplines and between the scientist and the non-scientist and has contributed to many films and television programmes. His main recreation is sailing.

Chemistry 1968

LARS ONSAGER

«for the discovery of the reciprocal relations bearing his name, which are fundamental for the thermodynamics of irreversible processes»

Chemistry 1968

Presentation Speech by Professor S. Claesson of the Royal Swedish Academy of Sciences

Your Majesty, Your Royal Highnesses, Ladies and Gentlemen.

Professor Lars Onsager has been awarded this year's Nobel Prize for Chemistry for the discovery of the reciprocal relations, named after him, and basic to irreversible thermodynamics. On hearing this motivation for the award one immediately gets a strong impression that Onsager's contribution concerns a difficult theoretical field. A closer study shows this indeed to be the case. Onsager's reciprocal relations can be described as a universal natural law, the scope and importance of which becomes clear only after being put in proper relation to complicated questions in border areas between physics and chemistry. A short historical review emphasizes this.

Onsager presented his fundamental discovery at a Scandinavian scientific meeting in Copenhagen in 1929. It was published in its final form in 1931 in the wellknown journal *Physical Review* in two parts with the title «Reciprocal relations in irreversible processes». The elegant presentation meant that the size of the two papers was no more than 22 and 15 pages respectively. Judged from the number of pages this work is thus one of the smallest ever to be rewarded with a Nobel Prize.

One could have expected that the importance of this work would have been immediately obvious to the scientific community. Instead it turned out that Onsager was far ahead of his time.

The reciprocal relations, which were thus published more than a third of a century ago, attracted for a long time almost no attention whatsoever. It was first after the second world war that they became more widely known. During the last decade they have played a dominant role in the rapid development of irreversible thermodynamics with numerous applications not only in physics and chemistry but also in biology and technology. Here we thus have a case to which a special Rule of the Nobel Foundation is of more than usual applicability. It reads: «Work done in the past may be selected for the award only on the supposition that its significance has until recently not been fully appreciated.»

The great importance of irreversible thermodynamics becomes apparent if we realize that almost all common processes are irreversible and cannot by themselves go backwards. As examples can be mentioned conduction of heat from a hot to a cold body and mixing or diffusion. When we dissolve a cold lump of sugar in a cup of hot tea these processes take place simultaneously.

Earlier attempts to treat such processes by means of classical thermodynamics gave little success. Despite its name it was not suited to the treatment of dynamic processes. It is instead a perfect tool for the study of static states and chemical equilibria. This science was developed during the nineteenth and the beginning of this century. In this work many of the most renounced scientists of that time took part. The Three Laws of Thermodynamics gradually emerged and formed the basis of this science. These are among our most generally known laws of nature. The First Law is the Law of Conservation of Energy. The Second and the Third Laws define the important quantity entropy which among other things provides a connection between thermodynamics and statistics. The study of the random motion of molecules by means of statistical methods has been decisive for the development of thermodynamics. The American scientist J. Willard Gibbs (1839–1903) who made so many important contributions to statistical thermodynamics, has his name attached to the special professorship which Onsager now holds.

It can be said that Onsager's reciprocal relations represent a further law making possible a thermodynamic study of irreversible processes.

In the previously mentioned case with sugar and tea it is the transport of sugar and heat during the dissolution process which is of interest in this connection. When such processes occur simultaneously they influence each other: a temperature difference will not only cause a flow of heat but also a flow of molecules and so on.

Onsager's great contribution was that he could prove that if the equations governing the flows are written in an appropriate form, then there exist certain simple connections between the coefficients in these equations. These connections – the reciprocal relations – make possible a complete theoretical description of irreversible processes.

The proof of the reciprocal relations was brilliant. Onsager started from a statistical mechanical calculation of the fluctuations in a system, which could be directly based on the simple laws of motion which are symmetrical with regard to time. Furthermore he made the independent assumption that the return of a fluctuation to equilibrium in the mean occurs according to the transport equations mentioned earlier. By means of this combination of macroscopic and microscopic concepts in conjunction with an extremely skilful mathematical analysis he obtained those relationships which are now called Onsager's Reciprocal Relations.

Professor Lars Onsager. You have made a number of contributions to physics and chemistry which can be regarded as milestones in the development of science. For example, your equation for the conductivity of solutions of strong electrolytes, your famous solution of the Ising problem, making possible a theoretical treatment of phase changes, or your quantisation of vortices in liquid helium. However, your discovery of the reciprocal relations takes a special place. It represents one of the great advances in science during this century.

I have the honour to convey to you the congratulations of the Royal Academy of Sciences and to ask you to receive the Nobel Prize for Chemistry for 1968 from the hands of His Majesty the King.

Lars Onsager

The motion of ions: principles and concepts

Nobel Lecture, December 11, 1968

Today I shall try to help you grasp the significance of a fairly general principle which applies to diverse types of irreversible processes. After last night it will be just as well if we do not go into all fine points of definitions or survey all possible applications. Rather, I want to talk about progress over a period of time in one field of research where much has happened (some of this relevant to the principle I mentioned) and intriguing problems still remain. Before we survey the progress in our understanding of electrolytes since the days of Arrhenius, let us take a quick look at what went before.

Gay-Lussac's rule of combining volumes (1808) led Avogadro to surmise that under corresponding conditions of temperature and pressure equal volumes of different gases contain equal numbers of molecules (1811). This principle was to become the chemist's primary means to determine molecular weights, but it was long debated and not in general use until after 1860. By that time Cannizzaro could muster enough evidence for a strong argument at the first international congress in Karlsruhe, and within a few years Avogadro's principle gained wide acceptance.

We may at least speculate that contemporary developments in the kinetic theory of gases encouraged the chemists' change of attitude, although they rarely if ever admitted that; they preferred to maintain an inductive point of view in their publications. In 1860, Maxwell obtained his distribution law for molecular velocities, which implies equipartition of kinetic energy; Avogadro's principle is an automatic consequence. In the following year Boltzmann founded a more general theory of specific heats, explaining the empirical rule of Dulong and Petit; those results had to be exploited with semi-empirical modifications until the quantum theory accounted for the discrepancies much later. Guldberg and Waage (1864) formulated the mass-action law and supported it by experiments. After a while (1885) Van 't Hoff recognized a close analogy between solutions and gases, so that measurements of osmotic pressure or changes in vapor pressure or freezing point depression could substitute for vapor densities. Strangely, however, the observations on solutions of salts, acids and bases in water indicated the presence of more solute particles than there could be molecules by any reasonable interpretation of chemistry known then or now. Arrhenius[1,2] (1884) recognized that these electrolytes dissociate largely into free ions, and he could point to a pretty good correspondence between the Van 't Hoff «anomalies» and the degrees of ionization inferred from the electrical conductivity.

A greatly simplified picture of electrolyte solutions loomed. At fairly low but readily attainable concentrations solutions of readily dissociating compounds like hydrochloric acid, potassium hydroxide and a great many salts like sodium chloride would be completely dissociated and the properties of a solution would be additive, not just over molecules, but even over the constituent ions. At higher concentrations, admittedly, one would have to allow for combination to form molecules or compound ions according to the mass-action law, as suggested by Ostwald[3] (1888). Nernst developed appropriate simple theories for the diffusion of electrolytes and for the variation of an electrode potential with the concentration of the ion discharged.

Such was the simple picture presented to me as a freshman chemist in 1920. In spite of some idealization it sufficed for a great many purposes; it eased many tasks no end and we were eternally grateful for that. However, very soon the journals rather than the textbooks taught me about numerous observations which did not quite fit into the picture and of tentative explanations for the discrepancies. Whether the experimenters studied the electrical conductivities or the equilibrium properties like freezing point depressions and electromotive forces, significant deviations from the ideal additive behavior persisted to much lower concentrations than had been predicted according to the mass-action law from the measurements performed on more concentrated solutions. These phenomena became known as the «anomalies» of strong electrolytes. In many ways the anomalies displayed conspicuous regularities; if one compared salts of the same valence type like NaCl and KNO_3, the differences were typically small even at concentrations as high as 0.1 mole/liter. Suspicion centered on the long-range electrostatic forces between the ions.

Debye and Hückel[4] finally succeeded in predicting the effects of the electrostatic interaction from the general principles of kinetic theory. They pointed out that the electrostatic field around an ion must be screened by an average density of compensating charge. As had been found previously by Gouy (1913) in somewhat different context, the screening distance is given by the ionic strength (sum of concentrations multiplied by squares of charges), the dielectric constant of the solvent and the temperature; it varies inversely as the square root of the ionic strength. The resulting effects on the chemical potentials of electrolytes are proportional to the square root of the ionic strength; to compute the coefficient one has to know the magnitude of an elementary charge, and the measurements of Millikan[5] had already supplied that information (1917). These predictions agreed well enough with previous experiments, and the improved techniques of subsequent decades have only confirmed the agreement. The theory of Debye and Hückel was soon routinely exploited to great advantage. Those stubborn deviations from the laws of Van 't Hoff and Arrhenius and Guldberg and Waage became harmless because we could compute them, make proper allowance and extrapolate in comfort to exploit additive relations. To make matters even easier, many electrolytes turned out to be completely dissociated or nearly so.

Debye and Hückel also considered the conductivity of electrolytes, a most important source of information because the measurement is almost always feasible, and it takes only reasonable care to get accurate results. Kohlrausch had shown long since that the conductivities of strong electrolytes in water decrease linearly with the square root of the concentration. Debye and Hückel recognized that two effects contribute to this decrease. For one, while the external electric field exerts a force on the ion, an opposing force of equal magnitude is distributed over the screening cloud of compensating charge. As a result, every ion is driven against a countercurrent; the speed of this current is proportional to the charge of the central ion and independent of its own mobility. This is called the electrophoretic effect, and the theory is closely related to that of the effect so profitably utilized by Tiselius[6]; but there is a significant difference between small and large particles, and the meaning of the word «electrophoresis» varies according to context. The so-called «relaxation effect» depends on distortions of the screening clouds produced by the systematic motion of the ions in the external field. As it happened, Debye and Hückel overestimated that effect and concluded that in computing the «electrophoretic force» they had extrapolated the macroscopic hydrodynamics too far.

Fortunately, my own efforts in the summer of 1923 had produced a modest but firm result. The relaxation effect ought to reduce the mobilities of anion and cation in equal proportion. Much to my surprise, the results of Debye and Hückel did not satisfy that relation, nor the requirement that whenever an ion of type A is 10 Å West of a B, there is a B 10 Å East of that A. Clearly, something essential had been left out in the derivation of such unsymmetrical results. The model used was this: one particular ion is constrained to move at a constant speed along a straight line in the solution; neighboring ions respond to the fields in the distorted screening cloud, and in addition they mill around in random fashion according to the laws of Brownian motion. Recipe: Remove restraints on the central ion but retain an external force on it, let it execute its own thermal motion and respond to the fields of its neighbors, and recognize whatever external forces act on them. That done, the result for binary electrolytes became very simple: the relaxation effect reduces the migration velocity of every ion by a fraction which depends neither on its own mobility nor on that of the partner species (of opposite, numerically equal charge). Otherwise the effect is charge-dependent and proportional to the square root of the concentration – just like the corrections to the equilibrium properties, but with a different coefficient. As to the electrophoretic effect, it was easy to show that plausible variations of the hydrodynamics near the center of the countercurrent system driven by a widely distributed force could not matter enough to affect the limiting law; Debye and Hückel had unjustly impugned their own result.

As seen from Fig. 1, the general variation of conductivities with the concentration for 1–1 electrolytes was quite well explained; the divergence and individual variation at higher concentrations was foreseeable, but the theory was not so far developed that the significance of those features could be evaluated in detail.

Fig. 2 displays the difference between a strong acid (HCl) and a weaker one (HIO_3). Clearly, the concept of a dissociation equilibrium was still indispensable.

In Fig. 3 we see the conductivities of a few ternary electrolytes compared with the theoretical limiting formulas.

In Fig. 4 the curves with appropriate limiting tangents are extrapolated according to the theory; previous empirical extrapolations are indicated too. The point was that the new extrapolation for $MgSO_4$ agreed with the limiting value expected from the additivity rule while the old one did not. It became clear that $MgSO_4$ was incompletely dissociated (as well as $CdSO_4$), a conclusion confirmed by later studies of the chemical reaction kinetics (Eigen[9]).

In Fig. 5 we see some deviations from Kohlrausch's rule of independent mobilities, first computed by Bennewitz, Wagner and Küchler[10] (1929), then demonstrated by Longsworth (1930). Solutions containing HCl and KCl in varying proportion are compared at constant total concentration. The fast hydrogen ions are delayed as they overtake the slower potassium ions and detour around them; the potassium ions are speeded up by the same interaction.

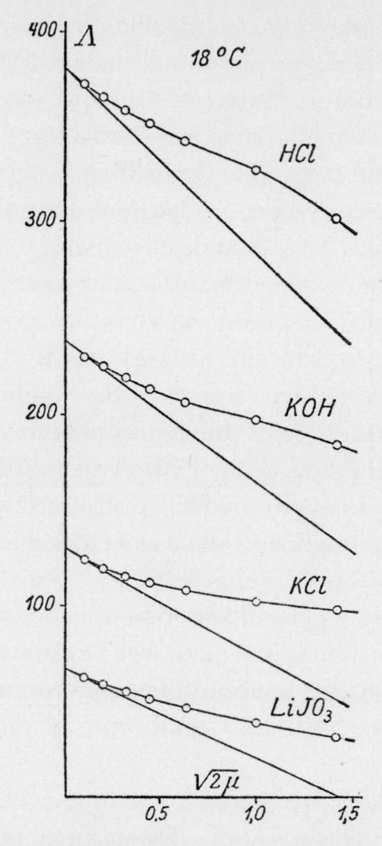

Fig. 1. Conductivities of 1–1 electrolytes in water. Λ, equivalent conductivity; μ, concentration in equiv./1. From *Physik.Z.*, 28 (1927) 277. Reproduced by permission of S.Hirzel Verlag K.G., Stuttgart.

The resultant net decrease of the total conductivity had been observed in a similar case before (Bray and Hunt, 1912) and pronounced a baffling mystery.

Going back to the time when I revised the theory of Debye and Hückel, the task was by no means easy. The key was a principle of superposition applied to the ion cloud around a pair. To begin with, it was a bit confusing that the force exerted by the external field on an ion as well as the interaction between the ions were proportional to the charge. In order to gain perspective I decided to ignore the relation between the charge and the driving force, and took a look at a more general problem. One constant field of force is acting on each kind of ion; what are the effects of the Coulomb interaction? The problem is in fact equivalent to that which arises in the most general case of diffu-

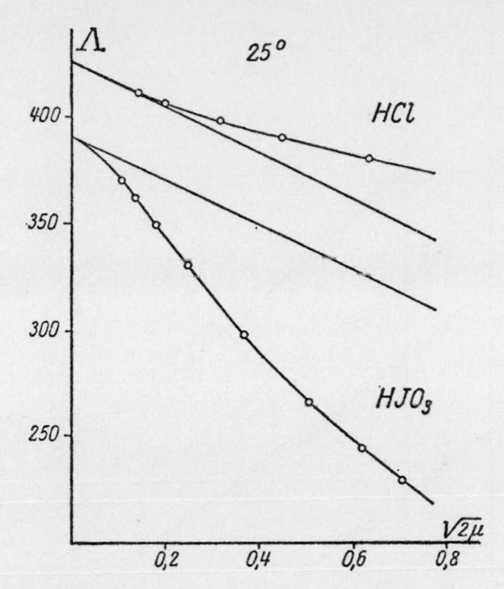

Fig. 2. Conductivities of hydrochloric acid and iodic acids in water. Same notation as Fig. 1. From *Physik. Z.*, 28 (1927) 277. Reproduced by permission of S. Hirzel K.G., Stuttgart.

sion and electrical conduction combined; the gradients of chemical potentials are equivalents of forces:

$$k_j = -\nabla\mu_j - e_j\nabla\varphi$$

where k_1, k_2... stand for forces, μ_1, μ_2... for chemical «potentials of ions», e_1, e_2... for charges and φ for the electrostatic potential. A measure of ambiguity in the definition of φ induces a corresponding ambiguity in μ_1, μ_2...; but the combination $(\mu + e\varphi)$, known as the «electrochemical potential», is uniquely defined for the purpose in hand. If the result of the computation was written in terms of transport J_1, J_2,...

$$J_j = \Sigma L_{ji} k_i$$

the coefficients L_{ji} were invariably symmetrical. It was soon evident that this did not depend on any mathematical approximations. For the relaxation effect I could depend on Newton's principle of action and reaction; for all the complications of hydrodynamics a «principle of least dissipation» derived by Helmholtz assured the symmetry. Admittedly, I did assume some consistent scheme of Brownian motion kinetics; but even that seemed not essential. The symmetry relation itself was equivalent to a principle of least dissipation; inverting the equations:

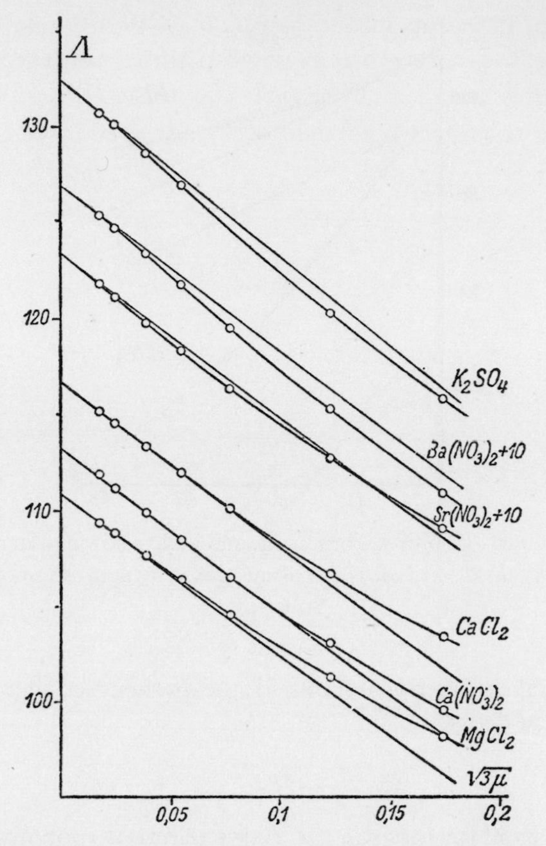

Fig. 3. Conductivities of 1–2 electrolytes in water. Same notation as Fig. 1. From *Physik. Z.*, 28 (1927) 277. Reproduced by permission of S. Hirzel Verlag K. G., Stuttgart.

$$k_j = \Sigma R_{j\,i} J_i$$

then

$$R_{j\,i} = R_{i\,j}$$

and the integral of the dissipation function

$$R(J, J) = \Sigma R_{j\,i} J_j J_i$$

equals the degradation of free energy, and it is a minimum in a case of stationary flow.

An unusual problem in chemical kinetics attracted my attention at the same time. C. N. Riiber was studying the mutarotation of various sugars by several precise methods: optical rotation, refractive index (interferometer) and volume changes (dilatometer). He discovered that there were (at least) three modifications of galactose, and the possibility that any one of these might transform into either of the others gave rise to a little problem in mathematics. In analyzing it I assumed, as any sensible chemist would, that in the state of equilibrium the reaction 1→2 would occur just as often as 2→1, etc., even

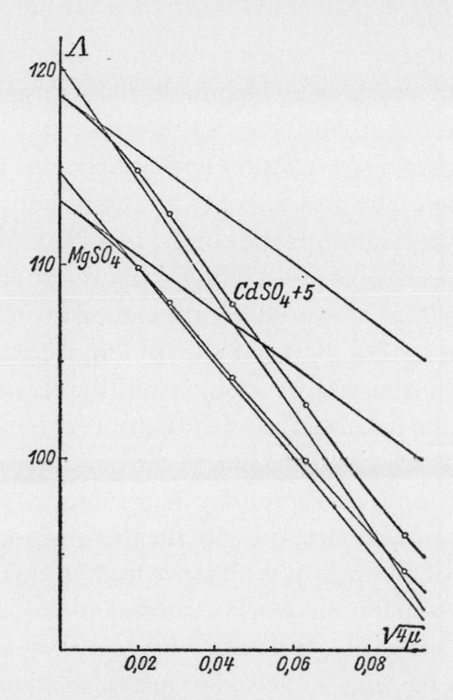

Fig. 4. Conductivities of 2–2 electrolytes in water. Same notation as Fig. 1. From *Physik. Z.*, 28 (1927) 277. Reproduced with permission of S. Hirzel Verlag K. G., Stuttgart.

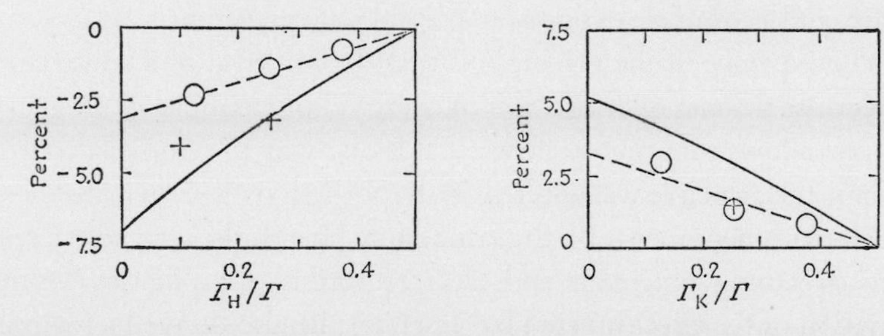

Fig. 5. The variation of cation mobilities with mixing ratio for aqueous solutions containing varying proportions of HCl and KCl at constant total concentration of 0.1 mole/liter. Measurements by Bennewitz, Wagner and Küchler[10] (1929). Figure from H. S. Harned and B. B. Owen, *The Physical Chemistry of Electrolyte Solutions*, ACS Monograph No. 95, Reinhold, New York, 1943, p. 143. Reproduced by permission of the authors and of Van Nostrand/Reinhold, New York.

though this is not a necessary condition for equilibrium, which might be maintained by a cyclic reaction – as far as the mathematics goes; the physics did not seem reasonable. Now if we look at the condition of detailed balancing from the thermodynamic point of view, it is quite analogous to the principle of least dissipation.

I developed a strong faith in the principle of least dissipation, and recognized that it had been used somehow by Helmholtz in his theory of galvanic diffusion cells and by Kelvin in his theory of thermoelectric phenomena. Some years later in Zürich in a conversation with P. Scherrer, I found that he had been strongly impressed by the ideas of G. N. Lewis about detailed balancing. This made me put the cart behind the horse. Now I looked for a way to apply the condition of microscopic reversibility to transport processes, and after a while I found a handle on the problem: the natural fluctuations in the distribution of molecules and energy due to the random thermal motion. According to a principle formulated by Boltzmann, the nature of thermal (and chemical) equilibrium is statistical, and the statistics of spontaneous deviations is determined by the associated changes of the thermodynamic master function; that is the entropy – or at constant temperature, equally well the free energy. Here was a firm connection with the thermodynamics, and we connect with the laws of transport as soon as we may assume that a spontaneous deviation from the equilibrium decays according to the same laws as one that has been produced artificially. When this reasoning was exploited by appropriate mathematics, the long-suspected reciprocal relations did indeed appear among the results, which were first announced in 1929. In view of the very general claims I felt that concepts and conditions ought to be defined with great care, and a complete exposition[11,12] did not appear until 1931.

One consequence of the principle is that the removal of a constraint will never decrease the rate of dissipation of energy. For example, closing an electric contact allows a current to flow; that is one way to remove a constraint. In this sense the principle was applied as a hypothesis by Kelvin in his theory of thermoelectric phenomena. By the same route Helmholtz arrived at a relation between streaming potentials and electrophoresis in capillaries (an inside-out variation of the effect utilized by Tiselius); he also derived a formula for the electromotive force of a concentration cell, which was later generalized by MacInnes and Beattie[13] (1920). The most important application of the dissipation principle not yet suggested in 1931 was a general relation between the cross-coefficients for diffusion of different solutes. This was announced for electrolytes in a joint paper with Fuoss[14] (1932), where of course the relation of MacInnes and Beattie was implied as well. By now there is a fairly extensive literature on the subject. A comprehensive review of varied applications and significant experimental tests was given some years ago (1960) by D. C. Miller[15]; he concluded that the relations are generally confirmed within the limits of error of the measurements.

Possibly the most important as a tool of research is the relation of Helmholtz, MacInnes and Beattie. The thermodynamic properties of electrolyte solutions can be determined from the measurements of the voltage between electrodes reversible to both ions. Largely through the efforts of H. S. Harned, methods of preparing reversible electrodes for several kinds of anions (halide, sulfate) and cations (hydrogen, silver, alkali metals and some others) have been perfected; but for a great many ions this has not been achieved and the prospects look poor.

Following MacInnes and Brown[16] (1936), the voltage of a concentration cell is measured between electrodes reversible to the same ion, and when in addition the transference number (fraction of the current carried by one ion) is known, the free energy of dilution can be computed. MacInnes and Longsworth[17] had shown (1932) how the transference numbers can be determined quite accurately by observing the displacement of a boundary between two solutions (with one common ion) by the passage of an electric current.

In 1932 Fuoss and I[14] computed the effects of the interaction between the ions on transport processes (conduction and diffusion, even viscosity) in mixtures of general composition. The algebraic techniques which enabled us to cope with a complicated system of equations were improved many years later (Onsager and Kim[18], 1957). Precision methods for the study of diffusion were not developed until the decade 1940–50. Then Kegeles and Gosting[19] (1947) showed that Gouy's optical fringe method gives excellent results when the principles of physical optics are properly applied; meanwhile Harned and co-workers[20,21] developed a relaxation method which depends on measurements of electrical conductivity for analysis *in situ*. The two methods supplement each other very nicely: at low concentrations of electrolytes, where Gouy's method fails for lack of fringes, the resistivities of the solutions suffice for easy measurement. Thus at long last Nernst's relation between the coefficient of diffusion and the electrolytic mobilities was verified to about 0.1%.

I have indicated already that the theory of long-range interaction by no means eliminated the need to consider a mass-action equilibrium with undissociated species. As was pointed out by Bjerrum[22] (1927), when the ions are highly charged or very small, or where the dielectric constant is not 80 but just 20 or less, the electrostatic interaction at close range will be so strong that pairs of ions will stay together for a long time and act pretty much like ordinary molecules. For that matter, recent kinetic studies have revealed (Eigen[9], 1967) that replacements in the innermost shell of solvent molecules and anions around a cation may be fairly infrequent – once in a microsecond, say, or even longer – so that molecules are reasonably well defined in the sense of chemical kinetics. However, even when the recombination kinetics is too fast for a sharp definition, it is often convenient to distinguish between «free ions» and «associated pairs» by some arbitrary but reasonable convention. Bjerrum[21] suggested (1927) that we draw the line at a distance where the work of separation against the coulomb force is twice the thermal energy (kT) per molecule; in water that distance is 3.5 Å for KCl, 7 Å for $MgCl_2$, 14 Å for $MgSO_4$, etc., and in a solvent of dielectric constant 20 at room temperature the «Bjerrum distance» is 14 Å for KCl. In solvents of very low dielectric constant only the salts of big complex ions dissolve and exhibit appreciable conductivity. Fig. 6 exhibits the effect of the dielectric constant. Fuoss[23] (with Kraus, 1933)

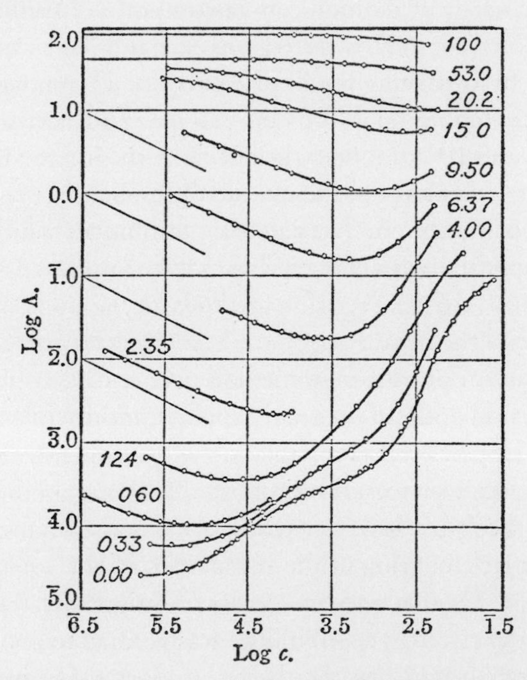

Fig. 6. The equivalent conductivity Λ of tetraisoamylammonium nitrate in mixtures of water and dioxane, as a function of the salt concentration *c*. From C. A. Kraus and R. M. Fuoss, *J. Am. Soc.*, 55 (1933) 21. Reproduced by permission of the surviving author and of the American Chemical Society.

measured the conductivities of solutions of tetraisoamyl ammonium nitrate over a wide range of concentrations in mixtures of water and dioxane, covering a range of dielectric constants from 78 to 2.25.

The descending branches of the curves represent a mass-action equilibrium between neutral pairs and individual ions. The minima and the increasing branches indicate that at higher concentrations the current is carried mostly by charged aggregates of several ions in mass-action equilibrium with smaller neutral aggregates and simple pairs, inflexions in the rising branches suggest ring-shaped neutral aggregates. Tentative estimates indicated that the coulomb forces could be held largely responsible for the variations of the various equilibrium constants. The long-range effects entail relatively small corrections compared to the enormous range of variation displayed in Fig. 6. Similar results are found quite often in solvents of low dielectric constants, but by no means always; we know a good many examples where strong specific interactions of the ions with each other or with the solvents are indicated. The accumulation of more and better data have motivated efforts to refine our original computations. Fuoss and I undertook such a computation[24] (1955); in Fig. 7 the predictions are compared with Shedlovsky's excellent measurements[25] (1932). In form our computed results agreed substantially with those of Pitts[26] (1953); but certain differences in the models entail differences of interpretation in terms of short-range interactions. In this context we might seek at least a partial answer to the question how closely the effects of short-range interactions on the conductivity may correspond to the effects on the thermodynamic properties in the sense of Arrhenius. That task begins to look feasible.

The theoretical developments of the nineteen twenties inspired a search for additional symptoms of long-range interaction, and several were found. For example, in a rapidly alternating field the ion can be caused to change its direction of motion before the relaxation force is fully developed. As a result the conductivity increases somewhat through a range of frequencies which corresponds to the (Maxwell) relaxation time, and the phase of the voltage lags a little behind that of the current. Alternatively, a very strong field causes an ion to move so fast that a screening cloud of the normal type has no time to form; in the limit of high speeds the screening is performed by a deficiency of other ions moving with the same speed (M. Wien; Onsager[29], 1934; Wilson[30], 1936; Eckstrom and Schmelzer[31], 1939). Attempts to exploit this effect as a means to eliminate the complications of long-range interactions for weak electrolytes met with a surprise (Schiele[32], 1932).

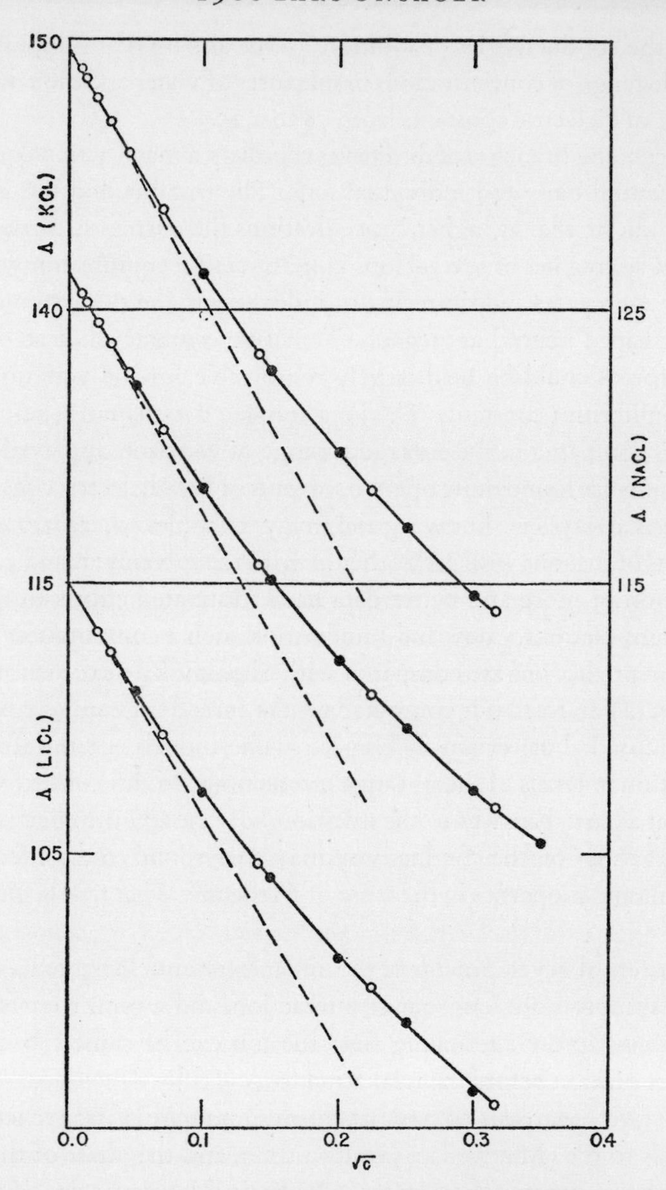

Fig. 7. Conductivities of chlorides in water. Measured points by Shedlovsky[25] (1932); computed curves by Onsager and Fuoss[24] (1955). Reproduced by concurrence of co-author.

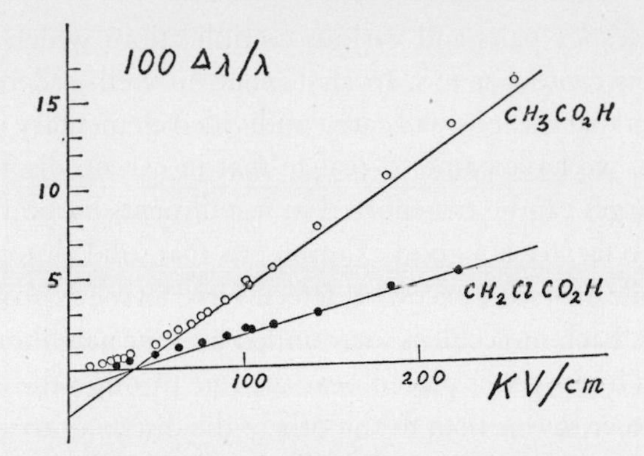

Fig. 8. Deviations from Ohm's law in aqueous solutions of weak acids. Points from Schiele[32].

Fig. 8 actually displays the excess field effects for two weak acids over the field effect for a strong one (HCl). The straight lines represent my own computations[29] (1934). The field disturbs the dissociation equilibrium because it helps pairs of ions to separate for good once they have reached the fringes of the coulomb field. The assistance is nearly proportional to the absolute value of the field. The negative intercepts represent mainly a decrease in the rate of recombination by the screening clouds of ions, effective in the absence of a strong external field. In the light of such analysis, the Wien effect seemed to hold promise as a good tool for the study of fast recombination kinetics; recent work – particularly by Eigen and DeMaeyer – has shown that this was not a vain hope (Eigen[9]).

Many solids exhibit electrolytic conduction, and symptoms of reaction kinetics, Wien effect and so forth have been observed. Impurities and other defects often play a decisive role, and these factors are none too readily controlled, so that the standard of precision has to be rather modest; but it is often possible to divine the mechanism. Arrhenius had to fight for his faith; but those days are long past. We now realize quite clearly that it takes excess charges moving somewhere to produce an electrolytic conductor. In a salt crystal such an excess charge can be an additional ion in an abnormal «interstitial» position (Frenkel defect) or a vacancy (Schottky defect) at a place normally occupied by an ion. The position of the vacancy is changed whenever a neighboring ion moves in to fill it. In KCl, for example, Schottky defects predominate as «ions» of both signs; but in AgCl some silver ions leave their normal sites for interstitial positions to produce positive Frenkel and negative Schottky defects. Schottky defects of opposite signs can combine to form neutral vacancy pairs, and various complications which involve more extensive defects can occur too. In any event, in well-ordered crystals we generally expect that the ions will carry undivided elementary charges.

Nevertheless, we have come to realize that in certain disordered crystals elementary charges can be transported in installments by point defects. We don't have to go far. Ice is a good example! In that solid almost all current is carried by mobile protons – excess or defect. First, let me explain the essentials of the structure. Each molecule is surrounded by four neighbors at a distance of 2.76 Å. Each hydrogen is placed near the line through the centers of two oxygens and closer to one than to the other; the distances are about 1 Å and 1.76 Å. Two neutrality rules are normally satisfied: each oxygen carries two near hydrogens, so that the water molecules are intact and neutral. The other neutrality rule requires that one and only one hydrogen connects any two neighboring oxygens: the hydrogen bonds are intact. Any violation of either neutrality rule produces an electrically active defect.

(Animated cartoon)

A chance rotation of a molecule produces a pair of bonding defects; these separate and move through the crystal by successive rotations of the participating molecules. Other bonding defects enter the picture and wander through it. A chance transfer of a proton from one molecule to its neighbor produces a pair of ionic defects: positive hydronium H_3O^+, negative hydroxyl OH^-. The positive ion moves by donating a proton, the negative by stealing one. The motion of the ions leave molecules oriented against the field; the drift of the bonding defects turns them into the field again.

(End of cartoon)

Estimates of the ionic mobilities vary over a considerable range; but in any event the positive ionic defect is much more mobile in the solid than in the liquid, and its mobility varies very little with the temperature. The mobilities of the bonding defects are more like those of ordinary ions in the liquid, and the temperature coefficients are similar. Nevertheless, the bonding defects determine the direction of polarization in pure ice, because they are much more numerous, possibly several pairs for each million molecules. As to the task of transporting a direct current, that is about equally distributed between bonding and ionic defects; each type carries about half an elementary charge. One kind of current may get out of step with the other for a short time; but this produces a polarization which equalizes the number currents.

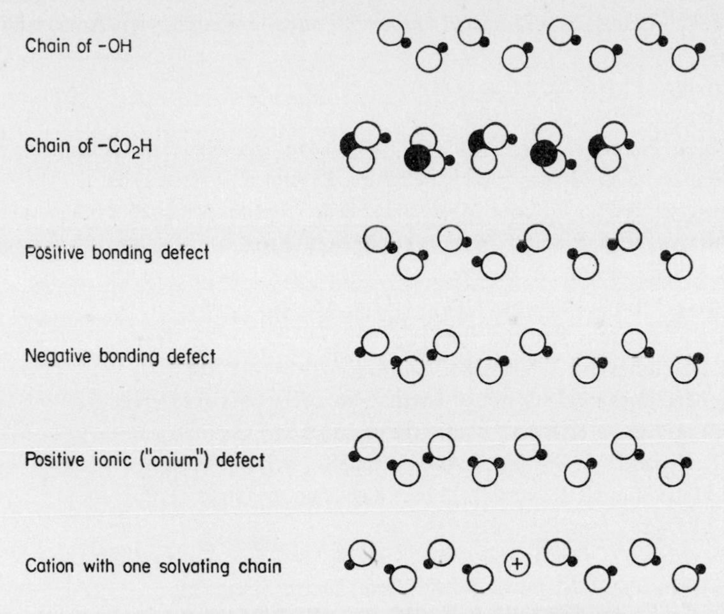

Fig. 9. H-Bond chains and electrically active defects.

If you apply the principle of least dissipation to this kind of coupling you may be stretching some thermodynamic concepts just a little bit; but it is a good safeguard against greater errors, Jaccard[33] (1964) found it quite helpful.

Many of the things I have told you have a bearing on problems in biology. For example, how do ions get through in cell membrane? Observations on poisoning suggest fixed facilities for such transport. Let me just toss on the screen what I think might be an essential element of such a facility (Onsager[34], 1967; Fig. 9).

This is a speculation, but one which is not yet refuted by observations and seems generally compatible with physical principles. The hope that it might be right adds interests to the exploration of ice and other protonic semiconductors.

1. S. Arrhenius, *Svenska Vetensk. Akad. Handl. Bih.*, 8 (1884) 13–14.
2. S. Arrhenius, *Z. Physik. Chem. (Leipzig)*, 1 (1887) 631.
3. W. Ostwald, *Z. Physik. Chem. (Leipzig)*, (1888).
4. P. Debye and E. Hückel, *Physik. Z.*, 24 (1923) 185, 305.
5. R. A. Millikan, *Nobel Lectures, Physics 1922–1941*, Elsevier, Amsterdam, 1965, p. 54.

6. A.W.K. Tiselius, *Nobel Lectures, Chemistry, 1942–1962*, Elsevier, Amsterdam, 1964, p.195.
7. L. Onsager, *Physik. Z.*, 27 (1926) 382.
8. L. Onsager, *Physik. Z.*, 28 (1927) 227.
9. M. Eigen, *Nobel Lectures, Chemistry, 1963–1970*, Elsevier, Amsterdam, 1972, p.170.
10. K. Bennewitz, C. Wagner and K. Küchler, *Physik. Z.*, 30 (1929) 623.
11. L. Onsager, *Beret. 18e Skand. Naturforskermøde, Copenhagen*, 1929, pp. 440–441.
12. L. Onsager, *Phys. Rev.*, 37 (1931) 405; 38 (1931) 2265.
13. D. A. MacInnes and J. A. Beattie, *J. Am. Chem Soc.*, 42 (1920) 1117.
14. L. Onsager and R. M. Fuoss, *J. Phys. Chem.*, 36 (1932) 2689.
15. D. G. Miller, *Chem. Rev.*, 60 (1960) 15.
16. D. A. MacInnes and A. S. Brown, *Chem. Rev.*, 18 (1936) 335.
17. D. A. MacInnes and L. G. Longsworth, *Chem. Rev.*, 9 (1932) 171.
18. L. Onsager and S. K. Kim, *J. Phys. Chem.*, 61 (1957) 215.
19. G. Kegeles and L. J. Gosting, *J. Am. Chem. Soc.*, 69 (1947) 2516.
20. H. S. Harned and R. L. Nuttall, *J. Am. Chem. Soc.*, 69 (1947) 736.
21. H. S. Harned, *Chem. Rev.*, 40 (1947) 461.
22. N. Bjerrum, *Kgl. Danske Videnskab. Selskab, Mat.-Fys. Medd.*, 7 (1927) 9.
23. C. A. Kraus and R. M. Fuoss, *J. Am. Chem. Soc.*, 55 (1933) 21.
24. L. Onsager and R. M. Fuoss, *Proc. Natl. Acad. Sci. (U. S.)*, 41 (1955) 274.
25. T. Shedlovsky, *J. Am. Chem. Soc.*, 54 (1932) 1411.
26. E. Pitts, *Proc. Roy. Soc. (London), Ser. A*, 217 (1953) 43.
27. P. Debye and H. Falkenhagen, *Physik. Z.*, 29 (1928) 401.
28. H. Falkenhagen, *Rev. Mod. Phys.*, 3 (1931) 412.
29. L. Onsager, *J. Chem. Phys.*, 2 (1934) 599.
30. W. S. Wilson, *Dissertation*, Yale, 1936.
31. H. Eckstrom and C. Schmelzer, *Chem. Rev.*, 24 (1939) 367.
32. J. Schiele, *Ann. Physik*, [5] 13 (1932) 811.
33. C. Jaccard, *Physik. Kond. Materie*, 3 (1964) 99.
34. L. Onsager, *The Neurosciences*, Rockefeller University Press, New York, 1967, p.78.

Biography

Lars Onsager was born in Oslo, Norway, November 27, 1903 to parents Erling Onsager, Barrister of the Supreme Court of Norway, and Ingrid, née Kirkeby. In 1933 he married Margarethe Arledter, daughter of a well-known pioneer in the art of paper making, in Cologne, Germany. They have sons Erling Frederick, Hans Tanberg, and Christian Carl, and a daughter Inger Marie, married to Kenneth Roy Oldham.

After three years with the experienced educators Inga and Anna Platou in Oslo, one year at a deteriorating private school in the country and a few months of his mother's tutoring, he entered Frogner School as the family returned to Oslo. There he was soon invited to jump a grade, so that he was able to graduate in 1920.

Admitted to Norges tekniske høgskole in the fall of that year as a student of chemical engineering, he entered a stimulating environment; the department had attracted outstanding students over a period of years. Among the professors particularly O. O. Collenberg and B. Holtsmark encouraged his efforts in theory and helped him in the evaluation of background knowledge.

After graduation in 1925 he accompanied Holtsmark on a trip to Denmark and Germany, then proceeded to Zürich, where he remained for a couple of months with Debye and Hückel and returned the following spring, for a stay of nearly two years. There he organized his results in the theory of electrolytes for publication, broadened his knowledge of physics and became acquainted with a good many leading physicists.

In 1928 he went to Baltimore and served for the spring term as Associate in Chemistry at Johns Hopkins University. The appointment was not renewed; but C. A. Kraus at Brown University engaged him as an instructor, and he remained in that position for five years. During this time he gave lectures on statistical mechanics, published the reciprocal relations and made progress on a variety of problems. Some of the results were published at the time, one with the able assistance of R. M. Fuoss; others formed the basis for later publications. In 1933 he accepted a Sterling Fellowship at Yale University, where he remained to serve as Assistant Professor 1934–1940, Associate Professor 1940–1945 and Josiah Willard Gibbs Professor of Theoretical Chemistry 1945–1972. Incidentally, he obtained a Ph.D. degree in Chemistry from Yale in 1935; his dissertation consisted of the mathematical background for his interpretation of deviations from Ohm's law in weak electrolytes.

Over the years, the subjects of his interest came to include colloids, dielectrics, order–disorder transitions, metals and superfluids, hydrodynamics and fractionation theory. In 1951–1952 he spent a year's leave of absence as a Fulbright Scholar with David Schoenberg at the Mond Laboratory in Cambridge, England, a leading center for research in low temperature physics. In the Spring of 1961 he served as Visiting Professor of Physics at the University of California in San Diego. Of his sabbatical leave 1967–1968 he spent the first three months as Visiting Professor at Rockefeller University and the last three as Gauss Professor in Göttingen. In 1962, at the suggestion of Manfred Eigen, he joined Neuroscience Associates, a small interdisciplinary group organized by F. O. Schmitt in Cambridge, Massachusetts.

Lars Onsager holds honary degrees of Doctor of Science from Harvard University (1954), Rensselaer Polytechnic Institute (1962), Brown University (1962), Rheinisch-Westfählische Technische Hochschule (1962), the University of Chicago (1968), Ohio State University (Cleveland, 1969), Cambridge University (1970) and Oxford University (1971), and Doctor Technical from Norges tekniske høgskole (1960).

In 1953 he received the Rumford Medal from the American Academy of Arts and Sciences, in 1958 The Lorentz Medal from The Royal Netherlands Academy of Sciences, in 1966 the Belfer Award in Science from Yeshiva University, in 1965 the Peter Debye Award in Physical Chemistry from the American Chemical Society, in 1962 the Lewis Medal from its California Section, the Kirkwood Medal from the New Haven Section and the Gibbs Medal from the Chicago Section, in 1964 the Richards Medal from the Northeastern Section.

In 1969 he received the National Science Medal, and he became an honorary member of The Bunsen Society for Physical Chemistry. During Spring 1970 he was Lorentz Professor in Leiden (The Netherlands).

Onsager is a Fellow of the American Physical Society and The New York Academy of Sciences, a member of The American Chemical Society, The Connecticut Academy of Arts and Sciences, The National Academy of Sciences, The American Academy of Arts and Sciences and The American Philosophical Society, a Foreign Member of the Norwegian Academy of

Sciences, The Royal Norwegian Academy of Sciences, The Norwegian Academy of Technical Sciences, the Royal Swedish Academy of Sciences and The Royal Science Society in Uppsala, and an Honorary Member of The Norwegian Chemical Society.

Chemistry 1969

DEREK H. R. BARTON

ODD HASSEL

«for their contributions to the development of the concept of conformation and its application in chemistry»

Chemistry 1969

Presentation Speech by Professor Arne Fredga, member of the Nobel Committee for Chemistry of the Royal Swedish Academy of Sciences

Your Majesty, Your Royal Highnesses, Ladies and Gentlemen.

One of the fundamental conditions for life on Earth is the ability of carbon atoms to bind each other to a practically unlimited extent. They form chains, often very branched, but also rings and net-works. The number of carbon compounds is thus very large–some years ago I saw the number two million–and many new ones are discovered or prepared every day. It is obvious that a multitude of different substances are required to build up a living organism and make it function.

The structure of carbon compounds, often called organic compounds, is, however, governed by rather simple principles. To describe an organic molecule, we have first the *constitution*, which can be said to represent the ground-plan. Next we have the *configuration*, which deals with the question of right or left. In the case of unsymmetrical objects like gloves or shoes there must exist a right form and a left form and the same is true for unsymmetrical molecules. What is then the *conformation*, which is of interest here to-day?

A molecule is not, in general rigid. There is a certain flexibility, which may, perhaps, be called limpness or floppyness. Certain distances and angles are invariable and the chain must not be broken, but it may bend, turn or twist in different ways. In ring-shaped molecules, the flexibility is more restricted. Small rings of three, four of five atoms are rather rigid and planar. Six carbon atoms permit a certain flexibility and large rings may be rather floppy. Complicated molecules with net-works of several rings are often more rigid. The rings check or lock each other. The *conformation* is the shape, which the molecule really assumes, utilizing the flexibility. It may be said that conformational analysis deals with the mode of behaviour of floppy molecules.

Metaphorically one could say that the molecule tries to arrange itself in the most comfortable way. It will avoid crowding and strain and must consider that certain groups may attract or repel each other.

Often a great number of conformations are possible, but some are more stable than others. These are statistically favoured. A ring of six carbon atoms can have two conformations, known as the chair and boat forms, which easily interchange. At room temperature, a molecule changes its conformation about a million times in a second. One of the conformations is, however, strongly predominant (about 99%). Professor Hassel has carried out fundamental investigations on this system and shown how heavy or bulky groups, attached to the carbon atoms, take up their positions relative to the ring and to each other.

The conformation is of great importance for the mode of reaction of the molecules. Reactive groups may be easily accessible, or they may also be more or less blocked by other groups. Knowledge of the conformation is therefore of great importance for explaining or predicting the mode of reaction of a certain molecule. It is always a good thing to know if an experiment has any chance of success.

Geometry in three dimensions is not very popular. I suppose that no one will mind if I refrain from discussing special cases in detail and from describing the physico-chemical methods used in conformational analysis.

In the development of scientific ideas it is generally possible to trace contributions, elements of thought, from many sides. But often the decisive advances, the intellectual syntheses from different thoughts and suggestions can be attributed to one or two scientists, who stand out from the others. Professor Hassel's elegant work on six-membered rings, carried out with ever increasing precision, has laid a solid foundation for a dynamic chemistry in three dimensions. Professor Barton has generalized, opening wider perspectives and deducing the consequences for many complicated ring systems, which play an important role in living nature. Let me only mention the ring system of the steroids, which is found in the bile acids, necessary for digestion, in sex hormones, cortisone, digitalis glycosides and cholesterol but also in the lather-forming saponins and in the special venoms of potato-tops and toads.

Professor Barton. In the classical work «The Conformation of the Steroid Nucleus» you have advanced the leading principles of conformational analysis. In this paper you have also drawn attention to the notable researches of Hassel, which have thrown considerable light on these more subtle aspects of stereochemistry. Your ideas were soon accepted and they play a fundamental role in organic chemistry of today. According to a prominent fellow scientist, your paper represents the first real advance in stereochemistry since the theory of Van 't Hoff and Le Bel, *i.e.* since 1874. I have no objections.

In recognition of your services to Chemical Science, the Royal Academy has decided to confer upon you the Nobel Prize. To me has been granted the privilege of conveying to you the most hearty congratulations of the Academy.

Professor Hassel. Not very long ago, organic chemists spoke of free rotation. Gradually they found that the rotation is restricted and that this fact is important. Many workers have contributed to the development, but in our opinion your work on the cyclohexane system is of outstanding importance. In your work on decaline, you have also taken the important step to polycyclic systems and pointed the way for coming development. In recognition of your services to Chemical Science, the Royal Academy has decided to confer upon you the Nobel Prize. To me has been granted the privilege of conveying to you the most hearty congratulations of the Academy.

Professor Barton. On behalf of the Academy I invite you to receive your prize from the hands of His Majesty the King.

Professor Hassel. On behalf of the Academy I invite you to receive your prize from the hands of His Majesty the King.

D. H. R. BARTON

The principles of conformational analysis

Nobel Lecture, December 11, 1969

> *« Il y a trois périodes dans l'histoire de toute découverte.*
> *Quand elle est annoncée pour la première fois, les gens pensent que ce n'est pas vrai.*
> *Puis, un peu plus tard, quand son exactitude leur paraît si flagrante qu'ils ne peuvent plus la nier, ils estiment que ce n'est pas important.*
> *Après cela, si son importance devient assez manifeste, ils disent : en tout cas, ce n'est pas nouveau. »*
>
> *William James*

The importance of conformational analysis in Chemistry became manifest during the decade immediately after the last World War. This lecture is, therefore, more an account of chemical history than of recent advances. In order to appreciate the significance of conformational analysis, a short introduction describing the development of structural theory in Organic Chemistry is necessary.

In the second half of the nineteenth century it became possible, thanks to the theories of Kekulé and others, to assign a constitution to each organic substance. The constitution is simply the specification of which atoms are bonded to which in the molecule and, in the great majority of cases, is unambiguous. A constitutional formula has no stereochemistry. The necessity to consider the stereochemistry of molecules became self-evident when two or more distinctly different substances were found to have the same constitution. It was Le Bel and, especially, Van 't Hoff who, also in the nineteenth century, introduced the idea of configuration. One can consider that the configurations of a molecule of a given constitution represent the specification of the order of the bonds in space about the atoms, or groups in the molecule which give rise to stereoisomers. For the great majority of substances this means the specification of the order of the bonds about an asymmetric «double bond» or about a «centre of asymmetry», normally a carbon atom substituted by four different groups. The first type of stereoisomerism is called «geometrical isomerism», the latter «optical isomerism». The number of possible stereoisomers of a given configuration then becomes $2^n \times 2^m = 2^{n+m}$ where n is the number of «asymmetric double bonds» and *m* is the number of «centres of asymmetry». This formula of Van 't Hoff, the first winner of a Nobel Prize, is still of fundamental importance today, nearly a century after it was first written down. Organic chemists have been very busy in demonstrating the truth of constitutional and configurational theory. Between one and two million different organic compounds have so far been prepared and we can prepare as many more millions as are required.

Van 't Hoff had a very clear idea of the reason for the success of the 2^{n+m} formula. It was based on the concept of «restricted» rotation about double bonds and of «free» rotation about single bonds. The latter concept was necessary to explain why most «optical isomers» did not have a myriad of isomers themselves. We shall not be concerned in this lecture with «geometrical isomerism» about double bonds. Optical isomerism is based on the idea of chirality, or the non-superposability of object and mirror image.

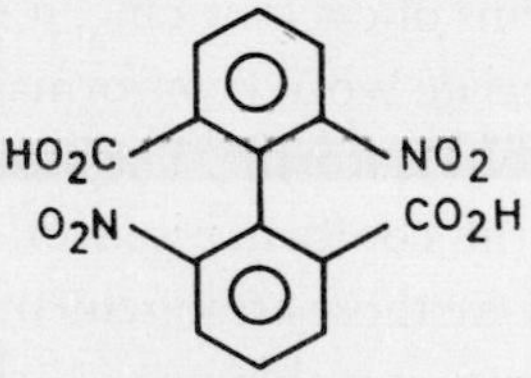

(I)  (II)  (III)  (IV)

The first indication that rotation about single bonds was not always free came in 1922 when Christie and Kenner[1] were able to resolve 2,2′-dinitrodiphenyl-6,6′-dicarbonylic acid (I) into optically active forms. This resolution is possible because the four bulky groups at the 2,2′-, 6- and 6′-positions prevent rotation about the central carbon–carbon single bond. Many analogous examples of restricted rotation of this type were later investigated[2]. It does not seem, however, that organic chemists were much worried about barriers to rotation in organic molecules in general at that time because there was no technique available to demonstrate the phenomenon experimentally.

In the decade starting in 1930, chemical physicists noted a discrepancy between the observed and calculated entropy of ethane. It was clear that this could only be explained by a barrier to free rotation about the two methyl groups, but whether this barrier was attractive or repulsive in origin with respect to the hydrogen atoms of the two methyl groups was a subject of considerable argument. However, the existence of such a barrier to free rotation in ethane implied that such barriers existed for aliphatic and alicyclic compounds in general.

The problem was clarified by studies on the simple alicyclic hydrocarbon cyclohexane (II) and on derivatives of this compound. It had been appreciated for many years that cyclohexane molecules could be constructed in two forms, the boat (IV) and the chair (III), both free from angle strain. This distinction had, however, no meaning to organic chemists since there was no reason to know which form was preferred, nor was it understood that the preference would have any chemical consequences. It was the fundamental electron-diffraction work of Hassel[3,4] that established clearly that the chair conformation (III) was always preferred. In this chair conformation the hydrogen atoms are as far apart as possible and correspond to the «staggered» form of ethane and of aliphatic compound in general. Similar considerations apply to medium and large alicyclic rings[5]. One can conclude, therefore, that the barrier to rotation in such substances is repulsive rather than attractive in character.

Rather than use the vague terms boat and chair forms of cyclohexane, it is convenient to have a general term. In fact, the appropriate word «conformation» had already been used in sugar chemistry by W. N. Haworth[6]. The most general definition of conformation is as follows[7]: «the conformations of a molecule (of defined constitution and configuration) are those arrangements in space of the atoms of the molecule which are not superposable upon each other». Such a definition includes arrangements of atoms in which angle strain has been introduced, as well as bond extension and compression. It replaces an earlier definition[8,9] which excluded angle strain and bond extension or compression. Thus all molecules have theoretically an infinite number of conformations. It is fortunate that the complexities which might arise from such a definition are minimised by the fact that, in general, only a few of the possible conformations are energetically preferred. One may therefore consider «chair», «boat» and «twist-boat» (a conformation half-way between two boats; see ref. 10) conformations of cyclohexane all of which are free from angle strain. The stability order is chair > twist-boat > boat.

It is obvious that in the chair conformation of cyclohexane two geometrically distinct types of carbon–hydrogen bonds are present[3,11]. Six of the C–H bonds are parallel to the three-fold axis of symmetry (V) and are called ‹axial› (ref. 12). The other six (VI) are approximately in an equatorial belt around the three-fold axis and hence are called ‹equatorial›. As soon as a substituent is introduced into a cyclohexane ring the molecule may adopt a preferred chair conformation with the substituent either axial or equatorial. Owing to repulsive non-bonded interactions between axial groups the equatorial conformation is, in general favoured[3,11]. In the case of multiply substituted cyclohexanes the preferred conformation is, provided dipolar interactions are not dominant, that with the maximum possible number of substituents equatorial.

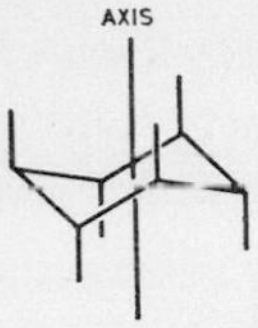

(V)  (VI) 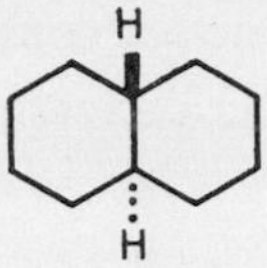 (VII) 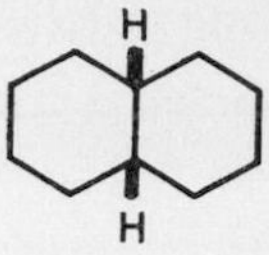 (VIII)

When two cyclohexane rings are fused together as in the configurational isomers *trans*- (VII) and *cis*- (VIII) decalin, a unique two-chair conformation, (IX) and (X) respectively, can be written for both molecules. Bastiansen and Hassel[13] showed that, as expected from consideration of non-bonded interactions, these two conformations were indeed the preferred ones.

(IX)  (X) 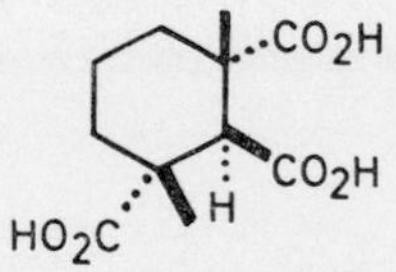 (XI) 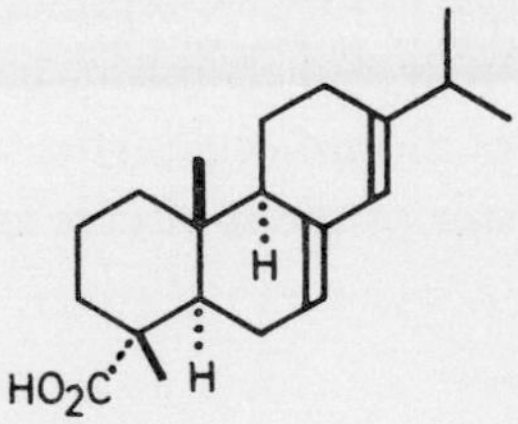 (XII)

At about this time the first semi-empirical, semi-quantitative calculations of non-bonded interactions were appearing in the literature[14]. Application of these methods to ethane, to cyclohexane and to the *trans*- and *cis*-decalins[15] gave results in qualitative agreement with the findings of Hassel and others mentioned above. In order to carry out these calculations, which at that time, in the absence of computers, were exceedingly arduous, special models were constructed[16] which later proved to be very useful in working out the principles of conformational analysis. The same models were also useful in understanding, in conformational terms, the dissociation constants of the tricarboxylic acid (XI) obtained by the oxidative degradation of abietic acid (XII)[17]. This was, in fact, an early example of the use of conformational analysis.

The stage was now set for a much fuller appreciation of the meaning of conformational analysis. At the time (1950) when our paper in *Experientia*[8] was written, steroid chemistry was already a major branch of science[18] which had just received a strong additional stimulus from the discovery of the utility of cortisone. There was an enormous literature of stereochemical fact which had not been interpreted properly in its three-dimensional aspects. Most steroids contain three six-membered rings fused *trans* to a five-membered ring and the two commonest configurational arrangements can be represented as in (XIII) and (XIV). Accepting that the three six-membered rings will in

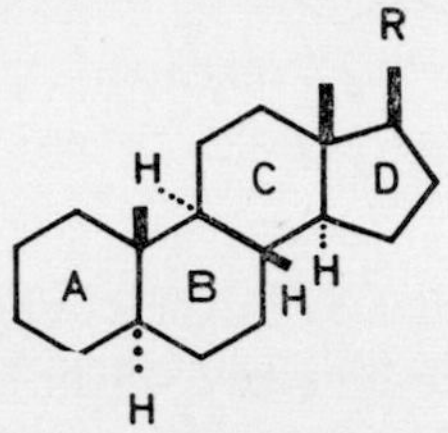

(XIII) 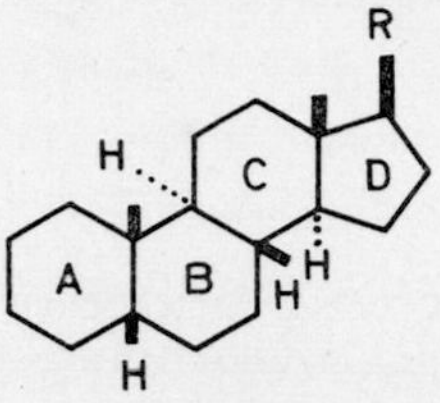 (XIV)

both cases adopt the preferred and unique three-chair conformation then (XII) may be represented in three dimensions as (XV) and (XIV) as (XVI). In steroid chemistry it is convenient to designate substituents on the same side of the molecule as the two methyl groups as β-oriented. Those on the opposite side of the molecule are then said to be α-oriented. Substituents attached to

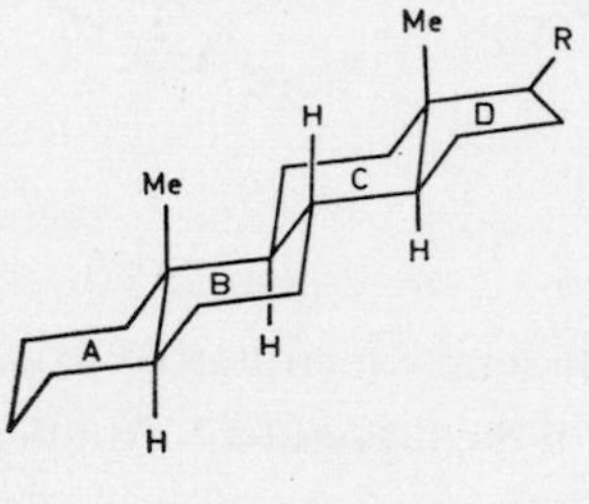

(XV) 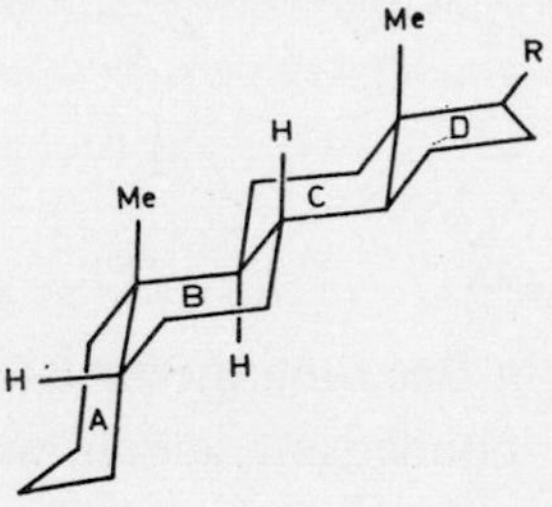 (XVI)

the steroid nucleus thus have a configuration which is α or β and which can be determined by the classical methods of stereochemistry (ring formation, ring fission, etc.). The key to the application of conformational analysis is that the ring fusions of the steroid nucleus fix the conformation of the whole molecule such that a substituent necessarily has both a configuration (α or β) *and* a conformation (equatorial or axial). Since, at a given carbon atom in the steroid nucleus, a substituent will be more stable equatorial than axial it follows that one can at once predict the more stable configuration between a pair of isomers. Thus a 3β-substituted (equatorial) *trans* A/B steroid (XVII) should be more stable than the corresponding 3α-substituted (axial) compound

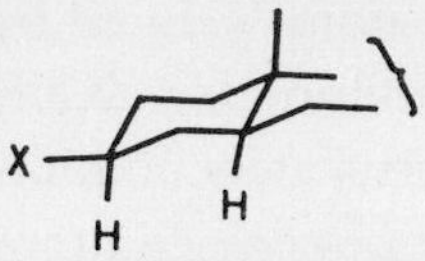

(XVII) 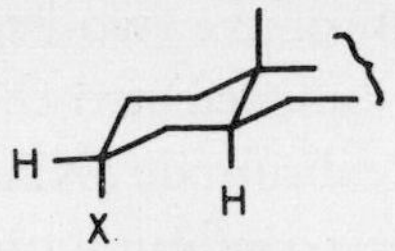 (XVIII) 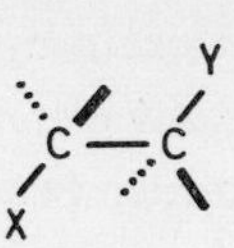 (XIX) 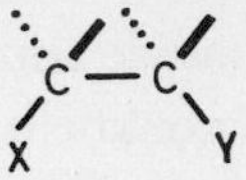 (XX)

(XVIII). This is in agreement with experiment. The same argument applies to all the other substitutable positions in the steroid nucleus in the six-membered rings, and, in general, good agreement with experiment is seen. The same applies for all molecules, for example triterpenoids, where fused all-chair conformations are present.

The relative stability of substituents is determined by repulsive non-bonded interactions (steric compression). Therefore, not only can one predict which isomer (α- or β-) will be formed in a chemical reaction which for mechanistic reasons gives the more stable product, but also any reaction which involves steric compression can be understood better. Thus in the alkaline hydrolysis of esters where the transition state for the reaction is more space-demanding than the initial state, one can predict that an equatorial isomer will hydrolyse faster than an axial isomer attached to the same carbon atom. This is, in general, true, and the principle aids in the prediction of selective hydrolysis reactions.

Every chemical reaction has a transition state. Many transition states have a well-defined preferred geometrical requirement. Thus in an E_2 type reaction[19], where two substituents attached to α-carbons are eliminated simultaneously by the attack of a reagent, the preferred geometry is that where the two carbons and the two substituents (X and Y) are coplanar. The two possible arrangements are *anti* (XIX) and *syn* (XX). The latter geometry is not available in steroids without distortion of a chair conformation. The former geometry (XIX) is, however, inevitably present in *trans*–1:2–diaxially substituted compounds (*e.g.* XXI), but not present in the corresponding *trans*–1:2-diequatorially substituted isomer (*e.g.* XXII). Such geometrical relationships are clearly shown if one looks along the C_5–C_6 bond. Thus in (XXI) the projection (XXIII) is seen and in (XXII) the projection (XXIV).

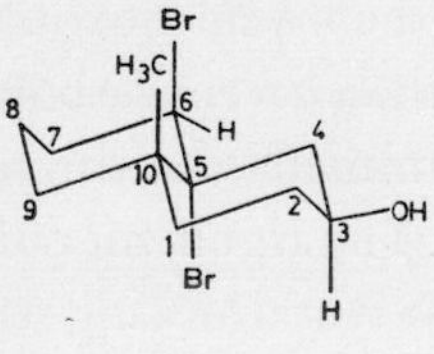

(XXI) 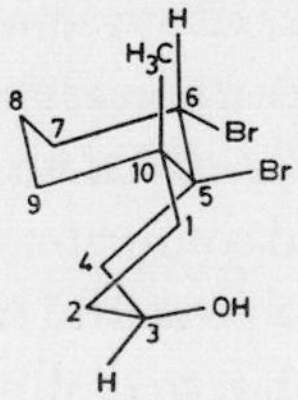 (XXII) 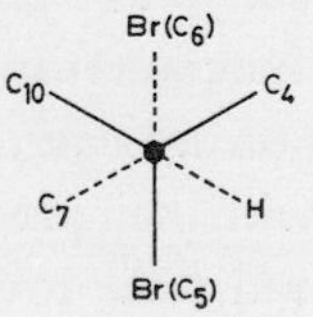 (XXIII) 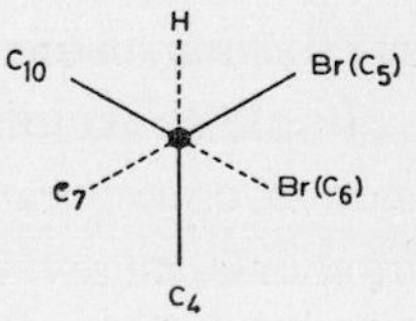 (XXIV)

Thus for a bimolecular E_2 reaction of the two bromine atoms induced by iodide ion (see XXV)[20], it could be predicted that the dibromide (XXI) would eliminate much faster than dibromide (XXII). Fortunately both dibromides could be prepared and their configurations determined[21]. As anticipated, the rate of elimination of bromines for the dibromide (XXI) was several powers of ten faster than the rate of the corresponding elimination from the dibromide (XXII)[21]. This was the first example of a phenomenon that was demonstrated later to be quite general for dibromides and many other types of eliminatable functions[22,23]. In general the geometrical requirements of the transition states of all chemical reactions are advantageously examined in conformational terms using steroids or other molecules with locked conformations.

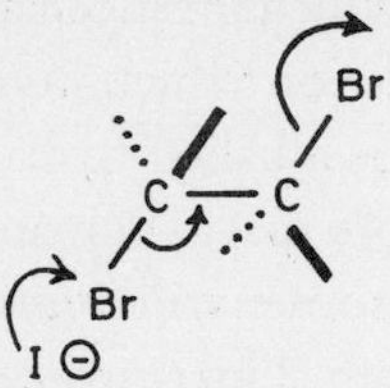

(XXV) 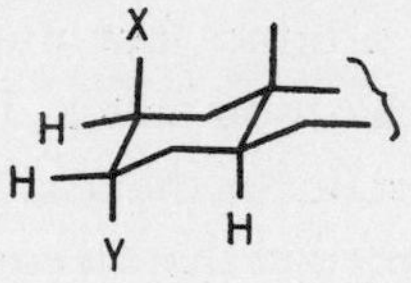 (XXVI) 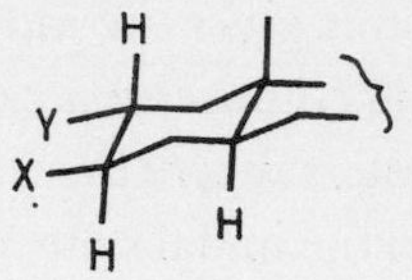 (XXVII)

The phenomenon of neighbouring group participation demands a conformational interpretation (diaxial participation) which is well exemplified in steroids[24]. Similarly, the opening of small membered rings like the halonium ion[24,25] or the epoxide group[9,26] gives predominantly diaxial rather than diequatorial products.

When the diaxial dibromide (XXI) is kept in solution at room temperature it rearranges spontaneously to an equilibrium with the more stable diequatorial dibromide (XXII). In effect two axial C–Br bonds are exchanged for two equatorial C–Br bonds. This is a general reaction[22,23] for 1,2-dibromides. We conceived that this rearrangement should be part of a generalised diaxial $\rightleftarrows$ diequatorial rearrangement process as in the scheme (XXVI) $\rightleftarrows$ (XXVII). By an appropriate choice of X and Y the truth of the proposition was demonstrated[27-29]. In practise, this reaction is a convenient route for shifting an oxygen function from one carbon atom to the adjacent carbon.

In the above discussion we have mainly correlated asymmetric centres of known configuration with their predicted conformations. The argument can, of course, be reversed and then provides a powerful method for deducing the configurations of compounds where a preferred all-chair conformation can be postulated. The first serious applications were in triterpenoid chemistry.

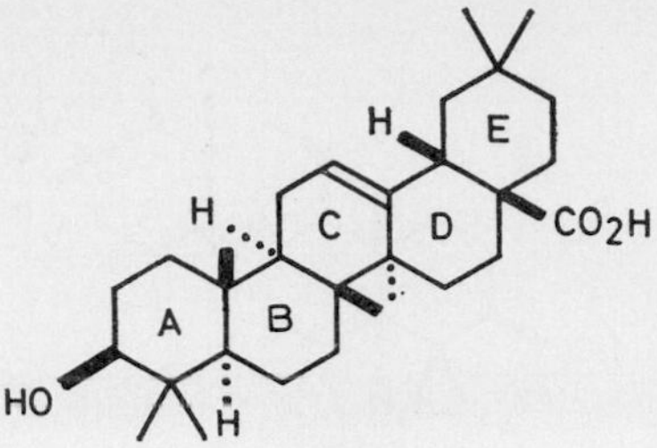

(XXVIII) 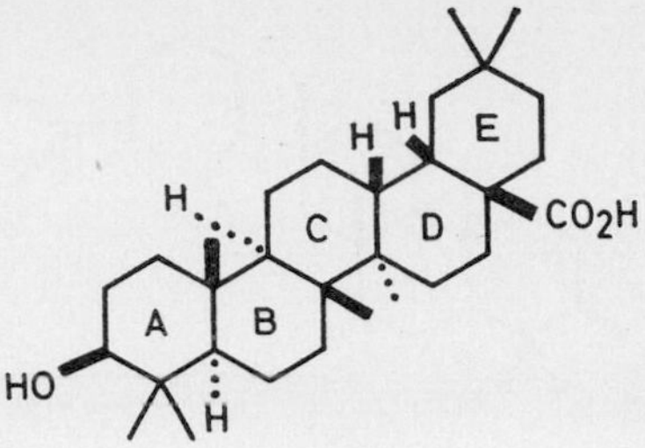 (XXIX)

The natural compound oleanolic acid (XXVIII) has 8 centres of asymmetry and can exist in principle in $2^7 = 128$ racemic configurations. Only one configuration [(+) or (−)] is synthesised by Nature. Because the analysis of conformations is easier with saturated six-membered rings we studied the saturated oleananolic acid (XXIX) which corresponds [(+) or (−)] to one racemate from a possible $2^8 = 256$ racemic configurations. Such is the power of the conformational method that the problem of configuration was reduced by chemical procedures only to a choice between two configurations (XXX

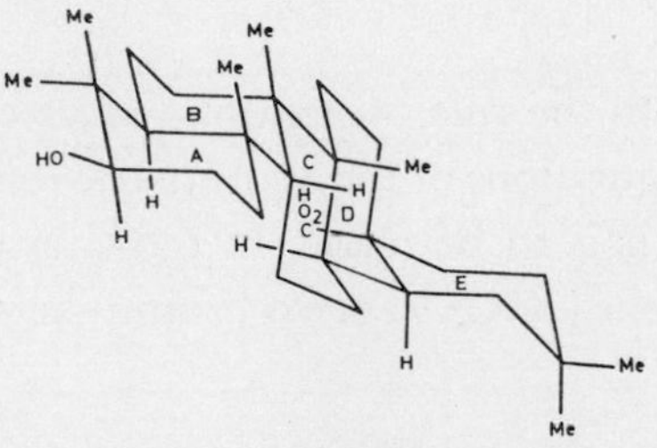

(XXX) 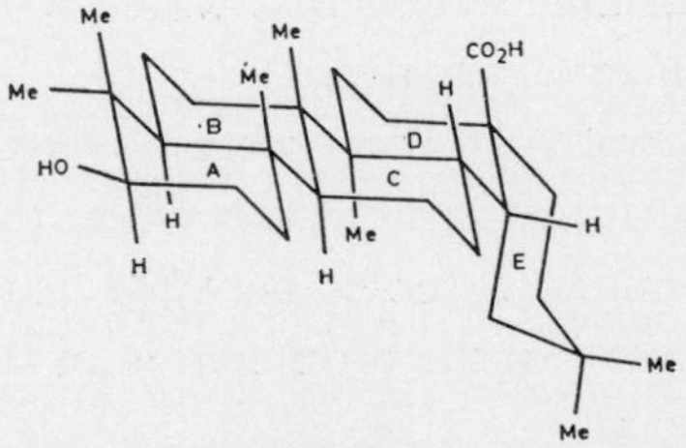 (XXXI)

and XXXI)[30]. The latter was shown to be correct by a later X-ray crystallography study[31]. It corresponds to the planar formula (XXIX).

By conformational analysis configurations could be assigned to typical triterpenoids like lanosterol (XXXII)[32], euphol (XXXIII)[33], cycloartenol

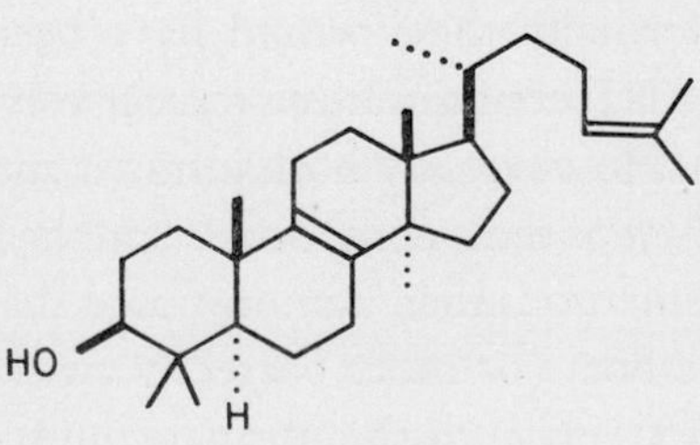

(XXXII) 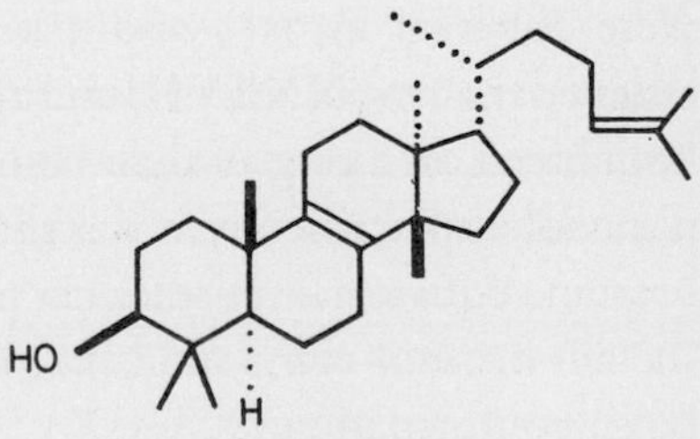 (XXXIII)

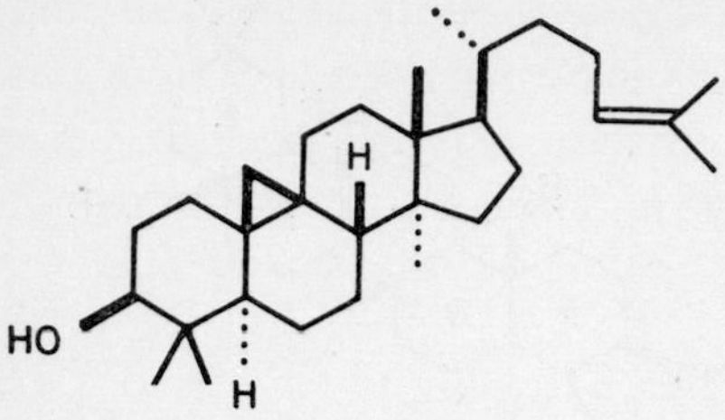

(XXXIV) 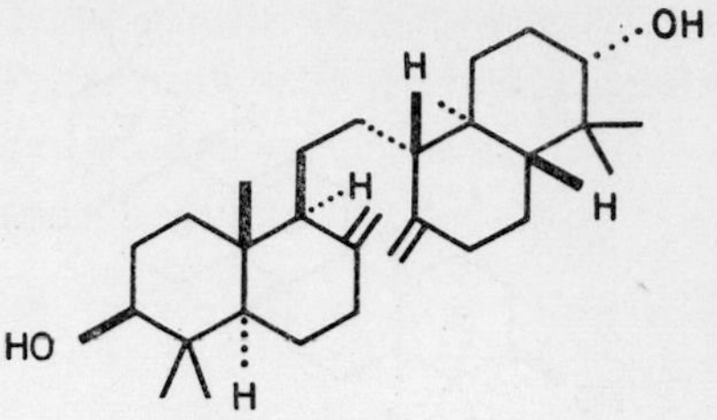 (XXXV)

(XXXIV)[34], and onocerin (XXXV)[35] at the same time as their constitutions were determined. Nowadays, of course, X-ray crystallographic analyses are done so easily and speedily that there is no special merit in the conformational method of determination of configuration. But it was important in the early 1950's and is useful today when the investigator does not have ready access to X-ray facilities. The X-ray method has the overall advantage that it determines constitution, configuration and preferred (in the crystalline state) conformation all at the same time. In general the preferred conformation in the crystalline state is that which would be predicted by the principles of conformational analysis.

Until 1957 there were no exceptions to the rule of preferred chair conformations in molecules where the configurations of the asymmetric centres permitted a choice between boat and chair to be made. In the course of studies[36] of the bromination of lanostenone (XXXVI) two bromo-ketones

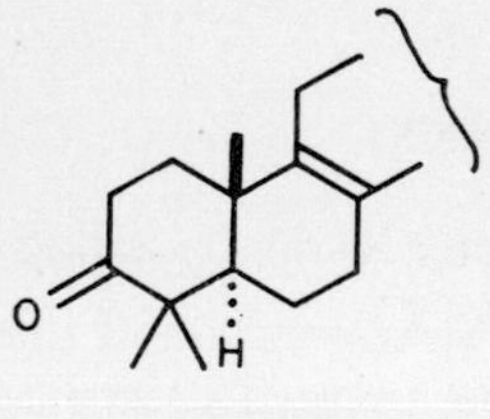

(XXXVI) 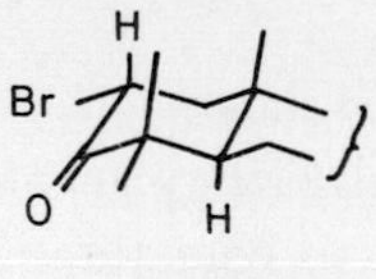 (XXXVII) 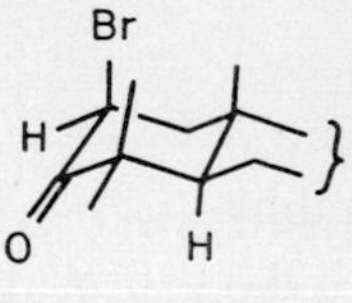 (XXXVIII)

were obtained in 95% and 5% yields. Normally these would have been assigned the 2α-(XXXVII) and 2β-(XXXVIII) configurations respectively, both based on a ring A chair conformation. However, by both infrared and ultraviolet spectroscopy it was shown that *both* of these compounds had their bromine equatorial. Further chemical investigations then demonstrated that the 2β-bromo-compound has, in fact, the boat (or more correctly twist-boat) conformation (XXXIX). This first exception to the normal conformational preference is due to the large methyl-methyl-bromine 1:3-diaxial interactions in the conformation (XXXVIII) and to the fact that the A ring contains one trigonal atom (the carbon of the carbonyl group). Later this conformational anomaly was extensively investigated[37] and its reality has been further confirmed by nuclear magnetic resonance studies[38].

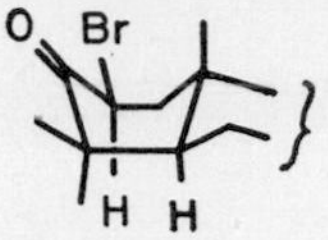

(XXXIX)

Conformational analysis was put on a quantitative basis by Winstein and Holness[39] and especially by Eliel[40]. The latter author, a recognised authority in the field, has made a thorough study[41] of the differences in free energy between axial and equatorial substituents in six-membered rings. There remain, however, certain subtle aspects of the conformations of molecules such as steroids and triterpenoids which still demand an adequate explanation from quantitative theory. Thus we have shown[23,42] that if benzaldehyde condenses with lanostenone (XXXVI) under mildly basic conditions at a rate

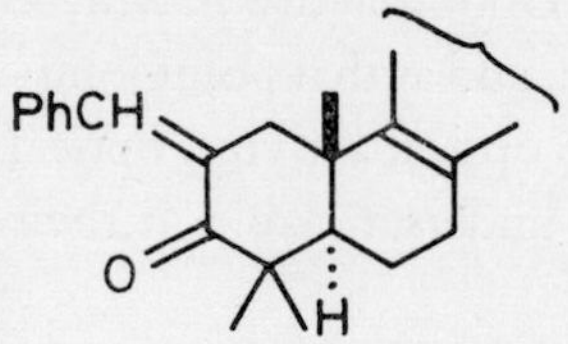

(XL)

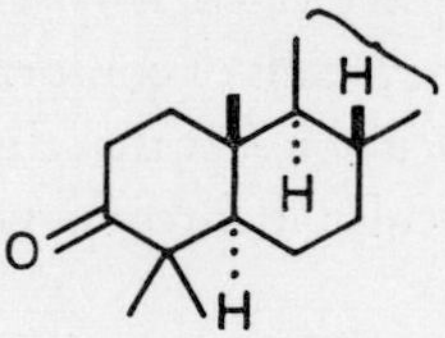

(XLI)
(55)

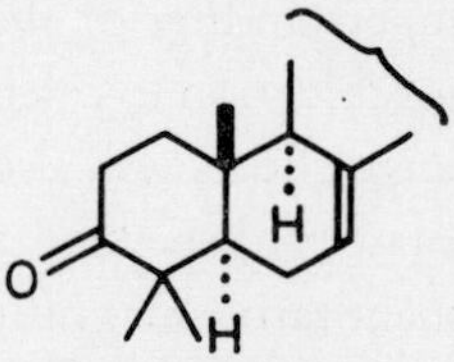

(XLII)
(17)

of (say) 100 to give the 2-benzylidene derivative (XL), then the simple non-polar derivative lanostanone (XLI) condenses at a rate of 55 and the isomeric olefin (XLII) at a rate of only 17. Similarly, cholestanone (XLIII, R = C_8H_{17}) give the 2-benzylidene derivative (XLIV) at a rate of 182. Simple side-chain derivatives (see XLIII)) condense at the same rate. However, the isomeric olefins (XLV) and (XLVI) condense at rates of 645 and 43, a difference of nearly thirty-fold for only a change in position of a (relatively) remote double bond. We have attributed these long-range effects to «conformational transmission» implying a distortion of bond angles by substituents that is transmitted through molecules to much greater distances than hitherto suspected. Such phenomena are beginning to receive adequate explanation, at least in qualitative and semi-quantitative terms[43].

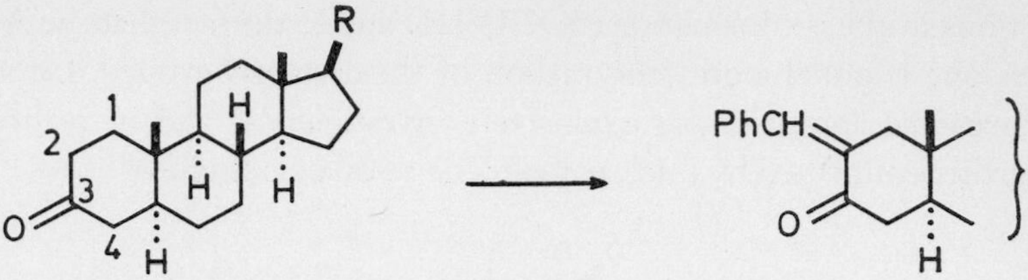

(R = C_8H_{17} ; 182)
(R = $C_{10}H_{21}$; 180)
(R = OH ; 188)
(R = C_9H_{17} ; 188)

(XLIII)

(XLIV)

As already mentioned above, conformation preferences can be calculated by semi-empirical methods. Now that computers have taken away the arduous arithmetic involved it has been possible to make rapid progress. For example, preferred conformations have been calculated for alicyclic rings larger than six-membered. These calculations provide valuable clues to an understanding of the chemistry of such systems[44]. Undoubtedly it will be possible soon to calculate fine details of conformation, and at that point long-range effects will also be calculable. At that time also optical activity, optical rotatory dispersion and circular dichroism will be understandable in their quantitative magnitudes[45].

Although the principles of conformational analysis are most clearly demonstrated in saturated six-membered cyclohexane ring systems, nevertheless the same basic approach is useful in understanding the reactions of unsaturated and of heterocyclic compounds. For example, cyclohexene can be assigned the conformation (XLVII) with equatorial and axial hydrogens as indicated. The hydrogens marked with a prime can then be called quasi-equatorial and quasi-axial[46]. The conformation of cyclohexene is more easily deformed than

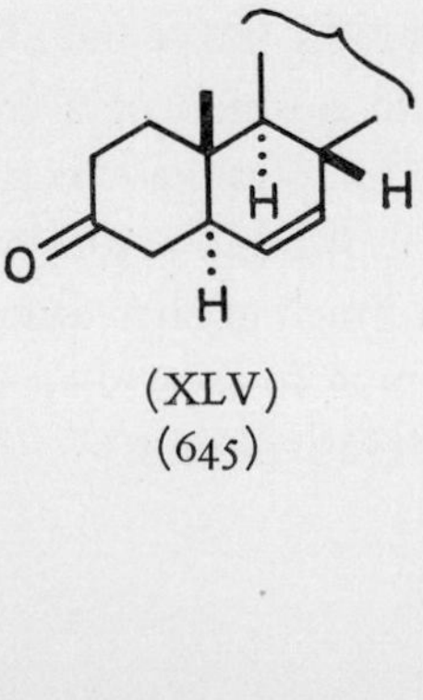

(XLV)
(645)

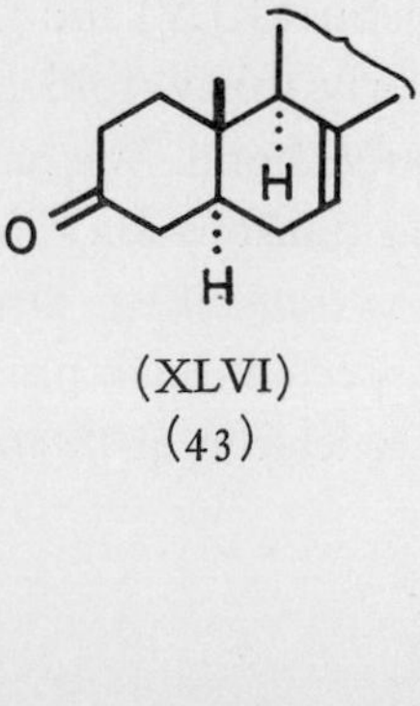

(XLVI)
(43)

that of cyclohexane even when fixed to other ring systems. Nevertheless the symbol (XLVII) has found general favour in conformational analysis as an expression of reality.

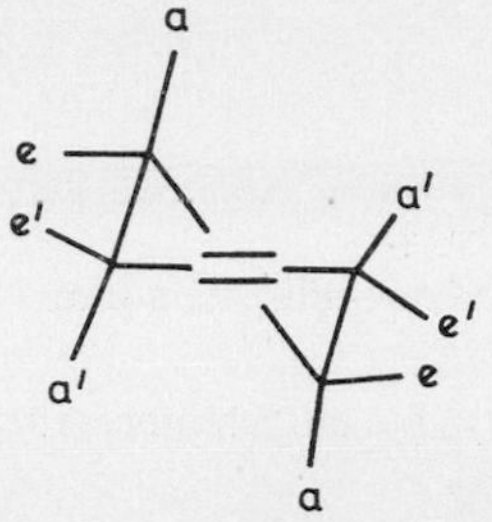

(XLVII) 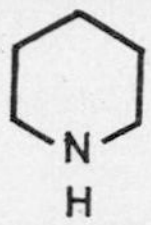 (XLVIII) 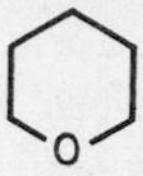 (XLIX)

The introduction of heteroatoms into a cyclohexane ring as in piperidine (XLVIII) or pyran (XLIX) raises conformational problems of considerable interest and sophistication. Thus one must consider if pairs of p-electrons have bulk or not and if so, is it greater or less than the hydrogen when attached to the heteroatom[7]. In this way a new field of conformational analysis has rapidly developed[41,47].

As has already been discussed, the choice of a preferred conformation was originally made on the basis of inference from the electron-diffraction work on simple compounds and from other physical evidence. X-Ray crystallography is nowadays an accurate and rapid method of determining conformation in the crystal lattice, which conformation usually corresponds to the preferred conformation in solution. There is, however, another physical method, nuclear magnetic resonance, which in the last decade has become– with every justification–predominant in the determination of conformation in solution. In many cases extremely detailed conformational analysis can be carried out. A simple example, which had great consequences for carbohydrate chemists, is the work of Lemieux and his colleagues[48].

An enzymatic reaction involves a large molecule–the enzyme–and a relatively small molecule – the substrate. Any complete understanding of the mode of action of an enzyme will require a knowledge of the conformations of the substrate and of the enzyme, as well as of the functional group reactivity. We are now in a position with most substrates to specify the preferred conformation involved in the reaction. In the case of steroidal substrates the conformation can be described in a detail which will soon be quite exact. This knowledge must soon have important consequences in biology.

Conformational analysis may be said to have come of age in that two excellent monographs have now appeared[41,49] which review present knowledge in detail. It is interesting to observe how an acorn of hypothesis can become a tree of knowledge.

1. G.H. Christie and J. Kenner, *J. Chem. Soc.*, 121 (1922) 614.
2. E.L. Eliel, *Stereochemistry of Carbon Compounds*, McGraw-Hill, London, 1962, pp. 156 *et seq.*
3. O. Hassel and H. Viervoll, *Acta Chem. Scand.*, 1 (1947) 149, and references there cited.
4. O. Hassel and B. Ottar, *Acta Chem. Scand.*, 1 (1947) 929.
5. V. Prelog, *J. Chem. Soc.*, (1950) 420.
6. W.N. Haworth, *The Constitution of the Sugars*, Arnold, London, 1929, p. 90.
7. D.H.R. Barton and R.C. Cookson, *Quart. Rev. (London)*, 10 (1956) 44.
8. D.H.R. Barton, *Experientia*, 6 (1950) 316.
9. D.H.R. Barton, *J. Chem. Soc.*, (1953) 1027.
10. W.S. Johnson, V.J. Bauer, J.L. Musgrave, M.A. Frisch, J.H. Dreger and W.N. Hubbard, *J. Am. Chem. Soc.*, 83 (1961) 606.
11. C.W. Beckett, K.S. Pitzer and R. Spitzer, *J. Am. Chem. Soc.*, 69 (1947) 2488.
12. D.H.R. Barton, O. Hassel, K.S. Pitzer and V. Prelog, *Nature*, 172 (1953) 1096; *Science*, 119 (1954) 49.
13. O. Bastiansen and O. Hassel, *Nature*, 157 (1946) 765.
14. I. Dostrovsky, E.D. Hughes and C.K. Ingold, *J. Chem. Soc.*, (1946) 173; F.H. Westheimer and J.E. Mayer, *J. Chem. Phys.*, 14 (1946) 733; F.H. Westheimer, *J. Chem. Phys.*, 15 (1947) 252.
15. D.H.R. Barton, *J. Chem. Soc.*, (1948) 340.
16. D.H.R. Barton, *Chem. Ind. (London)*, (1956) 1136.
17. D.H.R. Barton and G.A. Schmeidler, *J. Chem. Soc.*, (1948) 1197.
18. L.F. Fieser and M. Fieser, *Natural Products related to Phenanthrene*, 3rd ed., Reinhold, New York, 1949.
19. M.L. Dhar, E.D. Hughes, C.K. Ingold, A.M.M. Mandour, G.A. Maw and L.I. Woolf, *J. Chem. Soc.*, (1948) 2093.
20. W.G. Young, D. Pressman and C. Coryell, *J. Am. Chem. Soc.*, 61 (1939) 1640; S. Winstein, D. Pressman and W.G. Young, *ibid.*, 61 (1939) 1646.
21. D.H.R. Barton and E. Miller, *J. Am. Chem. Soc.*, 72 (1950) 1066.
22. D.H.R. Barton and W.J. Rosenfelder, *J. Chem. Soc.*, (1951) 1048; D.H.R. Barton, *Bull. Soc. Chim. France*, (1956) 973; D.H.R. Barton, A. da S. Campos-Neves and R.C. Cookson, *J. Chem. Soc.*, (1956) 3500.
23. D.H.R. Barton and A.J. Head, *J. Chem. Soc.*, (1956) 932.
24. G.H. Alt and D.H.R. Barton, *J. Chem. Soc.*, (1954) 4284.
25. D.H.R. Barton, E. Miller and H.T. Young, *J. Chem. Soc.*, (1951) 2598.
26. D.H.R. Barton and G.A. Morrison, *Fortschr. Chem. Org. Naturstoffe*, 19 (1961) 165.
27. D.H.R. Barton and J.F. King, *J. Chem. Soc.*, (1958) 4398.

28. D.H.R.Barton, *Svensk. Kem. Tidskr.*, 71 (1959) 356; *Suomen Kemistilehti*, 32 (1959) 27.
29. J.F.King, R.G.Pews and R.A.Simmons, *Canad.J.Chem.*, 41 (1963) 2187.
30. D.H.R.Barton and N.J.Holness, *J.Chem. Soc.*, (1952) 78.
31. A.M.Abel el Rehim and C.H.Carlisle, *Chem. Ind.* (*London*), (1954) 279.
32. W.Voser, M.V.Mijovic, H.Heusser, O.Jeger and L.Ruzicka, *Helv.Chim.Acta*, 35 (1952) 2414; C.S.Barnes, D.H.R.Barton, J.S.Fawcett and B.R.Thomas, *J.Chem. Soc.*, (1953) 576.
33. D.H.R.Barton, J.F.McGhie, M.K.Pradhan and S.A.Knight, *Chem. Ind.* (*London*), (1954) 1325; *J.Chem. Soc.*, (1955) 876; D.Arigoni, R.Viterbo, M.Dunnenberger, O.Jeger and L.Ruzicka, *Helv.Chim.Acta*, 37 (1954) 2306.
34. D.H.R.Barton, J.E.Page and E.W.Warnhoff, *J.Chem. Soc.*, (1954) 2715; D.S.Irvine, J.A.Henry and F.S.Spring, *ibid.*, (1955) 1316.
35. D.H.R.Barton and K.H.Overton, *J.Chem. Soc.*, (1955) 2639.
36. D.H.R.Barton, D.A.Lewis and J.F.McGhie, *J.Chem. Soc.*, (1957) 2907.
37. J.E.D.Levisalles, *Bull. Soc.Chim.France*, (1960) 551; M.Balasubramanian, *Chem. Rev.*, 62 (1962) 591.
38. R.J.Abraham and J.S.E.Holker, *J.Chem. Soc.*, (1963) 806.
39. S.Winstein and N.J.Holness, *J.Am.Chem. Soc.*, 77 (1955) 5562.
40. E.L.Eliel, *Stereochemistry of Carbon Compounds*, McGraw-Hill, London, 1962.
41. E.L.Eliel, N.L.Allinger, S.J.Angyal and G.A.Morrison, *Conformational Analysis* Interscience, New York, 1965.
42. D.H.R.Barton, *Experientia*, Suppl.II (1955) 121; D.H.R.Barton, A.J.Head and P.J.May, *J.Chem. Soc.*, (1957) 935; D.H.R.Barton, F.McCapra, P.J.May and F.Thudium, *J.Chem. Soc.*, (1960) 1297; D.H.R.Barton, *Theoret. Org.Chem.*, *Papers Kekule Symp.*, *London, 1958*, Butterworth, London, 1959, p.127.
43. M.Legrand, V.Delaroff and J.Mathieu, *Bull. Soc.Chim.France*, (1961) 1346; R.Bucourt, *ibid.*, (1962) 1983, (1963) 1262; M.J.T.Robinson and W.B.Whalley, *Tetrahedron*, 19 (1963) 2123; R.Baker and J.Hudec, *Chem.Commun.*, (1967) 891.
44. J.B.Hendrickson, *J.Am.Chem. Soc.*, 83 (1961) 4537; see also N.L.Allinger, *ibid.*, 81 (1959) 5727; R.Pauncz and D.Ginsburg, *Tetrahedron*, 9 (1960) 40.
45. C.Djerassi, *Optical Rotatory Dispersion*, McGraw-Hill, New York, 1960; W.Moffitt, R.B.Woodward, A.Moskowitz, W.Klyne and C.Djerassi, *J.Am.Chem. Soc.*, 83 (1961) 4013; G.Snatzke (Ed.), *Optical Rotatory Dispersion and Circular Dichroism in Organic Chemistry*, Heyden, London, 1967; J.Hudec, *J.Chem. Soc.*, in the press.
46. D.H.R.Barton, R.C.Cookson, W.Klyne and C.W.Shoppee, *Chem. Ind.* (*London*), (1954) 21.
47. F.G.Riddell, *Quart. Rev.* (*London*), 21 (1967) 364.
48. R.U.Lemieux, R.K.Kullnig, H.J.Bernstein and W.G.Schneider, *J.Am.Chem. Soc.*, 80 (1958) 6098.
49. M.Hanack, *Conformation Theory*, Academic Press, New York, 1965.

Biography

Derek Harold Richard Barton was born on 8 September 1918, son of William Thomas and Maude Henrietta Barton. In 1938 he entered Imperial College, University of London, where he obtained his B.Sc.Hons.(1st Class) in 1940 and Ph.D.(Organic Chemistry) in 1942. From 1942 to 1944 he was a research chemist on a government project, from 1944–1945 he was with Messrs. Albright and Wilson, Birmingham. In 1945 he became assistant lecturer in the Department of Chemistry of Imperial College, from 1946–1949 he was I.C.I. Research Fellow. In 1949 he obtained his D.Sc. from the same University. During 1949–1950 he was Visiting Lecturer in the Chemistry of Natural Products, at the Department of Chemistry, Harvard University (U.S.A.). In 1950 he was appointed Reader in Organic Chemistry and in 1953 Professor at Birkbeck College. In 1955 he became Regius Professor of Chemistry at the University of Glasgow, in 1957 he was appointed Professor of Organic Chemistry at Imperial College, which position he still holds.

In 1950, in a brief paper in *Experientia* entitled «The Conformation of the Steroid Nucleus», Professor Barton showed that organic molecules in general and steroid molecules in particular could be assigned a preferred conformation based upon results accumulated by chemical physicists, in particular by Odd Hassel. Having chosen a preferred conformation, it was demonstrated that the chemical and physical properties of a molecule could be interpreted in terms of that preferred conformation. In molecules containing fixed rings, such as the steroids, there resulted a simple relationship between configuration and conformation, such that configurations could be predicted once the possible conformations for the products of a reaction could be analysed. Thus the subject «conformational analysis» had begun. Barton later determined the geometry of many other natural product molecules using this method. Conformational analysis is useful in the elucidation of configuration, in the planning of organic synthesis, and in the analysis of reaction mechanisms. It will be fundamental to a complete understanding of enzymatic processes.

Prof. Barton was invited to deliver the following special lectures: 1956, Max Tischler Lecturer at Harvard University; 1958, First Simonsen Memorial

Lecturer of the Chemical Society; 1961, Falk-Plaut Lecturer, Columbia University; 1962, Aub Lecturer at Harvard Medical School; Renaud Lecturer at Michigan State University; Inaugural 3 M's Lecturer, University of Western Ontario; 1963, Hugo Müller Lecturer of the Chemical Society; 3 M's Lecturer at the University of Minnesota; 1967, Pedler Lecturer of the Chemical Society; 1969, Sandin Lecturer at the University of Alberta; 1970, Graham Young Lectureship, Glasgow.

In 1958 Prof. Barton was Arthur D. Little Visiting Professor at Massachusetts Institute of Technology, Cambridge, Mass.; in 1959 Karl Folkers Visiting Professor at the Universities of Illinois and Wisconsin.

In 1954 Derek Barton was elected to Fellowship of the Royal Society, in 1956 he became Fellow of the Royal Society of Edinburgh; in 1965 he was appointed member of the Council for Scientific Policy of the U.K.; in 1969 he became President of Section B, British Association for the Advancement of Science, and President of the Organic Chemistry Division of the International Union of Pure and Applied Chemistry.

Professor Barton holds the following honours and awards: 1951, First Corday-Morgan Medal of the Chemical Society; 1956, Fritzsche Medal of the American Chemical Society; 1959, First Roger Adams Medal of the American Chemical Society; 1960, Foreign Honorary Member of the American Academy of Arts and Sciences; 1961, Davy Medal of the Royal Society; 1962, D. Sc. *h.c.* Montpellier; 1964, D. Sc. *h.c.* Dublin; 1967, Honorary Fellow of the Deutsche Akademie der Naturforscher «Leopoldina»; 1969, Honorary Member of Sociedad Quimica de Mexico; 1970, D. Sc. *h.c.* St. Andrews: Fellow of Birkbeck College; Honorary Member of the Belgian Chemical Society; Foreign Associate of the National Academy of Sciences; Honorary Member of the Chilean Chemical Society; D. Sc. *h.c.*, Columbia University, New York; 1971, First award in Natural Product Chemistry, Chemical Society (London); D. Sc. *h.c.*, Coimbra (Portugal); Elected Foreign Member of the Academia das Ciencias de Lisboa; 1972, D. Sc. *h.c.* University of Oxford; Longstaff Medal of the Chemical Society.

Derek Barton was first married to Jeanne Kate Wilkins but this marriage was later dissolved. He is now married to Christiane Cognet, a Professor of the Lycée français de Londres. He has one son, W. G. L. Barton, by his first marriage.

O. HASSEL

Structural aspects of interatomic charge-transfer bonding

Nobel Lecture, June 9, 1970

In the nineteenth century convincing experimental proof was collected of interactions in liquid systems between molecular species generally regarded as chemically «saturated», leading to more or less stable complexes. In several cases stoichiometric solid compounds could even be isolated from such mixtures. The number of intermolecular complexes more or less extensively studied during the last two decades mainly employing spectrophotometric methods is considerable, and it has become possible not only to classify the different types of complexes, but also to predict and subsequently to prepare and study new complexes. The understanding of the processes leading to the formation of intermolecular complexes remained rather incomplete, however, and about the atomic arrangements in the complexes almost nothing was known until direct interferometric experiments were carried out.

The two types of reversibly functioning intermolecular interactions for which a more precise structural picture was available already before the second world war were apparently only those leading to the formation of hydrogen bonding and to «dative covalent bonds» like those formed in the interaction between amine nitrogen atoms and boron atoms belonging to boron trihalide molecules. For both reaction categories it was suggested that an electron transfer from a «donor» to an «acceptor» atom takes place. Quantitative experimental work carried out by spectroscopists during the last two decades has substantiated the belief that the formation of molecular complexes may generally be attributed to a transfer of negative electrical charge from a *donor* molecule (Lewis or Brønsted base) to an *acceptor* molecule (the Lewis acid) and it had become natural to classify the process involved as a «charge-transfer» interaction.

Among the molecules acting as electron donors those which owe their donor properties to the presence of atoms possessing one or more «lone pairs» of electrons – the «*n* donors» – are of particular significance. Other types of donor molecules are also known, like unsaturated hydrocarbons, in particular aromatic hydrocarbons and some of their derivatives, molecules deriving their donor properties from the presence of comparatively loosely bound π electrons. Molecules exhibiting electron-accepting properties may be of widely differing nature ranging from aromatic molecules with strongly electronegative substituents to molecular species containing an atom which should normally act as an «*n* donor» but has, in the molecule in question, acquired a certain positive surplus charge.

During the later part of the 1940's the interest in complexes formed by donor and acceptor molecules was stimulated by quantitative spectroscopic work dealing first with solutions of iodine in benzene, later with a considerable number of liquid systems containing different combinations of donor and acceptor molecules dissolved in solvents regarded as «inert» with respect to the interacting species. From the spectroscopic data equilibrium constants and thermodynamic values associated with the formation of 1:1 complexes were evaluated. A quantum mechanical theory of the «complex resonance» was worked out by Mulliken which is of a very general nature and explains spectroscopical observations, but has not yet made possible reliable predictions regarding the atomic arrangements within the complexes.

Direct interferometric structure determinations in the vapour phase are regarded virtually impossible in most cases because of the extremely low concentration of the complex which may be expected to be present. X-Ray crystallographic structure determinations have made it possible, however, to draw conclusions regarding atomic arrangements not only in solids, but even in isolated 1:1 complexes, conclusions which should be of value for theoretical workers trying to establish a more elaborate theory of charge-transfer interaction.

Particular importance may be attributed to complexes in which direct bonding exists between one atom belonging to the donor molecule and another atom belonging to the acceptor molecule. Complexes of this kind are above all those formed by donor molecules containing atoms possessing «lone pair electrons» and halogen or halide molecules. A presentation and discussion of the general results obtained by X-ray analysis of solid adducts exhibiting charge-transfer bonding between such atoms might therefore be of some interest.

First, some of the conclusions drawn before direct structure determinations had been carried out, deserve to be recalled.

The considerations were based on the assumption that halogen atoms are directly linked to donor atoms with a bond direction roughly coinciding with the axes of the orbitals of the lone pairs in the non-complexed donor molecule. The oxygen atom of an ether or ketone molecule was supposed to form bonds with both halogen atoms in an isolated 1:1 complex, which would require the halogen molecule axis to run orthogonal to the COC plane in an ether complex, and to be situated in the

$$\begin{matrix} C \\ C \end{matrix}\!\!>\!=O$$

plane in a ketone complex. In both cases three-membered rings would then result, containing one oxygen and two halogen atoms. Even in an isolated 1:1 complex formed by pyridine both halogen atoms were supposed to be attached to the donor atom, although in different ways: one of them was expected to be situated in the plane of the pyridine molecule on the axis of the nitrogen lone pair orbital, the second (supposed to carry a certain negative surplus charge) outside this plane, under attraction of the (positively charged) nitrogen atom.

Early in the 1950's X-ray crystallographic investigations of halogen adducts were started in Oslo, beginning with the solid 1:1 1,4-dioxan–bromine compound. The most striking feature of the resulting crystal structure[1], the endless chains of alternating dioxan and bromine molecules depending on linear O—Br–Br—O arrangements running in a direction roughly equal to the «equatorial» direction in cyclohexane (Fig. 1) was rather unexpected. It proved that both atoms belonging to a particular halogen molecule may simultaneously be bonded to oxygen atoms, although probably not to the same oxygen atom. The existence of halogen molecule bridging between donor atoms contradicts previous assumptions according to which a charge-transfer bond formed by one of the atoms belonging to a particular halogen molecule creates a marked negative charge on the partner atom.

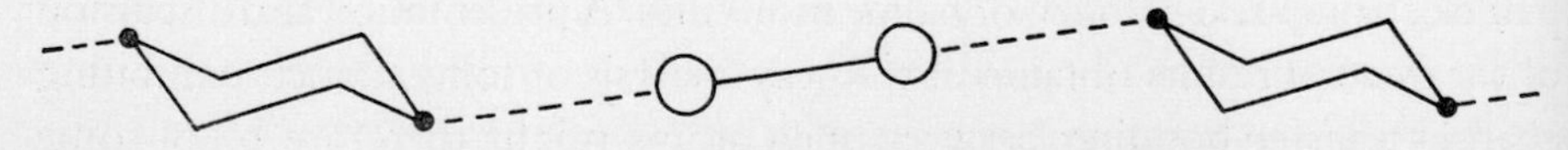

Fig. 1. Chains in the 1:1 adduct of 1,4-dioxan and bromine.

The oxygen–bromine distance in the dioxan–bromine adduct is 2.7 Å and thus considerably larger than the sum of the covalent radii of oxygen and bromine, but at the same time definitively shorter than the sum of the corresponding van der Waals radii. The type of «polymerisation» of simple complexes into endless chains observed in the crystalline dioxan–bromine compound has also been observed in the crystalline 1:1 adducts of dioxan and chlorine, resp. iodine. A comparison between the oxygen–halogen separations and of the interhalogen bond lengths in the three 1,4-dioxan adducts leads to the conclusion that the former increases rather slowly from chlorine to iodine, indicating a certain degree of compensation of the effect of larger halogen radius by the increase in charge-transfer bond strength. On the other hand, the halogen–halogen bond length increases, although slowly, from chlorine to iodine compared with that observed in «free» halogen molecules, an observation which must also find its explanation in the stronger charge-transfer interaction between oxygen and halogen.

In the solid 1:2 iodomonochloride 1,4-dioxan adduct, isolated complexes are observed in which iodine is bonded to oxygen in a linear oxygen–iodine–chlorine arrangement, again approximately pointing in the direction of an equatorial bond in cyclohexane. Similar results have been obtained for halogen complexes formed by 1,4-dithiane and 1,4-diselenane, with the exception that in diselenane an «axial» bond direction appears to be more favourable than an equatorial. The increased strength of the charge-transfer bonding is indicated by a more pronounced lengthening of the halogen–halogen bond. The fact that no example of halogen molecule «bridging» between donor atoms has, so far, been observed in dithiane or diselenane compounds also points to a relatively strong donor–acceptor interaction.

From the structures of the crystalline adducts of 1,4-dioxan it became clear that both atoms of a particular halogen molecule are able to form bonds to donor atoms, although apparently not to the same donor atom. The question then arose whether or not a particular donor atom may be involved in more than *one* bond to halogen. This would appear possible for *n* donor atoms like oxygen possessing *two* lone electron pairs. In the crystal structure of the 1:1 acetone–bromine adduct it was actually found that each keto oxygen atom is linked in a symmetrical way to two neighbouring bromine atoms, thus serving as a starting point for two bromine molecule bridges, both with a linear O—Br–Br—O arrangement and with an angle between the two bond directions of 110°.

In the case of amine adducts it would not be expected that a nitrogen atom might be capable of forming more than *one* single bond to halogen. The correctness of this anticipation has been borne out by the results of a considerable number of crystal structure determinations of addition compounds, usually choosing iodine or an iodine monohalide as the acceptor partner. Here again, the bond direction corresponds to that expected from simple considerations regarding the orbital of the lone electron pair on the amine nitrogen atom in the donor molecule. The nitrogen–halogen–halogen arrangement has always been found to be nearly linear, the bonds between the nitrogen atom and the carbon, resp. the iodine atom are tetrahedrally arranged in the case of aliphatic amines, essentially co-planar if the donor molecule is a heteroaromatic amine. The strength of the charge-transfer bond may be inferred from the short nitrogen–halogen bond distance which is only 2.3 Å in all complexes formed by tertiary amines and iodine or iodine monohalides, a value only about 0.25 Å larger than the sum of the covalent radii of nitrogen and iodine. A lengthening of the interhalogen bond by approximately 0.2 Å observed in these complexes is therefore not surprising. The fact that halogen molecule bridges have never been observed between amino nitrogen atoms also indicates that the nitrogen–halogen bond is rather strong. This does not imply, however, that such bridges can not be stable between other kinds of nitrogen atoms. Thus, in the isolated complex containing two molecules of acetonitrile and one bromine molecule such bridges are present, which appears very natural because nitriles are known from spectroscopical measurements to be weaker donors than are the amines.

The only crystal structure of an amine adduct so far investigated in which a cyanogen halide acts as the acceptor molecule is that formed by pyridine and cyanogen iodide[2]. The complex contains a linear arrangement N—I–C≡N which is symmetrically situated in the pyridine plane along the line drawn between the pyridine nitrogen and the γ-carbon atoms. The N—I bond distance is larger than 2.3 Å by about one quarter of an Å unit, in agreement with the spectroscopical finding that cyanogen iodide is a relatively weak acceptor.

Only one single addition compound containing alcohol and halogen molecules has been investigated crystallographically, the 2:1 methanol–bromine adduct[3]. Its structure may be, and probably is, characteristic of the type of structure to be generally expected for alcohol–halogen adducts in that oxygen atoms are linked together partly by hydrogen bonds, partly by linear O—Br–Br—O arrangements. Each oxygen atom is actually tetrahedrally surrounded by a methyl group, two hydrogen bonds and one bromine molecule bridge. This structure provides us in any case with a striking example of analogy between intermolecular charge-transfer and hydrogen bonding. Further examples will be presented below.

It would appear natural to search for a closer interdependence between interatomic distances (between donor and acceptor atoms and between the two halogen atoms) and the strength of the charge-transfer bond in crystalline adducts exhibiting the types of linear arrangements of donor and halogen atoms described above. This has, however, not turned out to be an easy task. It is, however, possible to evaluate, with considerably success, overall distances along the linear chains of donor and halogen atoms, either between the donor atom and the second halogen atom or, in the case of halogen molecule bridges, the distance between the two donor atoms. This is done by simple summation of the radii of individual atoms along the chain. Accepted covalent radii of the donor atoms are used, and for halogen atoms radii related to such atoms involved in charge-transfer bonds. For the latter the mean values of the covalent and the Van der Waals radius are chosen. The overall distances thus computed are generally in acceptable agreement with observed distances (Table 1). Such predictions of overall distances may no doubt be useful when working out new crystal structures, but considerations which may be expected to yield informations about the strength of the bond between a halogen and a donor atom must evidently involve the shortening of this bond and the lengthening of the adjacent halogen–halogen bond caused by the charge-

Table 1

Overall distances in adducts

Substance	*Observed distance*	*Calculated distance*
$N(CH_3)_3$–I_2	5.10	5.19
4-Picoline–I_2	5.14	5.19
$N(CH_3)_3$–ICI	4.82	4.85
Pyridine–ICI	4.77	4.85
Pyridine–IBr	4.92	5.00
Pyridine–ICN	4.68	4.63
1,4-Dioxan–2 ICI	4.92	4.81
Benzyl sulphide–I_2	5.60	5.53
1,4-Dithiane–I_2	5.66	5.53
1,4-Diselenane–I_2	5.70	5.66
	D—X–X—D distances	
1,4-Dioxan–Cl_2	7.36	7.34
1,4-Dioxan–Br_2	7.73	7.70
1,4-Dioxan–I_2	8.35	8.38
Acetone–Br_2	7.92	7.70
2 Acetonitrile–Br_2	8.00	7.84
2 Methanol–Br_2	7.85	7.70

transfer process. For this reason evaluation of «effective» radii R_I and R_2 of the centre halogen atom in the two opposite directions, towards the donor and towards the second halogen atom, were carried out simply by subtracting the covalent radii of the donor atom, resp. the second halogen atom from the observed distances. In Fig. 2 the values R_I and R_2 are plotted against each other for a number of adducts in which iodine is the centre atom[4]. Points representing the individual adducts show a marked tendency to concentrate near a straight line descending from the left to the right according to the equation:

$$R_2 = -1.92\,R_I + 4.59$$

Besides the points representing *n* donor–IX adducts the figure also contains points giving «effective» radii of iodine in solids in which trihalide ions are present with iodine as the centre atom and linked to two halogen atoms of the same kind. Such IX_2^- ions may be either symmetrical or non-symmetrical, in the latter the halogen atom closest to the centre atom is believed to carry a smaller negative charge than the more distant halogen atom. We assume that in the weakest IX complexes like those with oxygen as the donor atom, this atom corresponds to the halogen atom carrying the larger negative charge in

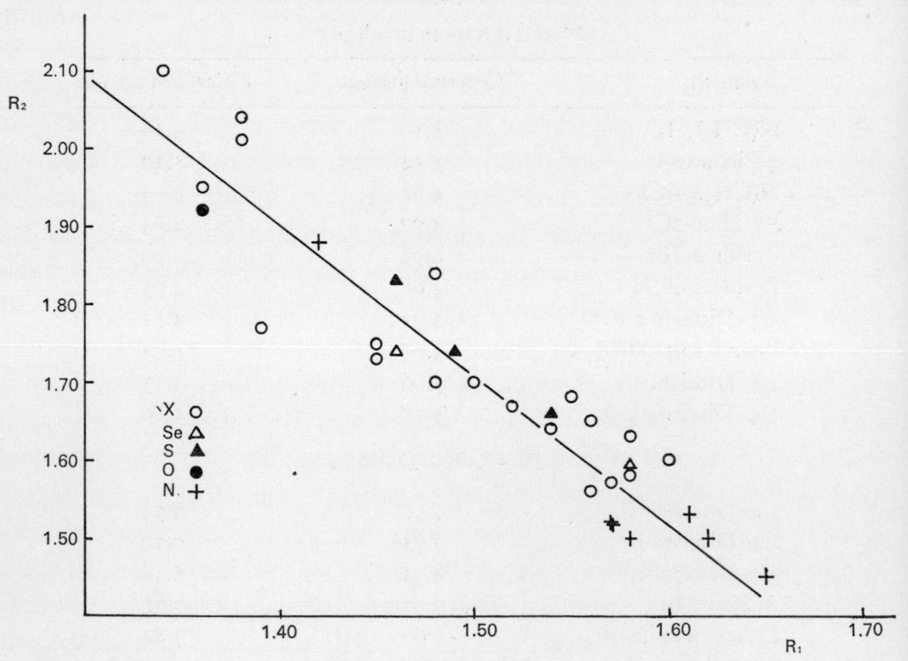

Fig. 2. Plot of «effective» radii R_I and R_2 of iodine in adducts and in trihalide ions.

non-symmetrical trihalide ions. Further, that the nitrogen atom in the strong IX adducts of tertiary amines takes the place of the other halogen atoms in trihalide ions which has a smaller negative charge and comes closer to the centre atom. The points in Fig. 2 representing the weakest complexes are situated in the upper left part of the diagram, those representing the strongest complexes bottom right. It appears significant that the point corresponding to the relatively weak pyridine–cyanogen iodine complex ($R_1 = 1.42$, $R_2 = 1.88$) is found in the upper middle of the diagram. Theoretically, it is difficult to assess if the relation between R_1 and R_2 should be expected to be strictly linear. It might perhaps be more adequately expressed by another curve, *e.g.* a hyperbola with an angle almost equal to 180° between its asymptotes, than by a straight line.

Turning now to complexes in which halide molecules act as electron acceptors it should be mentioned that the concept of a halogen atom linked to a non-halogen atom but still acting as an electron acceptor emerged at a rather recent date. The conclusion that halides may serve as acceptors in the formation of addition compounds formed with *n* donor molecules might have been drawn decades ago, but the idea was apparently not put forward until 1957[5]. It had in fact been known for some time that 1:3 adducts of iodoform and of antimony triiodide may exhibit a remarkable (macroscopic) crystallographic similarity and this was now so interpreted that the bonding between the two molecular species present in such adducts depended on bonds linking iodine atoms to *n* donor atoms. Crystal structure determinations carried out in the following years fully confirmed this suggestion and showed that the atomic arrangement within the complexes is such that the arrangement carbon (antimony)–iodine–*n* donor atom is linear or at least very nearly so. The crystal structure of a considerable number of other adducts formed by halogenated hydrocarbons and *n* donor molecules have now been determined with the result that one may be rather confident about the general principles governing the atomic arrangements in such adducts. In particular, a nearly linear arrangement carbon–halogen–donor atom has always been observed.

In most adducts so far investigated both participants contain more than one atom capable of taking part in charge-transfer bonds. When each oxygen, sulphur or selenium atom is linked to only one iodine atom in 1:1 addition compounds of iodoform with 1,4-dioxan or its analogues, structures exhibiting endless chains of alternating donor and acceptor molecules would be anticipated. Such chains are actually present in these adducts, chains analogous to those found in the sulphuric acid–dioxan compound. Figs. 3a and b il-

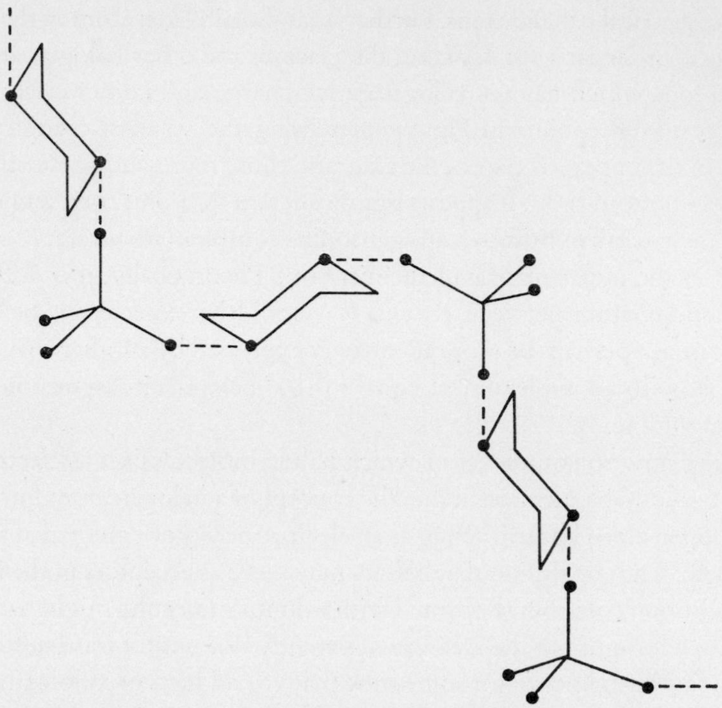

Fig. 3a. Chains of sulphuric acid and dioxan molecule in the 1:1 adduct.

lustrate the shape of the chains observed in the sulphuric acid–dioxan, resp. the iodoform–dithiane compound. The similarity between these chains again affords an example of analogy between hydrogen and charge-transfer bonding.

The possibility that an oxygen, sulphur or selenium atom may be involved in *two* charge-transfer bonds with halogen atoms should always be kept in mind, however, particularly when the acceptor molecule contains a large number of halogen atoms. Thus, in the 1:1 diselenane–tetraiodoethylene adduct[6] every selenium atom is bonded to *two* iodine atoms, the bond directions being roughly equatorial resp. axial and the selenium–iodine bond lengths almost identical. A «cross-linking» of the chains results, all iodine atoms are linked to selenium and each selenium atom to *two* iodine atoms.

In view of the moderate energies apparently involved in bonding between halide halogen atoms and *n* donor atoms it would appear natural to suggest that the Van der Waals interaction energy between acceptor molecules containing a sufficiently large number of the heaviest halogen species may contribute substantially to the lattice energy of a solid addition compound. Crystal

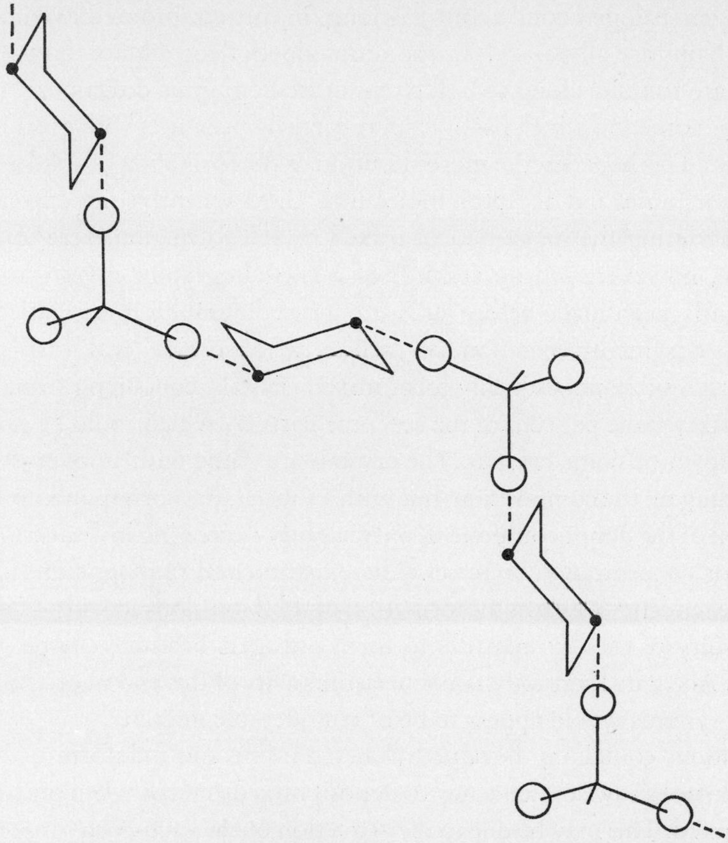

Fig. 3b. Chains of iodoform and dithiane molecules in the 1:1 adduct.

structure determinations of tetrabromo- and tetraiodoethylene and of their 1:1 pyrazine adducts actually give some support to this suggestion. The mutual arrangement of the halide molecules is virtually identical in the tetrahalogenoethylene crystals and in the corresponding crystalline addition compounds. The four crystals all belong to the space group $P2_1/c$ and their unit cells all contain four molecules which are in the tetrahalogenoethylene crystals practically identical in shape, but not crystallographically equivalent. The structures of the 1:1 adduct crystals may formally be derived from those of the tetrahalide crystal by removing one set of equivalent molecules and replacing them by pyrazine molecules. For both adducts this results in the formation of endless chains of alternating donor and acceptor molecules in which each nitrogen atom is bonded to a halogen atom situated near the plane of the pyrazine ring with a nearly linear nitrogen–halogen–carbon arrangement and a nitrogen–halogen bond about 3 Å long. In the tetrabromoethylene adduct these chains are all parallel, in the tetraiodoethylene adduct, however, the chains are running along two different crystallographic directions[7].

A phenomenon which has not yet apparently been met with great interest, but should perhaps deserve more attention, is the formation of solid solutions between donor and acceptor molecules. Until recently, the experimental facts favouring the suggestion of mixed crystal formation were somewhat meagre, however, and no attempt of a crystallographic investigation had apparently been made before an X-ray crystallographic investigation of the system hexamethylenetetramine–carbon tetrabromide was carried out[8]. These two substances actually form mixed crystals, containing from zero to about sixty mole per cent of the acceptor partner, which could be examined in the form of single crystals. The crystals are cubic with an over-structure depending on the composition, but with a subcell that corresponds to the true unit cell of the donor component, only slightly decreasing in dimension as the acceptor concentration increases. The experimental findings seem to prove that the tendency towards the formation of solid solutions actually depends on the faculty of the two partners to form nitrogen–bromine charge-transfer bonds. Accurate thermodynamic measurements of this and of certain related binary systems would appear to be of considerable interest.

Solutions containing hexamethylenetetramine and iodoform or bromoform do not show any tendency to deposit mixed crystals when the solvent is evaporated. This may be due to the attraction of the «active» hydrogen atom towards amino nitrogen. A solid 1:1 hexamethylenetetramine–iodoform adduct has been prepared and its crystal structure determined[9]. In this crystal (Fig. 4) every acceptor molecule is tetrahedrally linked to four neighbouring donor nitrogen atoms by three I—N bonds (2.94 Å long) and to a fourth nitrogen atom *via* the CH hydrogen atom. The C(H)—N distance is only 3.21 Å and the bond appears therefore to be energetically of some importance. It appears very probable that a 1:1 bromoform adduct of hexamethylenetetramine, isostructural with the iodoform adduct, may exist. In this adduct, the halogen–nitrogen bonds would be expected to be somewhat weaker, the C–H—N bonds, however, somewhat stronger than in the iodoform adduct.

Crystals having the 1:1 composition have so far not been isolated in the hexamethylenetetramine–bromoform system, but another compound, a solid 1:2 compound has been obtained in the form of crystals suitable for X-ray examination. It is obvious that in a solid of this composition all CH

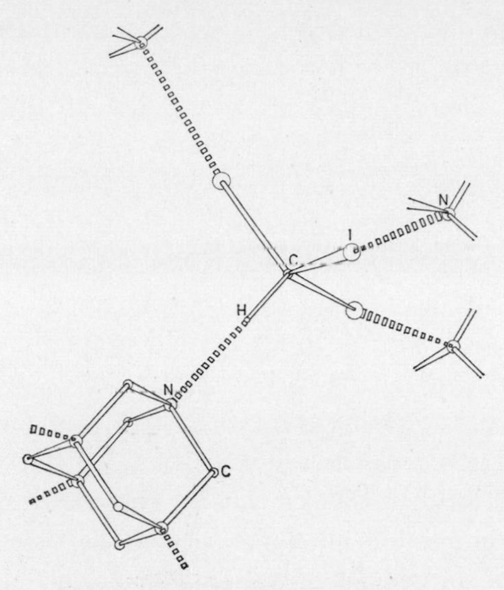

Fig. 4. Tetrahedral arrangement of hexamethylenetetramine molecules around an iodoform molecule in the 1:1 adduct.

groups and all bromine atoms cannot form bonds with nitrogen atoms. This means that a competition between the two kinds of intermolecular bonds may take place, the outcome of which may give indications regarding relative strengths of the bond types. In the crystal structure derived[10] from the X-ray data of the adduct containing hexamethylenetetramine and bromoform molecules in the proportion 1:2 it is obvious that two of the bromine atoms belonging to half the number of acceptor molecules form charge-transfer bonds to nitrogen atoms in neighbouring donor molecules, the other half none. Each donor molecule is tetrahedrally linked to two bromine atoms and to two CH-groups all belonging to neighbouring bromoform molecules. Fig. 5 visualizes the arrangement of bromoform molecules surrounding each hexamethylenetetramine molecule. These results seem to indicate that the CH··· N bonds are stronger than are the Br··· N bonds between hexamethylenetetramine and bromoform molecules.

The relative stability of C–H··· N bonding indicated by the results of X-ray investigations of adducts of hexamethylenetetramine and trihalogenomethanes make it appear possible to carry out chemical separations *via* addition compounds held together by such bonds. It may be suggested for example that the separation into optically active components of racemic haloform–HCFClBr–can be achieved *via* adduct formation with a properly chosen optically active amine.

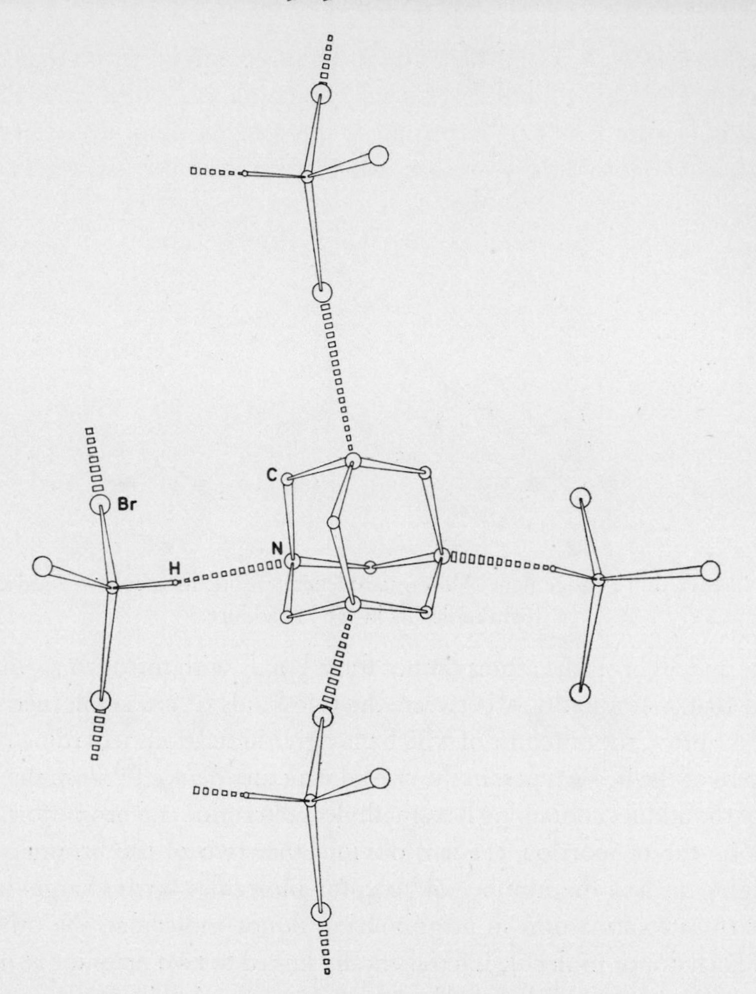

Fig. 5. Bromoform molecules surrounding each hexamethylenetetramine molecule in the 2:1 adduct.

Even in crystals containing only one molecular species intermolecular bonds between *n* donors and halogen atoms should be expected in particular cases. They were searched for in the crystals of carboxylic acid halides where such bonds between halogen and carbonyl oxygen atoms appeared to be possible since an earlier investigation had proved that oxgen–halogen bonding actually exists in 1:1 adducts formed by oxalyl chloride, resp. oxalyl bromide, and 1,4-dioxan. X-Ray investigations of oxalyl chloride and oxalyl bromide proved that this kind of bonding determines the crystal structure of the bromine compound but are absent in the chlorine compound.

In the former each oxalyl bromide molecule is linked to its four nearest neighbours by O··· Br bonds. That the two compounds have quite different crystal structures is not so surprising as the oxygen–chlorine interaction is expected to be considerably weaker than that between oxygen and bromine.

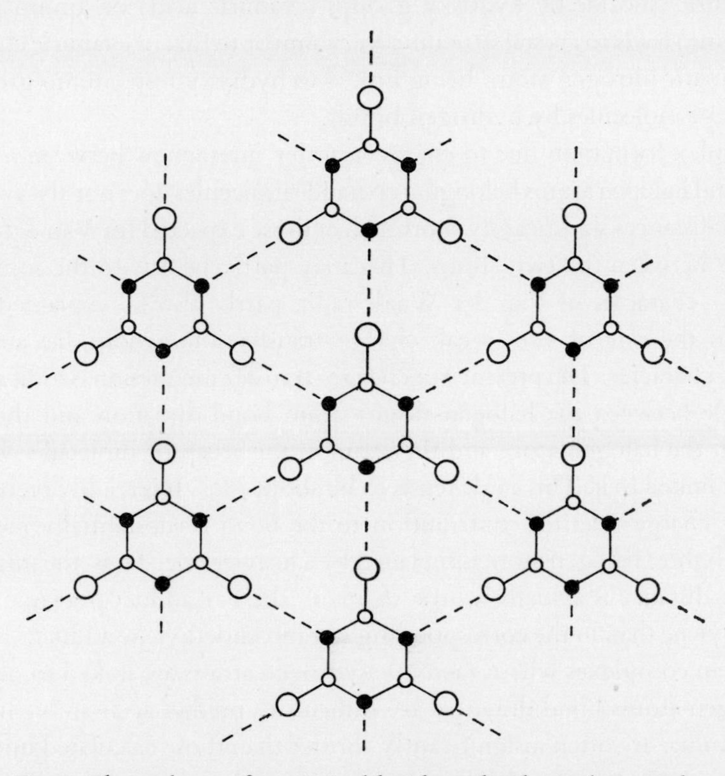

Fig. 6. Planar sheets of cyanuric chloride molecules in the crystal.

A particularly interesting example of charge-transfer interaction between nitrogen and halogen atoms is provided by the structure of solid cyanuric chloride – the trimeric form of cyanogen chloride. According to available X-ray data[11] the crystals are built up of planar layers of the type shown in Fig. 6, held together by Van der Waals forces. Within the layers each molecule is linked to six neighbour molecules by nitrogen–chlorine bonds and the C—Cl—N arrangement is probably strictly linear. At the time of publication of this paper no experimental facts about similar charge-transfer bonding were available, and it is not surprising that the authors failed to interpret their results by suggesting bonds between nitrogen and chlorine atoms to be present.

When *n* donor atoms like oxygen or nitrogen are involved in hydrogen bonds the resulting atomic arrangements are usually similar to those observed in charge-transfer complexes in which the same donor atom is linked to a halogen atom. It is therefore not surprising that a replacement of chlorine in cyanuric chloride by hydroxy groups (cyanuric acid) or amino groups (melamine) leads to crystal structures very similar to that of cyanuric chloride, the aromatic nitrogen atoms being linked to hydroxy resp. amino groups in neighbour molecules by hydrogen bonds.

Complex formation due to charge-transfer interaction between *n* donor atoms and halogen atoms belonging to halide molecules does not always result in bond distances significantly shorter than those expected for Van der Waals contacts between the two atoms. This may partly be due to the somewhat «diffuse» character of Van der Waals radii, partly also be explained if the bond, in the case of very weak charge-transfer interactions, has an intermediate character. The presence of charge-transfer interaction is indicated by the angle between the halogen–donor atom bond direction and the bond between the halogen atom and the atom in the acceptor molecule which is directly linked to it. This angle tends to be about 180°. It is readily recognized that the charge-transfer contribution to the bond is substantially increased when a lighter halogen atom is replaced by a heavier one. Thus, the nitrogen–halogen distance is actually a little *shorter* in the 1:1 adduct pyrazine–tetraiodoethylene than in the corresponding tetrabromoethylene adduct.

Even in complexes where «active» hydrogen atoms are linked to nitrogen or oxygen atoms bond distances are difficult to predict accurately, and observed values are often insignificantly shorter than those calculated under the assumption of a Van der Waals interaction. In such cases arguments in favour of a weak «hydrogen bond» between donor and acceptor molecule must to some extent be based on the actual geometry of the complex.

X-Ray investigations of solid adducts in which molecules containing oxygen, nitrogen, etc. act as donors, molecules containing halogen atoms or «active» hydrogen atoms as acceptors, have contributed substantially to our present knowledge about atomic arrangements in certain donor–acceptor complexes. Previous suggestions based on spectroscopic observations have in some cases been found incorrect. Simple rules for the arrangements of the bonds (to some extent even bond lengths) have been formulated. Perhaps most striking, and certainly not expected, is the farreaching analogy between atomic arrangements in complexes formed by the same category of donors with the two so different types of acceptors. It may perhaps be suggested that the results obtained from direct X-ray structure analysis may contribute to the theoretical understanding of both categories of donor-acceptor interaction.

1. O. Hassel and J. Hvoslef, *Acta Chem. Scand.*, 8 (1954) 873.
2. T. Dahl, O. Hassel and K. Sky, *Acta Chem. Scand.*, 21 (1967) 592.
3. P. Groth and O. Hassel, *Acta Chem. Scand.*, 18 (1964) 402.
4. O. Hassel and Chr. Römming, *Acta Chem. Scand.*, 21 (1967) 2659.
5. O. Hassel, *Chem. Soc. (London)*, Centenary Lecture, May 9th, 1957.
6. T. Dahl and O. Hassel, *Acta Chem. Scand.*, 19 (1965) 2000.
7. T. Dahl and O. Hassel, *Acta Chem. Scand.*, 22 (1968) 2851.
8. T. Dahl and O. Hassel, *Acta Chem. Scand.*, 22 (1968) 372.
9. T. Dahl and O. Hassel, *Acta Chem. Scand.*, 24 (1970) 377.
10. T. Dahl and O. Hassel, *Acta Chem. Scand.*, 25 (1971) 2168.
11. W. Hoppe, H. U. Lenné and G. Morando, *Z. Krist.*, 108 (1957) 321.

Biography

Odd Hassel was born in Kristiania (now Oslo), Norway, 17 May, 1897. His father was Ernst Hassel, a physician who specialized in gynaecology, his mother Mathilde née Klaveness.

In 1915 he entered the University of his native town where he studied mathematics and physics with chemistry as his chief subject and graduated as a cand. real. in 1920. After a year of leisure in France and Italy he went to Germany in the autumn of 1922 where he first spent more than half a year in Munich in the laboratory of Professor K. Fajans. Work on the sensibilisation of silver halides by organic dyes led to the detection of what is now termed the «adsorption indicators». After moving to Berlin Hassel worked at the Kaiser Wilhelm Institute in Dahlem, carrying out X-ray crystallographic work. During that time he obtained, on the proposal of Fritz Haber, a Rockefeller Fellowship. In 1924 he graduated as Dr. Phil. at the Berlin University. From 1925 to 1926 he worked at the University of Oslo in the capacity of «universitetsstipendiat», from 1926 to 1934 as «dosent» in physical chemistry and electrochemistry. From 1934 to 1964 he had the chair of physical chemistry in Oslo, the first of its kind in Norway, and headed the department of physical chemistry started in 1934.

Hassel's main interest during the first years of his teaching at the University of Oslo dealt with inorganic chemistry, but from 1930 onwards his work was concentrated on problems connected with molecular structure, particularly the structure of cyclohexane and its derivatives and other substances containing six-membered rings related to that of cyclohexane.

In order to supplement the experimental methods available two additional methods not previously used in Norway were introduced; the measurements of electric dipole moments and electron diffraction by vapours. Sufficient experimental material had been gathered by 1943 to allow more general conclusions regarding the possible configurations (conformations) and the transition between them to be drawn. A short paper had just been published in a Norwegian journal when Hassel was arrested by Norwegian Nazis and later taken into custody by the German occupants. Released in November 1944 he found the institute almost deserted. After the war experimental work could be taken up again and in particular electron-diffraction work based on the rotating sector method.

During the early 1950's Hassel opened a new field of structure investigation, namely that of the charge-transfer compounds. Compounds formed by organic electron-donor molecules like ethers and amines and electron acceptors as halogen molecules or organic halides had mainly been investigated by spectroscopic methods. Information on the steric structures was scarce, however, and a series of structure determinations was undertaken. After some years work he was able to set up rules for the geometry of this kind of addition compounds, and this field still remains his main interest in structural chemistry.

Hassel holds honorary degrees from the Universities of Copenhagen and Stockholm. He is an honorary Fellow of the Norwegian Chemical Society and of the Chemical Society, London.

He is a Fellow of the Norwegian Academy of Sciences, the Royal Danish Academy of Sciences, the Royal Swedish Academy of Sciences and the Royal Norwegian Academy of Science. In 1964 he received the Guldberg–Waage Medal from the Norwegian Chemical Society and the Gunnerus Medal from the Royal Norwegian Academy of Sciences.

He is a Knight of the Order of St. Olav.

From 1967 a lecture is given yearly by distinguished scientists from abroad to his honour at the University of Oslo («The Hassel Lecture»).

Chemistry 1970

LUIS F. LELOIR

«for his discovery of sugar nucleotides and their role in the biosynthesis of carbohydrates»

Chemistry 1970

Presentation Speech by Professor Karl Myrbäck, member of the Nobel Committee for Chemistry of the Royal Swedish Academy of Sciences*

Your Majesty, Your Royal Highnesses, Ladies and Gentlemen.

The 1970 Nobel Prize for chemistry has been awarded to Dr. Luis Leloir for work of fundamental importance for biochemistry. Dr. Leloir receives the prize for his discovery of the sugar nucleotides and their function in the biosynthesis of carbohydrates.

Carbohydrates, as everybody knows, form a comprehensive group of naturally occurring substances, which include innumerable sugars and sugar derivatives, as well as high-molecular carbohydrates (polysaccharides) like starch and cellulose in plants and glycogen in animals. A polysaccharide molecule is composed of a large number of sugar or sugar-like units.

Carbohydrates are of great importance in biology. The unique reaction, which makes life possible on Earth, namely the assimilation of the green plants, produces sugar, from which originate, not only all carbohydrates but, indirectly, also all other components of living organisms.

The important role of carbohydrates, especially sugars and starch, in human food and, generally, in the metabolism of living organisms, is well known. The biological break-down of carbohydrates (often spoken of as «combustion») supplies the principal part of the energy that every organism needs for various vital processes. It is not surprising, therefore, that the carbohydrates and their metabolism have been the subject of comprehensive and in many respects successful biochemical and medical research for a long time. While working on these problems, Leloir made the discoveries for which he has now been awarded the Nobel Prize.

Before these discoveries were made, our knowledge of carbohydrate biochemistry was rather one-sided. The biological processes which break down carbohydrates, including the so-called combustion, have been well known for several decades. Over the years many Nobel prizes have been awarded for chemistry and still more for physiology or medicine for discoveries about the reactions and catalysts involved. However, our knowledge about the innumerable corresponding synthetic reactions which occur in all organisms,

* The manuscript was read by Professor Arne Tiselius.

was fragmentary. We had to resort to doubtful hypotheses; it was usually assumed that the syntheses were a direct reversal of the well-known break-down reactions. The work of Leloir has indeed revolutionized our thinking about these problems.

In 1949 Leloir published the discovery which became the foundation for a remarkable development. He found that in a certain reaction, which results in the transformation of one sugar to another sugar, the participation of a so far unidentified substance was essential. He isolated the substance and determined its chemical nature. It turned out to be a compound of an unknown type, containing a sugar moiety bound to a nucleotide. Compounds of this type are now called sugar nucleotides. Leloir established that the transformation reaction does not occur in the sugars as such, but in the corresponding sugar nucleotides. To put it simply, one may say, that the linking with the nucleotide occasions an activation of the sugar moiety which makes the reaction possible.

The remarkable aspect of the discovery was not the explanation of a single reaction, but Leloir's quick comprehension that he had found the key which would enable us to unravel an immense number of metabolic reactions. He ingeniously realized that a path had been opened to a field of research containing an accumulation of unsolved problems. In the twenty years that have elapsed since his initial discovery he has carried on his research in this field in an admirable manner.

Other scientists were quick to grasp the fundamental importance of Leloir's discovery; they realized that a vast field was now accessible to worth-while scientific investigation and started research along the path which he had opened. There can be no doubt that few discoveries have made such an impact on biochemical research as those of Leloir. All over the world, his discoveries initiated research work, the volume of which has grown over since. Leloir has been the forerunner and guide throughout; he made all the primary discoveries which determined the path and the objectives of the ensueing research work.

Leloir soon found that besides the sugar nucleotide first isolated, several others of the same type occur in Nature, and many have also been found by other research workers. Today more than one hundred sugar nucleotides which are essential participants in various reactions are known and well characterized. Some of them have an action similar to that of the first isolated, namely in the transformations of simple sugars to other simple sugars or sugar derivatives.

Still more important was Leloir's discovery that other sugar nucleotides have another action which occurs in the biological synthesis of compounds which are composed of or contain simple sugars or sugar derivatives. Leloir showed that all these syntheses are essentially transfer reactions. Sugar moieties from sugar nucleotides are transferred to accepting molecules which thereby increase in size. Probably the most sensational discovery made by Leloir was that the synthesis of the high-molecular polysaccharides also functions in this manner. The first example of the fundamental role of the sugar nucleotides in polysaccharide biosynthesis was found by Leloir in 1959 in the case of glycogen. It became clear that the polysaccharide biosynthesis is not a reversal of the biological breakdown, as had doubtfully been assumed earlier. On the contrary, Nature uses different and quite independent processes for synthesis and breakdown. Later on the same extremely important principle was also shown to be valid with other groups of substances, for instance with proteins and nucleic acids.

Through Leloir's work and the work of others, who were inspired by his discoveries, knowledge of great significance has been gained in wide and important sections of biochemistry, which were previously obscure. It can be readily appreciated that Leloir's work has also had far-reaching consequences in physiology and medicine.

LUIS F. LELOIR

Two decades of research on the biosynthesis of saccharides

Nobel Lecture, 11 December, 1970

Our work on the biosynthesis of oligo- and polysaccharides started about in 1946 not by a deliberate selection of the subject but because it came to us. Due to the phenomenal progress of biochemistry our initial experiments seem to belong to the paleolithic period but fortunately there are also some very recent and exciting advances in the field.

After returning from Cambridge in 1936 I did some work with J. M. Muñoz on the oxidation of fatty acids in liver. We managed to prepare a cell-free system which was active when suitably supplemented and this was a novel result since the process of oxidation was believed to require the integrity of the cells. I suppose the young generation of biochemists finds it hard to understand many of the things which we believed at that time.

After that came an incursion into the field of renal hypertension with E. Braun Menéndez, J. C. Fasciolo and A. C. Taquini. This work was carried out quite rapidly and was rather successful.

Then I worked in Carl F. Cori's laboratory in St. Louis and with D. E. Green in Columbia University.

On returning to Buenos Aires in 1945 I started to work with R. Caputto and R. Trucco. Dr. Caputto had done some research on the mammary gland and had the idea that glycogen was transformed into lactose. At the time one had to rely on osazones for identification and we soon reached a dead end. On looking back I think that what we were observing was the degradation of glycogen by amylase.

We then decided to study lactose breakdown by a yeast *Saccharomyces fragilis* with the idea that this would give us information on the mechanism of synthesis. In fact it did give us information but only after a long and tortuous process.

First we studied the lactase, then the phosphorylation of galactose (Trucco, Caputto, Leloir and Mittelman[55], 1948) and the transformation of galactose 1-phosphate. What we measured was the increase in reducing power of the following reaction sequence:

Galactose 1-phosphate → glucose 1-phosphate → glucose 6-phosphate (1)

We soon found that a thermostable factor was required and set out to isolate it in collaboration with C.E. Cardini and A.C. Paladini.

At the time things were not so easy because we did not have the powerful methods which we have nowadays and because we were working under rather poor conditions.

The results of our experiments were very confusing because we did not know that we were dealing with two thermostable factors. Finally we realized what was happening and we concentrated on the purification of the factor involved in the second reaction. That is in the phosphoglucomutase reaction.

We sent a letter to the editors of *Archives of Biochemistry* (Caputto, Leloir, Trucco, Cardini and Paladini[10], 1948) describing a new cofactor and mentioned that Kendall and Strickland[29] (1938) had previously described an activation by fructose 1,6-diphosphate but that our cofactor was different. After we had sent the manuscript we happened to test again fructose 1,6-diphosphate and obtained a strong activation. Furthermore our purified preparations were loaded with fructose 1,6-diphosphate. We had decided to ask the letter back but as a consequence of much worrying we struck on the idea that the activator might be glucose 1,6-diphosphate. Since the latter compound has the reducing group blocked we reasoned that it should be alkali-stable. Strangely enough everything turned out as expected. If it had not been for this mistake we might still be talking of the allosteric effect of fructose 1,6-diphosphate on phosphoglucomutase.

When we finished working with glucose 1,6-diphosphate we continued with the other cofactor. The concentrates were found to absorb light at 260 mμ and had a spectrum similar to that of adenosine but with some differences. At the time the only soluble nucleotides known to be present in tissues were the inosine and adenosine nucleotides. It was an exciting day when Caputto came in one morning with a *Journal of Biological Chemistry* which showed the absorption spectrum of uridine. It looked identical to that of our cofactor. After measuring glucose and phosphate content and doing a titration curve the structure shown in Fig.1 was proposed (Cardini, Paladini, Caputto and Leloir[12], 1950; Caputto, Leloir, Cardini and Paladini[9], 1950). This first sugar nucleotide was named uridine diphosphate glucose: UDPG. Its structure was confirmed by synthesis some five years later by Todd and coworkers in Cambridge. The mechanism by which UDP-glucose acts as a cofactor in the galactose 1-phosphate → glucose 1-phosphate transformation became

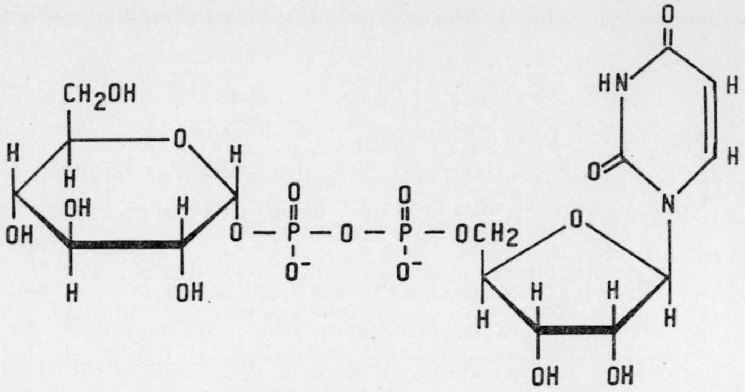

Fig. 1. Uridine diphosphate glucose (UDPG).

understandable when it was found that on incubation with yeast extracts part of the UDP–glucose was transformed into UDP–galactose (Leloir[31], 1951). After this we wrote the equations as follows:

$$\text{Galactose 1-phosphate} + \text{UDP–glucose} \rightleftharpoons \text{glucose 1-phosphate} + \text{UDP–galactose} \quad (2)$$

$$\text{UDP–galactose} \rightleftharpoons \text{UDP–glucose} \quad (3)$$

$$\text{Sum: galactose 1-phosphate} \rightleftharpoons \text{glucose 1-phosphate} \quad (4)$$

We used to call the whole system waldenase but Kalckar[27] (1958) suggested the names of uridylyl transferase and 4-epimerase for the enzymes corresponding to reaction 2 and 3 respectively.

After we found that yeast which was not adapted to galactose contained a lot of UDP–glucose we concluded that UDP–glucose should have some other function besides being a cofactor of galactose metabolism. I don't know if the reasoning was quite right but the facts were. For some years it was a joke in the laboratory because we were always asking: «What's the use of UDP–glucose?»

Since we had a method for estimating UDP–glucose with the galactose 1-phosphate → glucose 6-phosphate reaction, we began to measure UDP–glucose disappearance in different extracts and under different conditions. With yeast extracts it was observed that addition of glucose 6-phosphate increased UDP–glucose disappearance and finally this was found to be due to the formation of trehalose phosphate, a substance which had been isolated from yeast many years before by Robison and Morgan[50] (1930). The reaction is as follows:

$$\text{UDP-glucose} + \text{glucose 6-phosphate} \rightarrow \text{trehalose phosphate} + \text{UDP} \qquad (5)$$

This work which was carried out with Cabib (Leloir and Cabib[34], 1953), described the first case in which UDP-glucose was found to act as a glucose donor. Such a role had been suggested by Buchanan *et al.*[4] (1952), and by Kalckar[26] (1954).

Once we had found one transfer reaction we were soon able to detect another one using wheat germ extracts. Actually we found two enzymes, one which gave rise to the formation of sucrose (Cardini, Leloir and Chiriboga[11], 1955) and another which gave sucrose phosphate (Leloir and Cardini[35], 1955) as follows:

$$\text{UDP-glucose} + \text{fructose} \rightleftharpoons \text{sucrose} + \text{UDP} \qquad (6)$$

$$\text{UDP-glucose} + \text{fructose 6-phosphate} \rightleftharpoons \text{sucrose phosphate} + \text{UDP} \qquad (7)$$

This was a rather interesting finding because it explained the mechanism of sucrose synthesis in plants.

Another novel result of that period was the isolation of UDP-*N*-acetylglucosamine (Cabib, Leloir and Cardini[8], 1953). This substance was first detected as an impurity of UDP-glucose concentrates and we used to call it UDP-X until we were able to identify the sugar moiety as *N*-acetylglucosamine. It is now known to be involved in the biosynthesis of bacterial cell walls and mucoproteins.

Other members of the sugar nucleotide family were isolated in our laboratory. In 1954 (Cabib and Leloir[7]) GDP-mannose was found in yeast extracts, and later Pontis[47] (1955) detected UDP-*N*-acetylgalactosamine in liver. These substances are now known to be involved in the biosynthesis of mannan (Behrens and Cabib[1], 1968) and of some proteoglycans.

Other laboratories made important contributions. The identification of UDP-glucuronic acid as a donor for the formation of glucuronides (Dutton and Storey[13], 1953) was the first example of a transfer reaction from a sugar nucleotide.

Another important compound was detected by Park and Johnson[43] (1949) (Park[42], 1952) at about the same time that we isolated UDP-glucose. They found that a uridine containing compound accumulated in penicillin-treated Staphylococcus. This substance turned out to be difficult to identify because the sugar moiety was unknown at the time. This compound which kept biochemists in the dark behaved like a strange hexosamine and it was Strange and Dark[53] (1956) who first obtained a crystalline preparation. We now know that the sugar moiety is acetylglucosamine joined to lactic acid forming an ether linkage and the substance has been named muramic acid. The isolation of UDP–muramic acid was the starting point of the beautiful work carried out on bacterial cell wall synthesis which owes so much to Park and Strominger.

The number of known sugar nucleotides increased progressively for several years and in the 1963 census (Cabib[6]) they numbered more than 48. Furthermore, many enzymes involved in interconversion reactions have been studied. Herman Kalckar's group found that NAD is required in the UDP–glucose 4-epimerase reaction and it is believed that the glucose moiety of UDP–glucose is oxidized to a 4-keto intermediate which can then be reduced either to glucose or galactose.

Several other more complicated transformations have been carefully studied, for instance the transformation of GDP–mannose to GDP–fucose which requires a reduction at C-6 and inversions at -3 and -5 (Ginsburg[19], 1958). A similar case is the formation of TDP–rhamnose from TDP–glucose in which OH groups at C-3, -5 and -6 become inverted and a reduction at C-6 occurs (Glaser and Kornfeld[22], 1961; Pazur and Shuey[46], 1961.)

Polysaccharides

Many transfer reactions from sugar nucleotides have been detected. Thus Glaser and Brown[21] (1957) detected a transfer of *N*-acetylglucosamine from UDP–*N*-acetylglucosamine to chitin catalyzed by mold extracts. The formation of a β-1,3 glucan (callose) from UDP–glucose and of xylan from UDP–xylose was obtained by incubation with plant extracts (Feingold, Neufeld and Hassid[16,17], 1958, 1959).

A transfer from UDP–glucose to cellulose was also described by Glaser[20] (1957) working with *Acetobacter xylinum* which is a cellulose forming bacterium. Later it was found that the donor for cellulose formation in plants is GDP–glucose (Elbein, Barber and Hassid[14], 1964).

In our laboratory (Leloir and Cardini[36], 1957; Leloir, Olavarría, Goldenberg and Carminatti[38], 1959) we were able to detect the formation of glycogen from UDP–glucose (reaction 8) with liver and muscle enzymes:

$$\text{UDP-glucose} + G_n \rightarrow \text{UDP} + G_{n+1} \qquad (8)$$

In this equation G_n represents a glucogen molecule and G_{n+1} the same after addition of a glucosyl residue joined α-1,4.

The search for this enzyme glycogen synthetase or transferase was stimulated by reading a book by Herman Niemeyer[41] (1955) and its detection was a rather interesting finding since until then the synthesis of glycogen was believed to occur by reversal of the phosphorylase reaction (reaction 9):

$$(\text{Glucose})_{n+1} + \text{inorganic phosphate} \rightleftharpoons (\text{glucose})_n + \text{glucose 1-phosphate} \quad (9)$$

The same enzyme was thought to be involved in synthesis and in degradation. Another finding of considerable interest was that glucose 6-phosphate acts as an activator of glucogen synthetase.

Many years before, the Cori's had found that muscle phosphorylase has two forms which differ in their requirement for adenylic acid. Similarly J. Larner and C. Villar-Palasi[56] described two interconvertible forms of glycogen synthetase one active *per se* and another which requires glucose 6-phosphate. From then on a lot of work has been done on the regulation of glycogen metabolism.

Both phosphorylase and glycogen synthetase are regulated by the concentration of metabolites (adenylic acid and glucose 6-phosphate, respectively, as well as others such as ATP) and by reversible conversion of active to inactive forms. The latter changes are brought about by the action of several enzymes on one another. The picture which we have of the mechanism of glycogen regulation is too complicated to be shown here (for reviews see refs. 30, 32, 52, 54, 56).

Most of the studies on the biosynthesis of polysaccharides have consisted only in measuring the transfer of minute amounts of radioactive sugars. However, the studies should go further and we should be able to obtain *in vitro* polysaccharides identical to those made by cells. Some work of this type has been done with glycogen. One can obtain glycogen by incubating glucose 1-phosphate with phosphorylase or UDP-glucose with glycogen synthetase (in both cases with branching enzyme). The resulting products have been found to be of high molecular weight but different as judged by their pattern of degration by acid or alkali. The product formed with UDP-glucose and glycogen synthetase proved to be identical to that isolated from liver[39,40,44,45].

A logical extension of our work on glycogen was to investigate the formation of starch in plants. Enzymes were found which catalyzed the transfer of radioactivity from UDP-glucose labelled in the glucose moiety to starch (Fekete, Leloir and Cardini[18], 1960; Leloir, Fekete and Cardini[37], 1961). Studies on the specificity of the enzyme using synthetic nucleotides showed that ADP-glucose was used about ten times faster (Recondo and Leloir[49],

1961). This led to a search for ADP–glucose in natural sources which resulted in its isolation from corn (Recondo, Dankert and Leloir[48], 1963). An enzyme which can synthesize ADP–glucose was found by Espada[15] (1962).

Since then a lot of work has been done on the subject by several workers particularly by Carlos Cardini, Rosalía Frydman, Jack Preiss, T. Akazawa and others.

In Euglena the reserve polysaccharide is a β-1,3-linked glucan usually called paramylon. Its synthesis was studied by Goldemberg and Marechal[23] (1963) who found that it is formed from UDP–glucose.

Many more transfer reactions have been described so that the search was becoming monotonous.

Lipid Intermediates

From the data reported it may be concluded that most of the di-, oligo- and polysaccharides which occur in Nature in an amazing variety are synthesized from nucleotide sugars. However at least in some cases it seems that the transfer is not direct but mediated by lipid intermediates. This has been one of the most important findings of the last years and it is linked to the work of several groups (Osborn, Horecker, Strominger, Robbins, Lennartz and others). The structure of the first lipid intermediate detected in bacteria (Wright, Dankert, Fennesy and Robbins[57], 1967) is shown in Fig. 2.

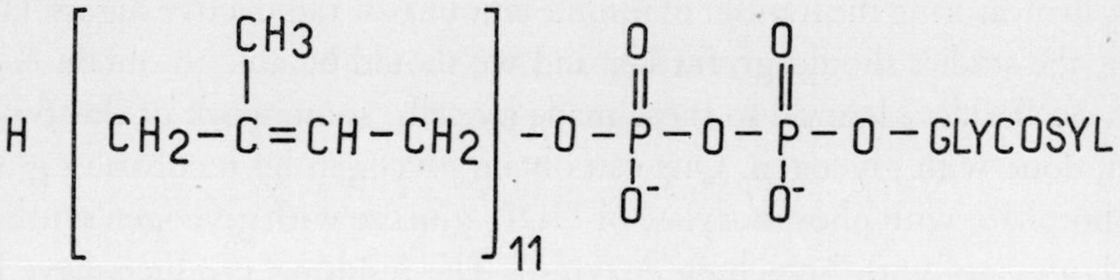

$$H\left[CH_2-\underset{}{\overset{CH_3}{\overset{|}{C}}}=CH-CH_2\right]_{11}-O-\overset{O}{\overset{\|}{\underset{O^-}{\underset{|}{P}}}}-O-\overset{O}{\overset{\|}{\underset{O^-}{\underset{|}{P}}}}-O-\text{GLYCOSYL}$$

Fig. 2. Antigen carrier lipid.

The structure of the compound was established in very small amounts mainly by mass spectroscopy. The compound, undecaprenol pyrophosphate, contains eleven isoprene residues one of them bearing an OH group joined to a pyrophosphate which in turn is linked to sugar residues.

The role of the carrier lipid in the formation of Salmonella lipopolysaccharide may be summarized in the following equations (where LP stands for the monophosphorylated lipid intermediate):

$$\text{LP} + \text{UDP–galactose} \rightarrow \text{LPP–galactose} + \text{UMP} \quad (10)$$

$$\text{LPP–galactose} + \text{TDP–rhamnose} \rightarrow \text{LPP–galactose–rhamnose} + \text{UDP} \quad (11)$$

$$\text{P–galactose–rhamnose} + \text{GDP–mannose} \rightarrow \text{LPP–galactose–rhamnose–mannose} + \text{GDP} \quad (12)$$

$$\text{LPP–galactose–rhamnose–mannose} \rightarrow \text{LPP(galactose–rhamnose–mannose)}_n + (n-1)\,\text{LPP} \quad (13)$$

$$\text{(galactose–rhamnose–mannose)}_n + \text{core} \rightarrow \text{(galactose–rhamnose–mannose)}_n \cdot \text{core} + \text{LPP} \quad (14)$$

$$\text{LPP} \rightarrow \text{LP} + \text{P} \quad (15)$$

In the first step (eqn. 10) there is a transfer of galactose 1-phosphate so that the lipid pyrophosphate and UMP are formed. Then rhamnose and mannose are added from the respective sugar nucleotides. Finally the trisaccharide units are transferred so as to form long chains (n = about 60) of galactose–rhamnose–mannose repeating units joined to the intermediate. In the next step (reaction 14) these would be transferred to the core of the lipopolysaccharide.

Undecaprenol pyrophosphate plays a similar role in the formation of bacterial cell walls in Staphylococci. The wall material, murein, is formed by alternating units of acetylglucosamine and muramic acid residues. These chains are cross-linked by peptides joined to the muramic acid residues.

The mechanism by which the cell wall is assembled has been elucidated mainly by the work of Strominger's group (Higashi, Strominger and Sweeley[25], 1967) and can be written as follows: (M = *N*-acetylmuramic acid joined to the following peptide: L-Ala–D-Glu–L-Lys–D-Ala–D-Ala; *N*-Ac stands for *N*-acetylglucosamine):

$$\text{UDPM} + \text{LP} \rightarrow \text{LPPM} + \text{UMP} \quad (16)$$

$$\text{UDP–}N\text{-Ac} + \text{LPPM} \rightarrow \text{LPPM–}N\text{-Ac} + \text{UDP} \quad (17)$$

$$\text{tRNA gly} + \text{LPPM–}N\text{-Ac} \rightarrow \text{tRNA} + \text{LPPM–}N\text{-Ac gly} \quad (18)$$

$$\text{LPPM–}N\text{-Ac gly} + \text{acceptor} \rightarrow \text{(M–}N\text{-Ac gly)–acceptor} + \text{LPP} \quad (19)$$

The first step (eqn. 16) is a transfer of muramyl peptide phosphate from the corresponding uridine nucleotide (one of the compounds isolated by Park) to undecaprenol monophosphate. Next (eqn. 17) *N*-acetylglucosamine is transferred from UDP–*N*-acetylglucosamine. After that (eqn. 18) five glycine residues are added (from a transfer ribonucleic acid) and then the whole disaccharide peptide is added to a part of the growing cell wall (referred to as acceptor in eqn. 19). After this the cross links are established between the peptide chains and the cell wall is complete.

Another piece of work dealing with lipid intermediates should be mentioned. This refers to the formation of mannan by *Micrococcus lysodeikticus* (Scher, Lennartz and Sweeley[51], 1968). The reactions are as follows:

GDP–mannose + undecaprenol-P → GDP + undecaprenol-P–mannose (20)

Undecaprenol-P–mannose + acceptor → mannose–acceptor + undecaprenol-P (21)

The difference with the previously mentioned cases is that in the first reaction (20) the sugar without the phosphate is transferred so that no pyrophosphate is formed.

While all this work was going on, Dankert who had been working with Robbin's group returned to Buenos Aires and transmitted to us his enthusiasm for polyprenols.

A Polyprenol Intermediate in Animal Tissues

A group working at the University of Liverpool formed by Morton, Hemming and others has studied carefully the different polyprenols found in nature. The general formula is shown in Fig. 3.

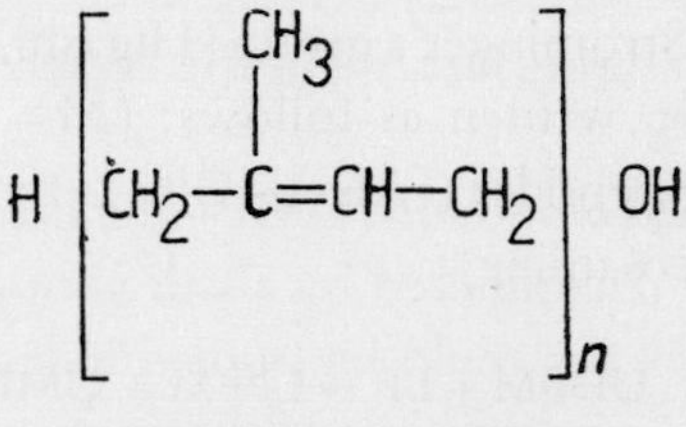

$$H\left[CH_2-\overset{\displaystyle CH_3}{\overset{|}{C}}=CH-CH_2\right]_n OH$$

Fig. 3. Polyprenols.

Many different types of compounds were detected which differ in the number *n* of isoprene residues, in the amount of *cis* or *trans* double bonds and also in that some of the double bonds may be saturated.

The compound isolated from animal tissues was named dolichol. In this substance the number of isoprene units is around 20 (it can vary from 16 to 23), and two of the double bonds are *trans*. Furthermore the double bond nearest to the alcohol group is saturated. Many other compounds were isolated from different sources (see Hemming[24], 1969).

With N. Behrens (Behrens and Leloir[2], 1970) we have studied a process occurring in liver in which it turned out that a phosphate of dolichol is involved. The reactions may be written as follows:

$$\text{UDP–glucose} + \text{DMP} \rightarrow \text{DMP–glucose} + \text{UDP} \tag{22}$$

$$\text{DMP–glucose} + \text{E} \rightarrow \text{glucose–E} + \text{DMP} \tag{23}$$

$$\text{glucose–E} \rightarrow \text{glucose} + \text{E} \tag{24}$$

In these equations DMP stands for dolichol monophosphate and E for an endogenous acceptor believed to be a protein.

The studies were carried out by incubating liver microsomes with radioactive UDP-glucose. It was found that a product soluble in organic solvents was formed. Further work showed that reaction 22 could be carried out so as to measure the lipid acceptor (DMP in eqn. 22). This allowed a purification process to be developed. The concentrates obtained gave infrared spectra having similarities with polyprenols. The compound had acidic character and was relatively stable to acid and alkali. It differed from undecaprenol phosphate in that the latter is acid-labile. It was reasoned that this difference could be due to the fact that in undecaprenol there is a double bond near the phosphate which is not present in dolichol. With this idea in mind the identification of our lipid acceptor was approached from another angle. Dolichol was prepared from liver (Burgos, Hemming, Pennok and Morton[5], 1963), phosphorylated chemically and the product tested for activity as lipid acceptor. The synthetic compound turned out to be identical, in all the properties tested, to that obtained from natural sources. For this reason we refer to it as dolichol monophosphate.

As to the glucosylated compound (DMP-glucose) it was found to be very labile to acid and to be decomposed by alkali giving 1,6-anhydroglucosan. The following reaction (23) could be studied independently from the first by using DMP-glucose prepared in a preliminary run. The optimal conditions for activity were determined. This step (reaction 23) does not require any cation in contrast to reaction 22 in which Mg^{2+} ions are necessary. Detergents are required in both steps.

The product formed from DMP-glucose indicated as glucose-E in eqn. 23 appears to be a glucosylated protein but work has just started on this point. There are very few glucose containing proteins. One of them is collagen which contains glucosyl, glactosyl hydroxylysine residues. However, the compound formed with liver microsomes seems to be clearly different from collagen. The last reaction (eqn. 24) has not been studied in any detail and could be brought by some of the glucosidases known to be present in liver.

The possibility that the glucosylation of ceramide, which is the first step in the formation of gangliosides, might be mediated by DMP-glucose has been investigated with results that are not quite conclusive but indicate that DMP-glucose is not involved.

Other sugar nucleotides have been tested and it was found that UDP-*N*-acetylglucosamine and GDP-mannose can serve as donors for the formation of the corresponding DMP-sugars. Other compounds such as UDP-*N*-acetylgalactosamine and UDP-galactose gave negative results (Behrens, Parodi, Leloir and Krisman, 1971).

The study of the lipid intermediates is becoming most interesting. The variety of polyprenols is large since they may vary in chain length, number of *cis* or *trans* double bonds, and degree of saturation. Furthermore they may have one or two phosphates and carry different sugars. The variety of polyprenol phosphate sugars may turn out to be as large as that of sugar nucleotides. It has been suggested that their role may be to provide a lipophylic moiety to sugars so as to allow them to permeate the lipid layer of membranes. Since in Salmonella polyprenol phosphates are involved in the formation of specific antigen it seems likely that in animal tissues they may be responsible for the formation of the surface carbohydrates which are so important in the behaviour of contacting cells. These external specific substances and interactions which Kalckar[28] (1965), in one of his penetrating essays calls «ektobiological», appear to be of great importance in the «social» behaviour of cells. Undoubtedly this may become a fascinating problem for future research. Fortunately even after two decades our field of investigation has not become dull or too fashionable.

Acknowledgements

My whole research career has been influenced by one person, Prof. Bernardo A. Houssay, who directed my doctoral thesis and who during all these years generously gave me his invaluable advice and friendship. I also owe very much to my friends, colleagues and coworkers the names of which are mentioned in the text.

The help of the «Fundación Campomar», Consejo Nacional de Investigaciones Científicas y Técnicas, Facultad de Ciencias, Exactas y Naturales, Universidad de Buenos Aires, the National Institutes of Health (U.S.A.) and the Rockefeller Foundation, which allowed us to carry out our work, is gratefully acknowledged.

1. N.H. Behrens and E. Cabib, *J.Biol.Chem.*, 243 (1968) 502.
2. N.H. Behrens and L.F. Leloir, *Proc.Natl.Acad.Sci.(U.S.)*, 66 (1970) 153.
3. N.H. Behrens, A.J. Parodi, L.F. Leloir and C.R. Krisman, *Arch.Biochem.Biophys.*, 143 (1971) 375.
4. J.G. Buchanan, J.A. Bassham, A.A. Benson, D.F. Bradley, M. Calvin, L.L. Daus, M. Goodman, P. Hayes, V.H. Lynch, L.T. Norris and A.T. Wilson, *Phosphorus Metabolism*, Vol. II, Johns Hopkins, Baltimore, 1952, p. 440.
5. J. Burgos, F.W. Hemming, J.F. Pennok and R.A. Morton, *Biochem.J.*, 88 (1963) 470.
6. E. Cabib, *Ann.Rev.Biochem.*, 32 (1963) 321.
7. E. Cabib and L.F. Leloir, *J.Biol.Chem.*, 206 (1954) 779.
8. E. Cabib, L.F. Leloir and C.E. Cardini, *J.Biol.Chem.*, 203 (1953) 1055.
9. R. Caputto, L.F. Leloir, C.E. Cardini and A.C. Paladini, *J.Biol.Chem.*, 184 (1950) 333.
10. R. Caputto, L.F. Leloir, R.E. Trucco, C.E. Cardini and A.C. Paladini, *Arch.Biochem.*, 18 (1948) 201.
11. C.E. Cardini, L.F. Leloir and J. Chiriboga, *J.Biol.Chem.*, 214 (1955) 149.
12. C.E. Cardini, A.C. Paladini, R. Caputto and L.F. Leloir, *Nature*, 165 (1950) 191.
13. G.J. Dutton and I.D.E. Storey, *Biochem.J.*, 53 (1953) xxxvii.
14. A.D. Elbein, G.A. Barber and W.Z. Hassid, *J.Am.Chem.Soc.*, 86 (1964) 309.
15. J. Espada, *J.Biol.Chem.*, 237 (1962) 3577.
16. D.S. Feingold, E.F. Neufeld and W.Z. Hassid, *J.Biol.Chem.*, 233 (1958) 783.
17. D.S. Feingold, E.F. Neufeld and W.Z. Hassid, *J.Biol.Chem.*, 234 (1959) 488.
18. M.A.R. de Fekete, L.F. Leloir and C.E. Cardini, *Nature*, 187 (1960) 918.
19. V. Ginsburg, *J.Am.Chem.Soc.*, 80 (1958) 4426.
20. L. Glaser, *Biochim.Biophys.Acta*, 25 (1957) 436.
21. L. Glaser and D.H. Brown, *Biochim.Biophys.Acta*, 23 (1957) 449.
22. L. Glaser and S. Kornfeld, *J.Biol.Chem.*, 236 (1961) 1795.
23. S.H. Goldenberg and L.R. Marechal, *Biochim.Biophys.Acta*, 71 (1963) 743.
24. F.W. Hemming, *Biochem.J.*, 113 (1969) 23P.
25. Y. Higashi, J.L. Strominger and C.C. Sweeley, *Proc. Natl. Acad. Sci., (U.S.)*, 57 (1967) 1878.
26. H.M. Kalckar, *The Mechanism of Enzyme Action*, Johns Hopkins, Baltimore, 1954, p. 675.
27. H.M. Kalckar, *Advan.Enzymol.*, 20 (1958) 111.
28. H.M. Kalckar, *Science*, 150 (1965) 305.
29. L.P. Kendall and L.H. Strickland, *Biochem.J.*, 32 (1938) 572.
30. E.G. Krebs and E.H. Fischer, *Vitamins Hormones*, 22 (1964) 399.
31. L.F. Leloir, *Arch.Biochem.Biophys.*, 33 (1951) 186.
32. L.F. Leloir, *Proc. 6th Panamerican Congress of Endocrinology, Mexico City, Excerpta Medica Intern.Congr. Ser.*, Vol. 112, Excerpta Medica, Amsterdam, 1965, p. 65.
33. L.F. Leloir, *Natl. Cancer Inst. Monograph* 27 (1966) 3.
34. L.F. Leloir and E. Cabib, *J.Am.Chem.Soc.*, 75 (1953) 5445.
35. L.F. Leloir and C.E. Cardini, *J.Biol.Chem.*, 214 (1955) 157.
36. L.F. Leloir and C.E. Cardini, *J.Am.Chem.Soc.*, 79 (1957) 6340.
37. L.F. Leloir, M.A.R. de Fekete and C.E. Cardini, *J.Biol.Chem.*, 236 (1961) 636.

38. L.F.Leloir, J.M.Olavarría, S.H.Goldemberg and H.Carminatti, *Arch.Biochem. Biophys.*, 81 (1959) 508.
39. J.Mordoh, C.R.Krisman and L.F.Leloir, *Arch.Biochem.Biophys.*, 113 (1966) 265.
40. J.Mordoh, L.F.Leloir and C.R.Krisman, *Proc.Natl.Acad.Sci.(U.S.)*, 53 (1965) 86.
41. H.Niemeyer, *Metabolismo de los Hidratos de Carbono*, University of Chili, 1955, p. 150.
42. J.T.Park, *J.Biol.Chem.*, 194 (1952) 877, 885, 897.
43. J.T.Park and M.J.Johnson, *J.Biol.Chem.*, 179 (1949) 585.
44. A.J.Parodi, C.R.Krisman, L.F.Leloir and J.Mordoh, *Arch.Biochem.Biophys.*, 121 (1967) 769.
45. A.J.Parodi, J.Mordoh, C.R.Krisman and L.F.Leloir, *Arch.Biochem.*, *Biophys.*, 132 (1969) 111.
46. J.H.Pazur and E.W.Shuey, *J.Biol.Chem.*, 236 (1961) 1780.
47. H.G.Pontis, *J.Biol.Chem.*, 216 (1955) 195.
48. E.Recondo, M.Dankert and L.F.Leloir, *Biochem.Biophys.Res.Commun.*, 12 (1963) 204.
49. E.Recondo and L.F.Leloir, *Biochem.Biophys.Res.Commun.*, 6 (1961) 85.
50. R.Robinson and W.T.Morgan, *Biochem.J.*, 24 (1930) 119.
51. M.Scher, W.J.Lennartz and C.C.Sweeley, *Proc.Natl.Acad.Sci.(U.S.)*, 59 (1968) 1313.
52. D.Stetten and M.R.Stetten, *Physiol.Rev.*, 40 (1960) 505.
53. R.E.Strange and F.A.Dark, *Nature*, 177 (1956) 186.
54. E.W.Sutherland, I.Øye and R.W.Butcher, *Recent Progr.Hormone Res.*, 21 (1965) 623.
55. R.E.Trucco, R.Caputto, L.F.Leloir and N.Mittelman, *Arch.Biochem.Biophys.*, 18 (1948) 137.
56. C.Villar-Palasi and J.Larner, *Vitamins Hormones*, 26 (1968) 65.
57. A.Wright, M.Dankert, P.Fennesey and P.W.Robbins, *Proc.Natl.Acad.Sci.(U.S.)*, 57 (1967) 1798.

Biography

Luis F. Leloir was born in Paris of Argentine parents on September 6, 1906 and has lived in Buenos Aires since he was two years old. He graduated as a Medical Doctor in the University of Buenos Aires in 1932 and started his scientific career at the Institute of Physiology working with Professor Bernardo A. Houssay on the role of the adrenal in carbohydrate metabolism. In 1936 he worked at the Biochemical Laboratory of Cambridge, England, which was directed by Sir Frederick Gowland Hopkins. There he collaborated with Malcom Dixon, N. L. Edson and D. E. Green. On returning to Buenos Aires he worked with J. M. Muñoz on the oxidation of fatty acids in liver, and also together with E. Braun Menéndez, J. C. Fasciolo and A. C. Taquini on the formation of angiotensin. In 1944 he was Research Assistant in Dr. Carl F. Cori's laboratory in St. Louis, United States and thereafter worked with D. E. Green in the College of Physicians and Surgeons, Columbia University, New York. Since then he has been Director of the Instituto de Investigaciones Bioquímicas, Fundación Campomar. With his early collaborators, Ranwel Caputto, Carlos E. Cardini, Raúl Trucco and Alejandro C. Paladini work was started on the metabolism of galactose which led to the isolation of glucose 1,6-diphosphate and uridine diphosphate glucose. The latter substance was then found to act as glucose donor in the synthesis of trehalose (with Enrico Cabib, 1953) and sucrose (with Carlos E. Cardini and J. Chiriboga, 1955). Other sugar nucleotides such as uridine diphosphate acetylglucosamine and guanosine diphosphate mannose were also isolated. Further work showed that uridine diphosphate glucose is involved in glycogen synthesis and adenosine diphosphate glucose in that of starch.

More recent investigations (with Nicolás Behrens) have dealt with the role of a polyprenol, dolichol, in glucose transfer in animal tissues.

Luis Leloir was married in 1943 to Amelia Zuberbuhler and has a daughter, Amelia. At present Leloir is Professor in the Faculty of Sciences, University of Buenos Aires. He is a member of the following academies; National Academy of Sciences, American Academy of Arts and Sciences, Academia Nacional de Medicina, American Philosophical Society, Pontificial Academy of Sciences, and Honorary Member of the Biochemical Society (England). He has received honorary degrees of the following universities: Granada (Spain), Paris (France), Tucuman (Argentina) and La Plata (Argentina). Prof. Leloir has received the following awards: Argentine Scientific Society, Helen Hay Whitney Foundation (United States), Severo Vaccaro Foundation (Argentina), Bunge and Born Foundation (Argentina), Gairdner Foundation (Canada), Louisa Gross Horowitz (United States), Benito Juarez (Mexico); and at present he is President of the Pan-American Association of Biochemical Societies.

Name Index

Subject Index

Index of Biographies